Environmental Science and Engineering

Series Editors

The ultimate goal of this series is to contribute to the protection of our environment, which calls for both profound research and the ongoing development of solutions and measurements by experts in the field. Accordingly, the series promotes not only a deeper understanding of environmental processes and the evaluation of management strategies, but also design and technology aimed at improving environmental quality. Books focusing on the former are published in the subseries Environmental Science, those focusing on the latter in the subseries Environmental Engineering.

Varenyam Achal · Ahmad Zuhairi Abdullah ·
Mohd Razif Harun
Editors

Protect the Water World: Water Pollution Treatment and Related Biotechnology

Selected Papers of 2024 International Conference on Environmental Protection and Pollution Control (EPPC 2024)

Springer

Editors
Varenyam Achal
Israel Institute of Technology
Guangdong Technion
Shantou, China

Ahmad Zuhairi Abdullah
School of Chemical Engineering
Universiti Sains Malaysia
Gelugor, Pulau Pinang, Malaysia

Mohd Razif Harun
Universiti Putra Malaysia
Serdang, Selango, Malaysia

ISSN 1863-5520 ISSN 1863-5539 (electronic)
Environmental Science and Engineering
ISBN 978-3-031-92122-3 ISBN 978-3-031-92123-0 (eBook)
https://doi.org/10.1007/978-3-031-92123-0

This Springer imprint is published by the registered company Springer Nature Switzerland AG
The registered company address is: Gewerbestrasse 11, 6330 Cham, Switzerland

Preface

It is with great pleasure that we present the proceedings of the 2024 International Conference on Environmental Protection and Pollution Control (EPPC 2024). The event was a resounding success, fostering meaningful discussions and collaborations that will undoubtedly contribute to the advancement of sustainable practices and technologies.

The primary objective of EPPC 2024 was to provide a platform for the exchange of innovative ideas, cutting-edge research, and practical solutions in the field of environmental science and engineering. The conference aimed to bridge the gap between academia and industry, encouraging interdisciplinary approaches to tackle complex environmental issues. By facilitating dialogue among experts from various disciplines, EPPC 2024 sought to inspire new perspectives and drive forward the global agenda for environmental sustainability.

This conference was conducted through knowledge exchange and networking opportunities in the form of keynote speeches, peer reviewed paper presentations, round table discussions and time set aside for traditional networking. Attendees include technology developers, researchers, policy makers, academics, students and community leaders.

A group of committee members was convened to offer strategic guidance on pivotal aspects of the conference. Their responsibilities included devising rigorous criteria for the evaluation of submissions, deliberating on and advising the formulation of the conference agenda, and identifying prospective keynote and closing speakers. Foo Keng Yuen served as the chair of the conference, leading its initiatives with expertise and vision. Zawawi Bin Daud contributed to the organization and coordination of the event. Their commitment to excellence and attention to detail were evident in every aspect of the event, from the meticulous selection of keynote speakers to the seamless organization of technical sessions.

We were honored to welcome a distinguished lineup of keynote speakers, each of whom brought unique insights and expertise to the conference. Among them were Varenyam Achal, whose groundbreaking work on environmental remediation and bio-based building materials has revolutionized our understanding of environmental sustainability; Zawawi Bin Daud from Universiti Tun Hussein Onn Malaysia, a pioneer in water and wastewater treatment; and Rongkui Su, whose research domains focus on environmental functional materials and the resource utilization of agricultural and forestry waste.

The conference proceedings of EPPC 2024 is a comprehensive collection of the research presented at the conference. Each paper underwent a rigorous peer-review process, ensuring the highest standards of quality and relevance. These contributions cover a wide range of topics, from advanced pollution control technologies to innovative strategies for environmental protection. We believe that this volume will serve as a valuable resource for researchers and practitioners alike, providing insights and inspiration for future work in the field.

In closing, we would like to extend our heartfelt gratitude to all those who contributed to the success of EPPC 2024. Our sincere thanks go to the organizing committee, keynote speakers, session chairs, reviewers, and participants for their invaluable contributions. Special recognition is due to the technical support team and the conference staff, whose behind-the-scenes efforts ensured the seamless execution of the event. Finally, we would like to acknowledge the support of our sponsors and partners, whose generosity made this conference possible.

As we look to the future, we are confident that the insights and collaborations fostered at EPPC 2024 will continue to drive progress in environmental protection and pollution control. We hope that this paper volume will inspire further research and innovation, contributing to a more sustainable and resilient world.

The Committee of EPPC 2024

Committee Member

Conference Chair

Foo Keng Yuen	Universiti Sains Malaysia (USM), Malaysia

Publication Chairs

Varenyam Achal	Guangdong Technion-Israel Institute of Technology, China
Ahmad Zuhairi Abdullah	Universiti Sains Malaysia, Malaysia
Mohd Razif Harun	Universiti Putra Malaysia, Malaysia

Technical Program Committee Chair

Guangming Li	Tongji University, China

Organizing Committee Chair

Zawawi Bin Daud	Universiti Tun Hussein Onn Malaysia, Malaysia

Organizing Committee Members

Jie Liu	Qingdao University of Technology, China
Jianhua Zhang	Qingdao University of Technology, China
Rongkui Su	Central South University of Forestry and Technology, China
Azhar Abdul Halim	Universiti Kebangsaan Malaysia, Malaysia
Mohd Anuar Kamaruddin	Universiti Sains Malaysia, Malaysia
Halizah Awang	Universiti Tun Hussein Onn Malaysia, Malaysia
Yiting Luo	Hunan First Normal University, China
Mohd Baharudin Ridzuan	Universiti Tun Hussein Onn Malaysia, Malaysia
Ley Juen Looi	Universiti Putra Malaysia, Malaysia

Technical Program Committee Members

Contents

Research on Ecology and Environmental Sustainability

Research on Ecology and Environmental Sustainability

The Interactions and Turning Points in the City-Water Relationship in the Guangdong-Hong Kong-Macao Greater Bay Area

Xinghui Liang, Jing He, and Li Li(✉)

South China University of Technology, Guangzhou 511400, Guangdong, China
2209420548@qq.com

Abstract. Amidst global climate change, the processes of urbanization and population growth have placed immense pressures on water resources. To this point, there has been limited research on long-term spatiotemporal evolution within this context. The Guangdong-Hong Kong-Macao Greater Bay Area, currently one of China's three major urban clusters, has employed the City-Water System Relationship (CWSR) methodology to explore the progression of its cities from early human settlements to the Qing Dynasty. Utilizing ArcGIS software to depict the spatial relationships between urban areas and water, the study significantly highlights the connections between river systems and urban patterns, as well as the impacts of water systems, environmental heterogeneity, and the political and economic factors influencing urban development. The findings also, to some extent, represent the historical concentration of economic centers in the Guangdong-Hong Kong-Macao Greater Bay Area and reveal the intensity of human activities in regional development.

Keywords: City-Water System Relationship · Guangdong-Hong Kong-Macao Greater Bay Area · Pearl River System · Urban Development · ArcGIS

1 Introduction

The City-Water System Relationship (CWSR) plays a critical role in regional sustainable development [1]. It exemplifies the mutual influence and feedback mechanisms between water and land, which are of significant importance to the sustainable development of a region.

The combined effects of population growth, economic advancement, and climate change have significantly strained water resources across numerous cities and regions globally. Urban expansion continues to impact aquatic environments and disrupt natural hydrological processes. The Pearl River Basin, recognized as China's third-largest river, is not only the cradle of ancient Chinese culture but also the birthplace of early civilizations, nourishing numerous ancient towns. It is one of the oldest rivers in Chinese history and has nurtured one of the nation's three principal urban agglomerations [2]. The Pearl

V. Achal et al. (Eds.): EPPC 2024, ESE, pp. 3–22, 2025.
https://doi.org/10.1007/978-3-031-92123-0_1

River Delta, as an estuarine alluvial plain, faces threats from extreme weather conditions and sea level rise linked to global climate change. To elucidate the City-Water System Relationship (CWSR) within this region, this research delves into the evolution of cities and changes in water systems within the Guangdong-Hong Kong-Macao Greater Bay Area. By examining historical records of cities along the Pearl River Basin, we aim to uncover the essence of human-environment relationships from a dynamic and continuous perspective, thereby promoting water resource management and sustainable development efforts in similar regions worldwide.

2 Principles and Research Gaps in City-Water System Relationship (CWSR)

Numerous studies and practices have contributed to the development of principles and insights concerning the City-Water System Relationship (CWSR). One core principle emphasizes that cities should be situated in close proximity to rivers to secure water supply, while simultaneously maintaining an appropriate distance to mitigate flood risks [3]. "Guanzi," a classic work on ancient Chinese urban planning, provides an in-depth discussion and dialectical analysis of the relationship between water and cities. This work laid the foundation for the spatial layout of ancient Chinese cities and the principles of building near water, reflecting the ecological wisdom of traditional agricultural societies in conforming to natural laws and managing water resources [4]. However, the CWSR mechanism remains underdeveloped due to several key factors, as discussed below:

1. The absence of extensive historical data is a significant issue. Human-environment interactions are dynamically evolving, necessitating long-term observation of regional geographical phenomena. For instance, before the Industrial Revolution, human settlements were closer to major rivers, but after 1870, human activities gradually moved away from these rivers [3]. In the context of industrial civilization, rapid industrialization and urbanization have led to an imbalanced development between cities and water bodies. The limited availability of long-term data makes it difficult to precisely track these evolving patterns.
2. There is a shortage of large-scale studies, particularly at the watershed level. Skinner's research on cities during the late Qing Dynasty shows that regional economic centers were typically located in the central areas of watersheds, while secondary cities were situated at the peripheries [5]. The site selection of ancient Chinese cities also emphasized the importance of factors such as water resource utilization, transportation, military defense, geology, climate, and disaster prevention [6]. However, such studies remain limited and typically focus on urban growth and density, with inadequate exploration of the city-water system relationship.
3. The complex nature of interactions between cities and water systems makes it challenging to establish clear causal relationships within intricate ecosystems. As a result, methods like the Granger causality test [7] and its enhanced versions, along with network analysis, have been suggested for examining complex socio-ecological interdependencies [8]. However, frequent historical changes in water systems and unpredictable events in human society further complicate these analyses.

3 Data and Methods

3.1 Research Area

The study area is located in the Guangdong-Hong Kong-Macao Greater Bay Area (Fig. 1). The Pearl River, a historically complex water system, consists of three main channels and numerous tributaries. These three main channels are situated in different geographical environments: the West River, North River, and East River. Each flows along distinct mountain ranges, forming unique landforms and converging in the Pearl River Delta before flowing into the South China Sea. The many tributaries of the Pearl River have a unique geographical pattern that has provided important references for the development of cities along the river. The distribution of cities along the Pearl River not only shares common characteristics with other major river cities in China but also exhibits significant differences [2].

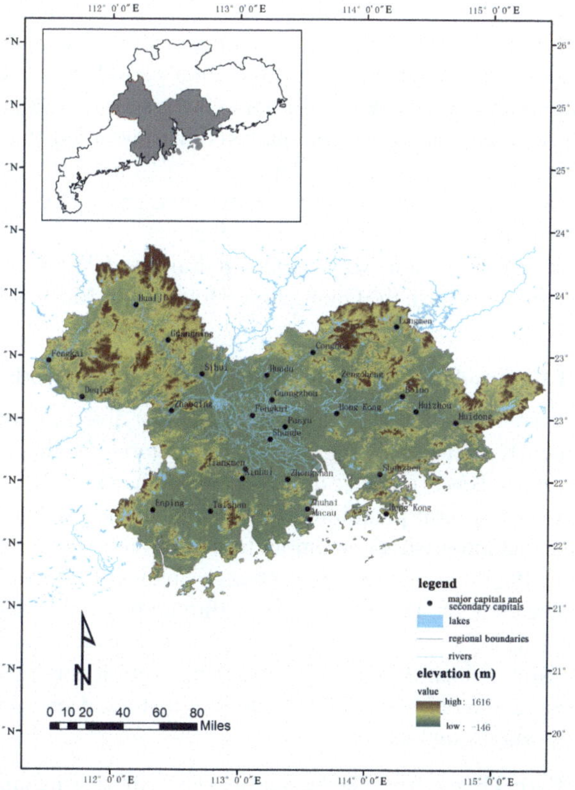

Fig. 1. Geographical location and environment in the Guangdong-Hong Kong-Macao Greater Bay Area (self-drawn).

As the Pearl River Delta plain continued to expand and agriculture developed, Neolithic cultures gradually transitioned to agrarian societies. This transformation led to a pattern of archaeological sites moving from the north to the south, evolving from older to newer. By the Ming and Qing dynasties, the urban development and landscape of the Pearl River Delta were essentially established. Consequently, there was significant interaction between the urban areas and water systems. This offers a valuable framework for studying the City-Water System Relationship (CWSR) at a watershed scale. As a major economic center in China, the Guangdong-Hong Kong-Macao Greater Bay Area faces ongoing challenges from population growth and environmental degradation. Gaining insights into the historical application of CWSR in this region could be essential for supporting its sustainable development now and in the future.

3.2 Methods

To explore the mechanisms and evolution of the City-Water System Relationship (CWSR) in this region, this study leverages historical data and applies ArcGIS 10.8 software to visualize their spatial relationships. The main analysis steps are as follows: (1) Determine Spatiotemporal Characteristics: Based on geomorphological features, the region is spatially divided, and the historical development of the Guangdong-Hong Kong-Macao Greater Bay Area is segmented into four time periods and spatial regions; (2) Describe Urban Development and Water System Changes from a Spatiotemporal Perspective: Through detailed spatiotemporal analysis, the historical trajectories of urban expansion and water system changes are depicted; (3) Conduct Correlation Analysis: Quantitative analysis is performed to reveal the dynamic processes of mutual influence between urban distribution and water system changes.

These steps aid in comprehensively understanding the CWSR characteristics and their evolutionary mechanisms in the Guangdong-Hong Kong-Macao Greater Bay Area.

3.2.1 Division of Geomorphological Structures and Time Periods

The three main streams of the Pearl River system create a diverse natural geographical environment, with each river possessing distinct characteristics: the West River, North River, and East River [2]. Based on these main streams of the Pearl River system and their associated formation of several piedmont alluvial plains, the piedmont alluvial plains can be further classified into distinct piedmont basin regions. These are divided into the areas through which the three main rivers flow each piedmont alluvial plain region and the final convergence area, the Pearl River Delta alluvial plain region.

To explore the long-term interaction between urban areas and water systems in the Guangdong-Hong Kong-Macao Greater Bay Area, the evolutionary process of cities and water systems in the region has been divided into four historical stages based on societal development. These stages include:

1. Remnants and Early Cities (from the Neolithic Age to the Spring and Autumn Period and the Warring States, before 221 BCE): This period marks the earliest interactions where settlements began forming along the rivers, relying on the natural resources provided by water systems.
2. Urban expansion from the Qin Dynasty to the Sui Dynasty period (221 BCE to 618 CE): During these dynasties, the consolidation of territories and the establishment

of administrative regions led to the growth of urban centers near water systems, facilitating trade and agriculture.

3. Urban Development from the Tang Dynasty to the Yuan Dynasty (618 CE to 1368 CE): This era saw significant advancements in technology and infrastructure, including the expansion of irrigation systems and the development of cities along these water networks.
4. Rapid Development during the Ming and Qing Dynasties (1368 CE to 1911 CE): The final stage marked by intensive urban expansion and sophisticated management of water resources to support large populations and high economic activities.

Each era is marked by distinct socio-economic situations and levels of technological advancement, reflecting the evolving relationship between urban areas and water systems throughout history.

3.2.2 Urban Expansion, Spatial Distribution, and Changes in the Hydrological System

To illustrate urban growth, we plotted the spatial distribution of the number of cities in each region and the entire study area for different years. To analyze the urban development trends in each subdivision more comprehensively, we used the Mean Center tool in ArcGIS to calculate the central tendency of city distribution patterns for each dynasty from 221 BCE to 1911 CE. The mean center represents the overarching spatial configuration of the urban network, and its calculation is expressed by the following formula:

$$\overline{X} = \frac{\sum_{i=1}^{n} x_i}{n}, \overline{Y} = \frac{\sum_{i=1}^{n} y_i}{n}$$

Where x_i and y_i represent the longitude and latitude of historical city points, respectively, $\overline{X}$ and $\overline{Y}$ represent the coordinates of the mean center for the city distribution pattern, and indicates the total number of cities within the subregion [1].

The plain was divided into a grid to calculate the city density for Periods 1–4. Subsequently, the density values were normalized to a range of 0–1, and changes in the water system were visualized based on city density. By employing the Kernel Density tool in ArcGIS, we calculated the kernel density of cities for different periods, determining the city densities for Periods 1–4. These densities were overlaid with river data to analyze the city-water relationship. Additionally, to quantify changes in the water system, we used grid division to calculate the frequency with which each grid cell was occupied by rivers (corridors) or oceans/lakes (patches), representing the number of changes.

3.2.3 Correlation Analysis of Spatiotemporal Trajectories

To compare the urban development trajectories across various subregions, the correlation coefficients between the city count time series were computed. The formula for the correlation coefficient is as follows:

$$\rho = \frac{\mathrm{Cov}(X, Y)}{\sigma_X \sigma_Y}$$

Where X and Y denote the time series, ρ is the Pearson correlation coefficient, Cov(X, Y) representing the covariance between X and Y, σ_X and σ_Y denote the standard deviations of Xand Y, respectively. [1]

To explore the relationship between urban density and variations in the water system across different periods, box plots were generated to illustrate urban density under varying degrees of water system changes. Additionally, both urban density and city-water relationship coefficients were normalized to a range between 0 and 1.

3.3 Sources of Data

3.3.1 Geographical and Historical Environmental Data

For the geographical data of the study area, digital elevation data for the Guangdong-Hong Kong-Macao Greater Bay Area with a resolution of 30 meters was obtained from the Shuttle Radar Topography Mission [9]. Historical changes in the water systems of the lower Pearl River were digitized from the Historical Atlas of China [10] .

3.3.2 Historical Urban Data

The urban data for the first period were digitized from the Historical Atlas of China [11], which includes sites from primitive societies and the urban development of the Zhou Dynasty from 770 BCE to 475 BCE. This era can be subdivided into the Spring and Autumn period and the Warring States period. Historical urban data for subsequent feudal periods were obtained from the *China Historical Geographic Information System* [12].

4 Results

4.1 Historical Population Evolution

The administrative structures and boundaries within the study area have undergone numerous changes over time, complicating efforts to accurately define the study area and gather population statistics for various periods. Considering that before the Song Dynasty, the main human activities in the Guangdong-Hong Kong-Macao Greater Bay Area were concentrated north of Guangzhou, with the area south of Guangzhou comprising bays and sandbanks, and that the water system characteristics only gradually stabilized during the Ming and Qing dynasties, the population figures for significant cities within the region are used to approximate the area's population. Historical population records span from the Han Dynasty (206 BCE to 220 CE) to 1911, and these figures are used to represent the socio-economic conditions of the corresponding dynasties [13].

According to specialized research on the history of ancient Chinese populations [13], recorded figures for the Guangdong-Hong Kong-Macao Greater Bay Area indicate significant fluctuations across different dynasties:

- During the Eastern Han period, there were 17,657 households with a population of 123,474.
- In the Western Han period, there were 91,731 households with 335,187 people.
- The Western Jin period saw 11,983 households.
- The Sui Dynasty recorded 485,775 households.
- During the Tang Dynasty, there were 77,560 households with a population of 370,955.
- The Northern Song period had 266,119 households.
- In the Yuan Dynasty, there were 255,472 households, housing 1,271,262 individuals.
- The Ming Dynasty had 218,240 households.
- The Qing Dynasty saw a dramatic increase to 12,845,822 households.

The missing specific population data for some dynasties and the non-standardization of household counts across periods introduce some variability, but these figures broadly reflect the population changes in the region.

From the aforementioned data, it is evident that the population in the Guangdong-Hong Kong-Macao Greater Bay Area developed gradually from the Han Dynasty. There was a dramatic decline in population from the Han Dynasty through the Western Jin period. Subsequently, from the Sui Dynasty to the Northern Song period, the population experienced rapid growth, reaching a minor peak during the Northern Song Dynasty. From the Northern Song to the Ming Dynasty, the population remained relatively stable. The population reached its highest peak during the Qing Dynasty.

4.2 Environmental Diversity Formed by Aquatic Systems

The geomorphological configuration of the region is classified into two primary types: the piedmont alluvial plains formed by the three main rivers and the delta alluvial plain where these rivers converge. The area is divided into six zones (Fig. 2): the Pearl River Delta alluvial plain south of Guangzhou (A) and five piedmont alluvial plains (P). Moreover, the P area is further divided into five sections, P1 through P5, each of which includes regions influenced by different rivers.

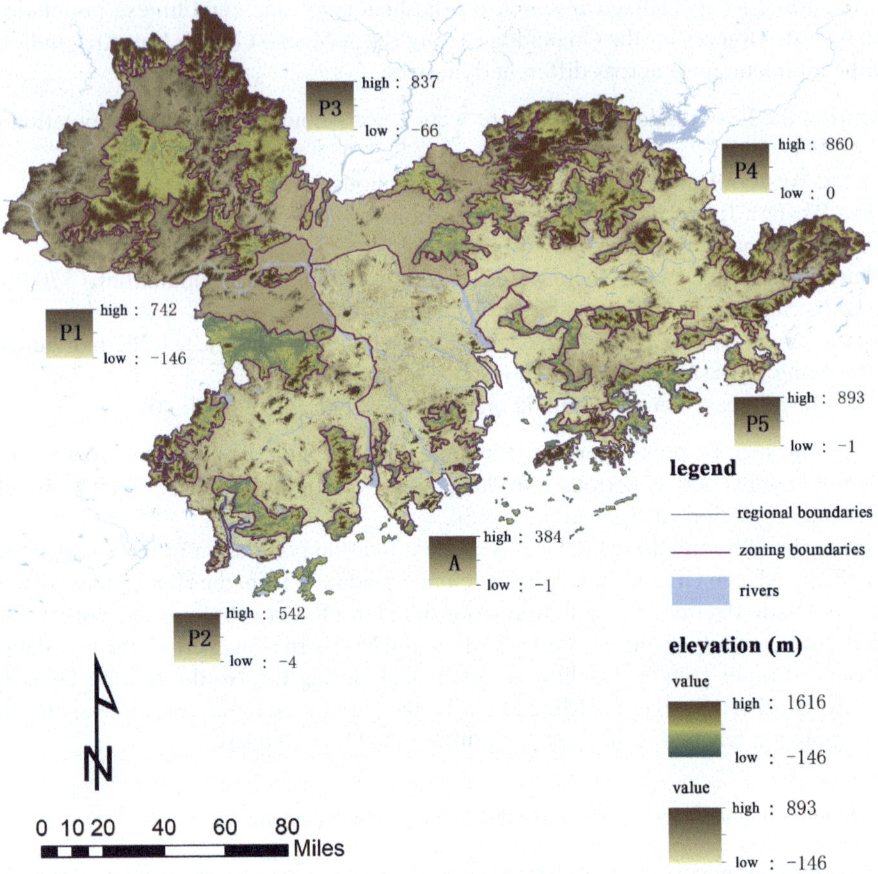

Fig. 2. illustrates the division of the geomorphological structures within the Pearl River Basin plains. P1 through P5 indicate regions directly affected by the rivers in the piedmont alluvial plains: P1 is influenced by the West River, P2 by the Tan River, P3 by the North River and Liuxi River, P4 by the Dong River, and P5 by branches of the Dong River. A represents the Pearl River Delta alluvial plain. Each area is accompanied by a color gradient legend indicating variations in elevation. (self-drawn).

Different subregions exhibit distinct natural characteristics during their formation processes, leading to environmental variability within the area (Table 1, Fig. 3). The alluvial plain, defined by the continuous accumulation and sedimentation of river deposits, is the flattest region where river confluences occur. Areas P1-P5 are formed by alluvial layers from mountain rivers, among which P5 experiences less intense river variations compared to other alluvial plains; thus, P5 is considered a foothill alluvium. From the paragraph below, it is evident that these subregions with varied natural characteristics provide different 'stages' for urban locations and profoundly influence their developmental processes.

Table 1. Geographic features of different areas

	P1	P2	P3	P4	P5	A	ALL
Average elevation (m)	94.05	31.83	74.53	62.81	54.86	13.13	130.3
Standard deviation of elevation	84.89	41.72	72.05	69.38	20.93	20.93	170.79
Area (km)	5861	6245	6287	8690	3065	7207	
Agricultural suitability	High	High	High	High	High	High	
Flood risk	Low	Low	High	Low	Low	Low	

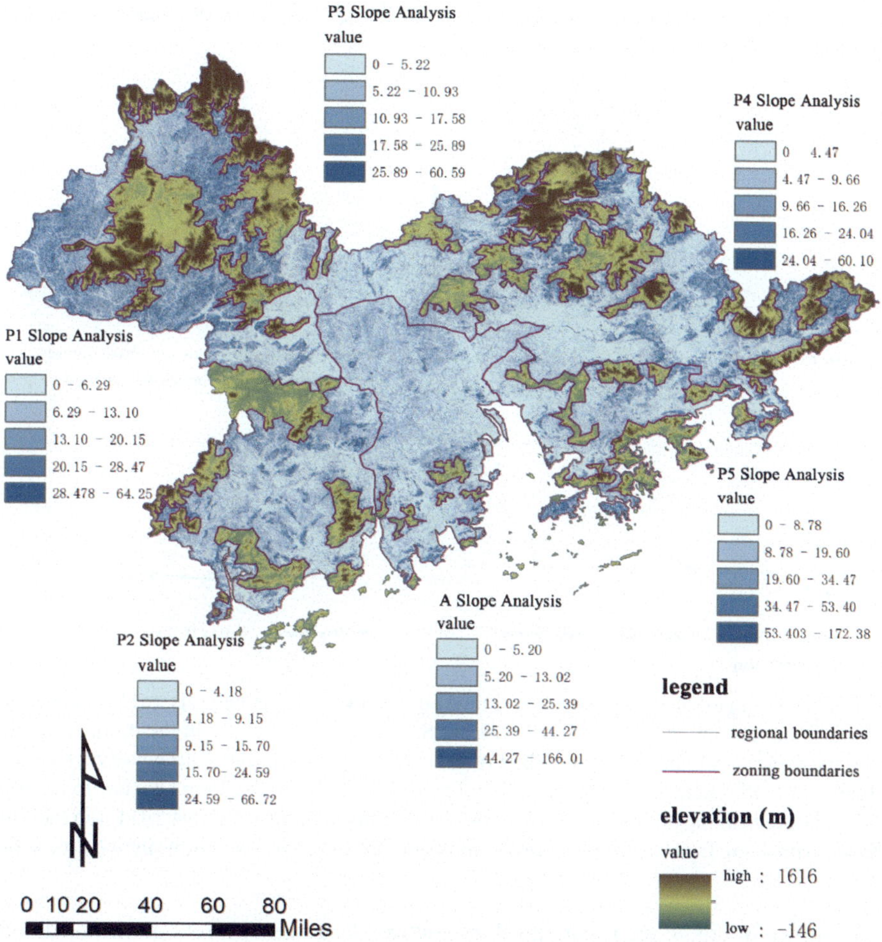

Fig. 3. Slope analysis of zones P1-P5, area A

4.3 Development and Changes in Urban and Water Systems

4.3.1 Variations in Urban Number and Density

The overall count of cities in the study area exhibits a gradual increasing trend, as shown in Fig. 4. Limited evidence from ancient sites indicates that the region has nurtured numerous primitive cultures. Historical records reveal that many cities from the Shang and Zhou dynasties were not fully recorded. With the unification under the Qin Dynasty, particularly with the implementation of the prefectural city system, most of the historical cities in the study area became systematically documented. However, due to Guangdong's distance from the political, cultural, and economic centers of the Central Plains, urban expansion progressed slowly from the Qin Dynasty through the Southern and Northern Dynasties [14].

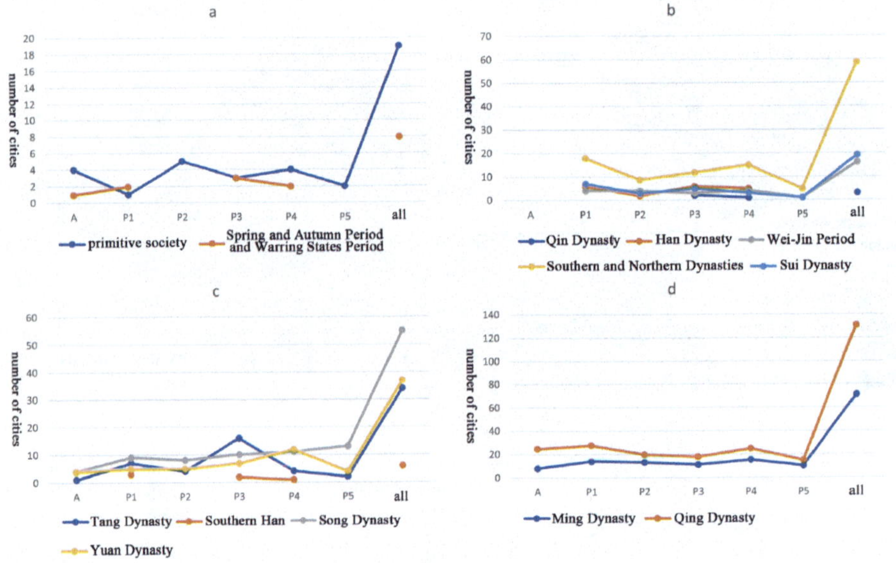

Fig. 4. Changes in urban development over time across study area subregions. (**a**) From primitive societies to the Spring and Autumn and Warring States periods, changes in the number of cities across six regions in two time series. (**b**) From the Qin Dynasty to the Sui Dynasty, changes in the number of cities across six regions in five time series. (**c**) From the Tang Dynasty to the Yuan Dynasty, changes in the number of cities across six regions in four time series. (**d**) From the Ming Dynasty to the Qing Dynasty, changes in the number of cities across six regions in two time series.

As shown in Fig. 4, the urban development trends across different partitions display distinct patterns. The foothill alluvial plain (P) and the alluvial plain (A) exhibit marked contrasts in city numbers compared to the overall regional trend, with a long historical period where no cities existed in the alluvial plain (A). Experiencing a lengthy period without any cities. Prior to the Qin Dynasty, the P region consistently maintained the highest urban density. With changes in the water systems leading to land expansion and driven by policy and economic factors during the Ming and Qing dynasties, the number of cities in the A region increased rapidly. Consequently, during this period, the urban density in the A region became comparable to that in the P region. Meanwhile, P retained the highest urban density across nearly all dynasties.

These results may indicate that the impact of environmental structures on urban distribution is more stable and enduring than anthropogenic factors. Additionally, within the foothill alluvial plain (P), there are variations in the number of cities and their densities. Larger plains and areas with tributaries tend to have more cities and higher urban density, potentially indicating a clustering effect of urban centers [15].

4.3.2 Spatial Changes in the River System

The Pearl River system includes three main rivers (the North River, West River, and East River) and several tributaries (such as the Tan River, He River, and Liu Xi River) (Fig. 5). Each main river has a distinct natural geographical environment. Historically, the delta river system maintained its main channels through siltation, a process considered to reflect the historical pattern of "divergence and convergence" and "strong trunk, weak branches." To this day, one of the characteristics of the Pearl River Delta's river network is the clear distinction between main and tributary rivers [14].

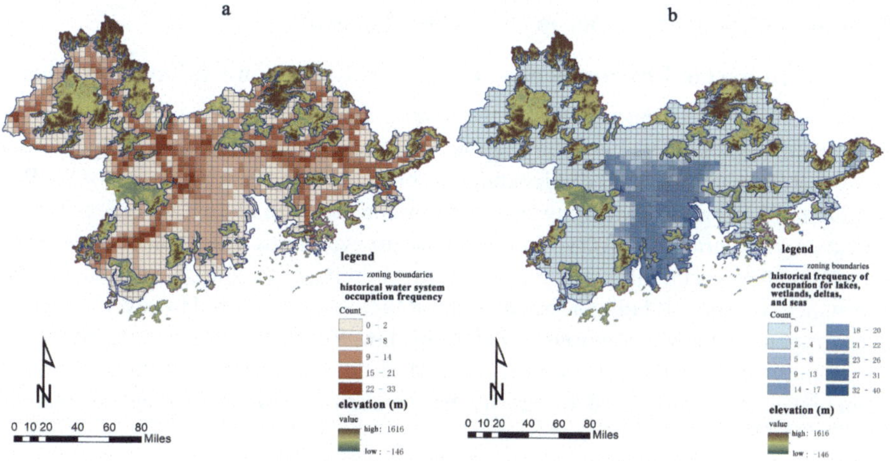

Fig. 5. Transformations in the Water System (**a**) Historical flow directions in the lower Pearl River basin, with the frequency of water flow recorded at each 5 km grid interval. (**b**) The number of times each 5 km grid was occupied by lakes, wetlands, or delta ocean. The data shows a noticeable reduction in the area occupied by the delta ocean over time.

In the early history, the Pearl River Delta plain, south of Guangzhou, was predominantly occupied by the sea (Fig. 5). The emergence of land in Guangzhou occurred during the Qin Dynasty, just over 2800 years ago [14]. The development of the Pearl River Delta from a bay—island state to extensive land formation—and estuarial sedimentation illustrates that the river network developed alongside the formation of land, typically following it. Many river sections near the estuary are actually only a few hundred or even just over a hundred years old, often presenting a stark contrast to the river's grand geomorphology with its brief history.

The formation of the Pearl River Delta can be specifically divided into three stages:

1. From the late Jumushi period to 4000 years ago, sea levels rose to today's baseline, transitioning from the river bays to the delta development stage. The widespread oyster shell layers and marine erosion and deposition landscapes south of Guangzhou are evidence that this area was once a sea. Historical records indicate that the bay existed until the Han and Jin dynasties;
2. During the Tang and Song dynasties, the delta still retained bay characteristics, which is evident from the transportation conditions at the time, the southward shift of salt fields, and administrative changes, indicating sea retreat and land expansion. The delta had completed extensive underwater accumulation and entered the distributary flow stage, displaying unique basin-island features;
3. The past 500 years during the Ming and Qing dynasties marked the most significant reversal period of water and land areas in the delta, transitioning from island-basin to plains and hills. The delta's landscape was characterized by widespread fields and scattered towns, with the modern river network beginning to emerge;

4.4 Impact of Water Systems on Urban Development

4.4.1 The Impact of Environmental Diversity as Evident in Regional Partitioning

The time series correlation analysis of the number of cities in different regions (Fig. 6) shows that the coefficients between Region A and other regions are relatively low, suggesting that Region A is largely independent. This reflects the historical state of the Pearl River Delta plain, which was long occupied by the sea with almost no cities. Tobler's First Law of Geography ("everything is related to everything else, but near things are more related than distant things") is also intuitively reflected in the correlation analysis. For instance, P5 and P4, being adjacent, exhibit higher correlation. However, geographical heterogeneity between regions can alter this law. Regions with similar geographical backgrounds (such as those located at the end of the main branch of the Pearl River) exhibit higher correlation, explaining why the correlation between P5 and P2 is relatively high.

4.4.2 The Average Central Trajectory of Urban Configurations in Various Sub-Regions

The spatial center (centroid) of urban distribution for each period reflects the trajectory of urban patterns and exhibits subregional characteristics (Fig. 7). All sections have undergone a "north-south" movement. In the foothill alluvial plains (P1-P3), the average center is generally located in the middle of each subregion, initially moving west, then north, and finally south. For the foothill alluvial plain (P4), the average center is also typically found in the middle of the subregion, moving first west and north, and ultimately south. In the foothill alluvial plain (P5), the average center, located in the western part of the subregion, first moves south, then north. In the alluvial plain (A), the average

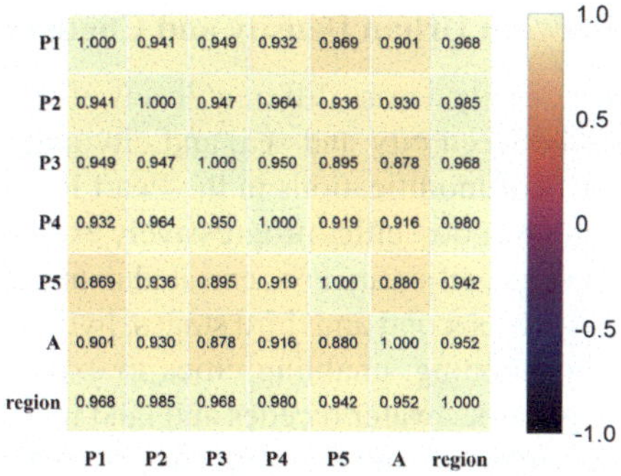

Fig. 6. Matrix of correlation coefficients for the time series of city counts across various subregions and the overall region (221 BCE to 1911 CE).

center's geographical change starts in the northern part of the subregion and then moves south.

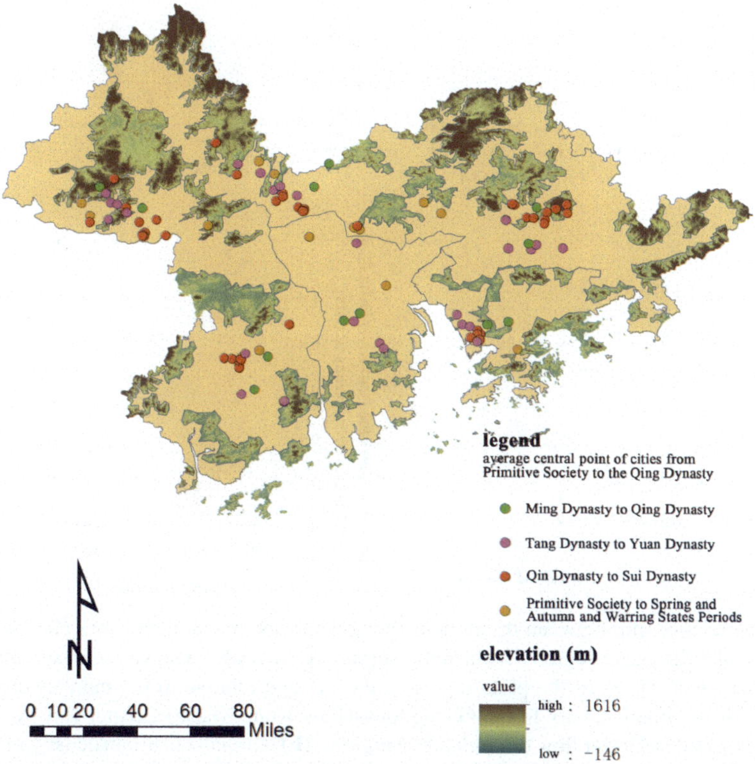

Fig. 7. Average Central Distribution Map of Urban Patterns from Primitive Society to the Qing Dynasty

4.5 The Relationship Between Urban Density and Changes in Water Systems

Box plots illustrating urban density across different levels of water system changes help elucidate the relationships between city and sea, and city and river in the Pearl River Delta. Regarding the impacts of modifications in the Pearl River, the prevailing pattern shows that as the river's water recedes, cities progressively expand into the areas adjacent to water bodies. In the first stage, cities are concentrated in areas influenced by the main channel of the Pearl River. In the second and third stages, hydraulic engineering projects are implemented to prevent flooding, enabling cities to expand into areas previously inundated. In the fourth stage, as seawater recedes and land increases, the distribution of cities shows a significant correlation with the frequency of river changes (Fig. 8a) and a stronger correlation with the expansion of the delta plain (Fig. 8b).

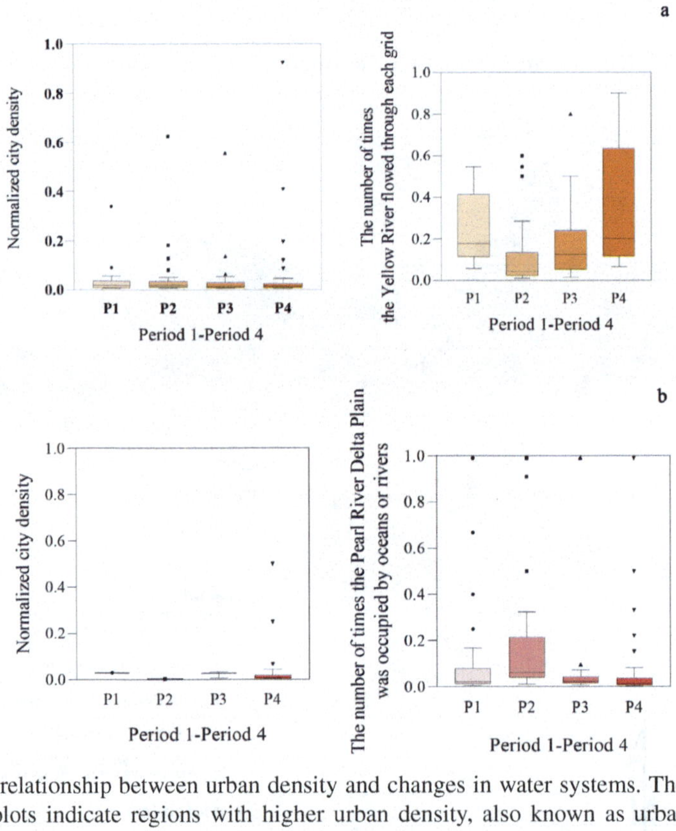

Fig. 8. The relationship between urban density and changes in water systems. The high values in the box plots indicate regions with higher urban density, also known as urban core areas, whereas the mean values in the plots represent the average relative to the number of changes. (**a**) Standardized urban density in the Guangdong-Hong Kong-Macao Greater Bay Area versus the frequency of Pearl River flow through urban regions. (**b**) Normalized urban density in the Pearl River Delta plain versus the number of times the area is occupied by the sea or lakes.

For the interactions between cities and water on the Pearl River Delta plain, larger changes in land area correspond to higher urban density, demonstrating an "urban advancement, water retreat" evolution pattern. This phenomenon may be due to cities in the study area prioritizing development along water bodies for reasons such as trade and land resources. Consequently, cities in the study area develop rapidly along water bodies, while water areas gradually recede with urban expansion.

4.6 Interaction Characteristics of CWSR with Water

Archaeological sites from the primitive society to the Spring and Autumn period are primarily located in the delta plains where two rivers converge or in the northern part of the delta plains (Fig. 9a). Early cities were generally founded in areas with favorable water and soil conditions, often referred to in historical texts as "urban seeds." During the Spring and Autumn period, cities underwent significant development and expansion. Particularly at the foothills, many cities with high population densities formed a core distribution, creating an urban belt along the foothills. In this period, although historical relics had already appeared, cities did not emerge in this region because most of the Pearl River Delta plain was still occupied by the sea.

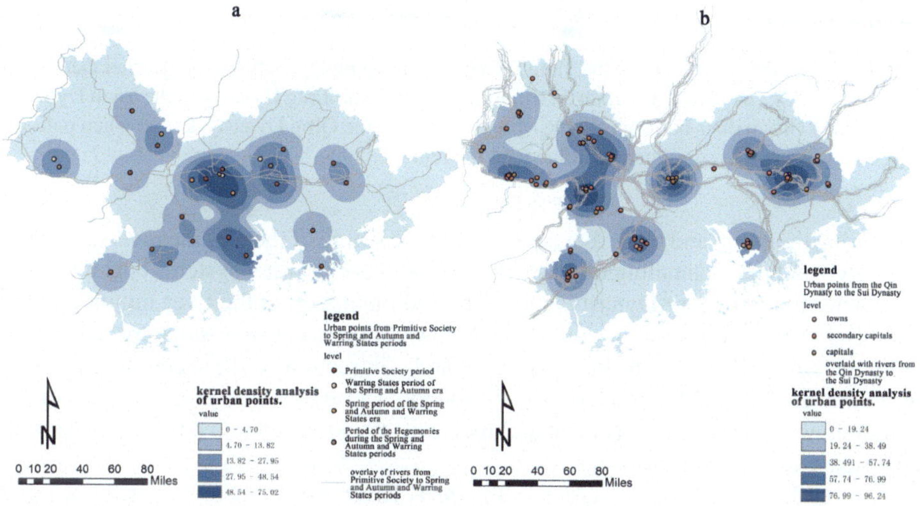

Fig. 9. Urban distribution from primitive societies to the Sui Dynasty and the Kernel Density in different periods. (**a**) Period 1: Main "urban seeds" prior to the Qin Dynasty. (**b**) Period 2: Urban distribution from the Qin Dynasty to the Sui Dynasty.

During the development from the Qin Dynasty to the Sui Dynasty, the Pearl River Delta region remained largely untouched by urban development (Fig. 9b). Throughout this period, the urban development in the Pearl River basin was influenced by a variety of factors including natural geography, social history, politics, and economics [14]. These factors contributed to the uneven urban development within the river basin, with cities along the river relatively underdeveloped. However, with the establishment of the Song Dynasty, the migration of populations from the north, and the further southward shift of the economic center, transportation in the Pearl River basin improved. Consequently, cities along the river gradually began to rise and develop (Fig. 10a).

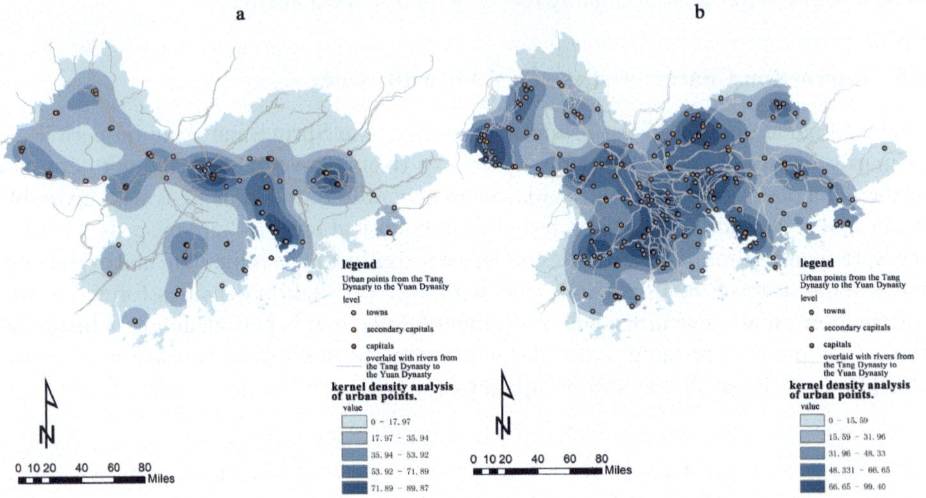

Fig. 10. Urban distribution from the Tang Dynasty to the Qing Dynasty and the Kernel Density in different periods. (**a**) Period 3: Urban distribution spanning the Tang Dynasty to the Yuan Dynasty. (**b**) Period 4: Urban distribution from the Ming Dynasty to the Qing Dynasty, a time when policies and economic growth significantly influenced the region.

During the Ming and Qing dynasties, due to the silting of the bay, the reversal of land and water, and the channeling of water flows, a network of water systems began to form [14]. After the late Qing Dynasty, due to the sedimentation at the river mouth and the extension of the river system, the gradient slowed down, causing sediment deposition along the river, which led to the narrowing and contraction of the water system. This eventually established the modern water network. Through a long process of historical evolution, what were once bays and archipelagoes transformed into a typical delta with a dense network of waterways [2].

During the Qing Dynasty, cities in the Pearl River basin entered a period of rapid development (Fig. 10b). With the growth of the commodity economy and improvements in transportation conditions, urban areas within the Pearl River basin experienced significant development, characterized primarily by an increase in the number of cities and the expansion of urban sizes.

After the founding of the People's Republic of China, numerous hydraulic engineering projects were implemented to stabilize river channels and mitigate the risk of urban flooding. Concurrently, rapid industrialization and urbanization facilitated the aggregation of large populations, industries, and buildings in cities. In the Guangdong-Hong Kong-Macao Greater Bay Area, many cities expanded urban construction land using large-scale, low-cost methods, such as filling in river channels and fish ponds, occupying waterfront spaces, and altering the hydrological environment. These practices have significantly impacted the coordination of urban production, living, and ecological spaces, thereby constraining sustainable economic, social, and environmental development (Fig. 11).

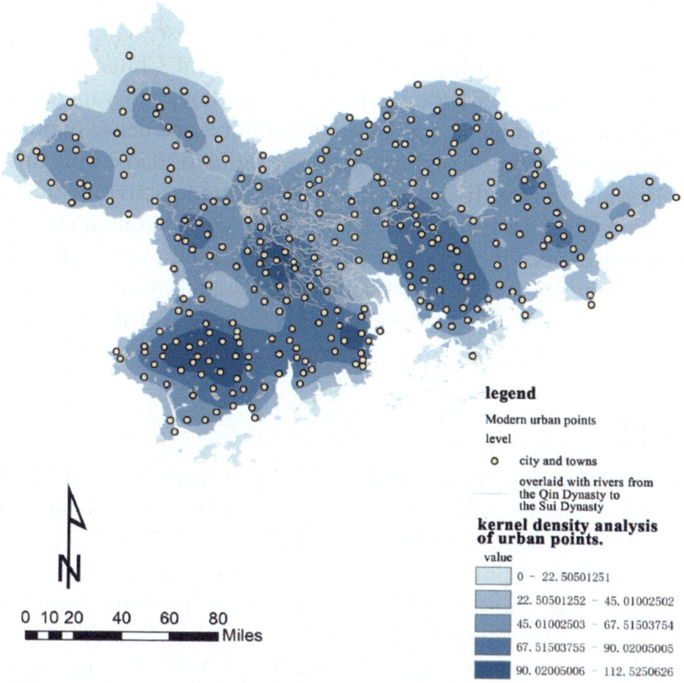

Fig. 11. Period 5: Urban Distribution in 2010, characterized by rapid industrialization and urbanization.

5 Discussion

Considering the aforementioned factors affecting the city-water system relationship, this research utilizes extensive historical urban data at the basin level to explore the City-Water System Relationship (CWSR) in the Guangdong-Hong Kong-Macao Greater Bay Area from an evolutionary perspective. The discussion attempts to address two questions:

In the context of the changing relationship between cities and water systems, this study discusses the evolutionary characteristics of the City-Water System Relationship (CWSR) in the Guangdong-Hong Kong-Macao Greater Bay Area over an extended period, including phase divisions and key driving factors. The Greater Bay Area is located in the downstream reaches of the Pearl River system, where the terrain consists largely of gentle alluvial plains. The downstream terrain of the Pearl River basin is flat and predominantly plain. The Tropic of Cancer passes through the center of the basin, and most of the area experiences a subtropical climate with ample sunshine, warm and humid conditions, abundant water resources, and rich natural products, making it highly suitable for agricultural development.

Water-adjacent regions are more expansive and offer superior conditions for city-building, with convenient water and land transportation. Therefore, the distribution of towns along the Pearl River exhibits two different patterns: one consists of towns connected by rivers, and the other consists of riverbank towns built along the river [16]. With the arrival of the Qianlong era during the Qing Dynasty, Guangzhou, as the sole port open to foreign trade and the foreign trade center of China at the time, and its surrounding key tributaries along the river offered excellent geographical conditions, a favorable ecological environment, developed agriculture, and dense population. Consequently, the urban agglomeration of the Pearl River Delta formed during this period. The open-trade policy and the bustling import and export trade activities further propelled the rapid development of the Pearl River Delta urban agglomeration [17]. Compared to the closely linked riverside cities of the Pearl River Delta, cities in the foothill urban belt have shrunk, indicating that changes in river channels and the development of economic and political systems have disrupted the original structure of the water system. This reflects the siphon effect of urban growth corridors.

6 Conclusion

In the foothill urban belt, early cities developed independently and rapidly adapted to the basin environment, resulting in the absence of a central city within the basin. The primary drivers of the City-Water System Relationship (CWSR) initially included natural factors (such as rivers and topography) and human efforts (including hydraulic engineering, transportation, economics, and trade). With a notable population surge starting in the sixteenth century and especially since the 1950s, there has been a significant increase in the demand for agricultural and industrial water. In the CWSR, water systems initially controlled the urban layout. However, as human impacts increased, urban development reshaped the water systems, although the original conditions of these systems still play a crucial role.

Regarding the evolutionary process of the Guangdong-Hong Kong-Macao Greater Bay Area, four interactive processes have been identified that illustrate the interconnected relationship between urban areas and water systems in the region. Their key characteristics are as follows:

The first process involves the geomorphological structure of the basin, which shaped the origins of cities and significantly influenced their subsequent development. During the early stages of human civilization, most settlements were established on the relatively higher elevations of the foothill alluvial plains. These higher foothill alluvial plains were formed by the accumulation of river flood deposits and alluvial fans, provided favorable soil conditions for agriculture. Additionally, these cities were located in the upper segments of river systems, where channels were comparatively stable, and changes in river courses were minimal. As a result, these regions became known as "urban seeds" for early cities. Early settlements sought out these geographic locations, where the soil was fertile, water supply was ample, and flooding was less frequent, making these areas breeding grounds for early urban development. These geographical positions were consistently occupied by densely distributed cities.

The second process involves the expansion of cities into broader areas and the construction of hydraulic engineering projects to prevent flooding and enhance irrigation. As human civilization progressed, especially with the increased capability to alter natural environments, the edge effects that once significantly influenced settlement patterns have become less pronounced. Consequently, humans have begun to explore and settle in areas with relatively poorer geographical conditions. This shift reflects a transition from purely adapting to the natural landscape to actively modifying it to suit human needs, allowing for the development of urban areas in less naturally favorable locations.

The third process involves the significant influence of river dynamics on urban distribution. The gradual stabilization of delta plains has influenced numerous cities, leading to shifts in their distribution over time. This trend is distinctly evident in the relocation of the average urban center following the Song Dynasty. As the delta plains stabilized, the emergence and development of cities migrated towards the southern parts of the region. This shift highlights the adaptability of urban planning to changing geographical conditions, with urban growth following the natural sedimentation and stabilization patterns of the deltas, enabling new areas to become suitable for development and settlement.

The fourth process is marked by economic development and policy implementation. During the Ming Dynasty, the rise of Cantonese merchants and the subsequent implementation of the Qing Dynasty's Canton System greatly elevated the economic and political status of Guangzhou. This, in turn, spurred rapid urban development across the stabilized delta plains. This can be seen as a pivotal turning point in the region's city-water relationships, representing a lengthy transition from the rapid rise of Cantonese commerce during the Ming Dynasty to the peak of urban development facilitated by the Canton System during the Qing Dynasty. This period highlights how economic policies and commercial growth directly influenced urban expansion, especially in strategically important locations like Guangzhou, which became central hubs due to their enhanced access to trade and resources.

In summary, the geomorphological structure of the study area significantly influences the origins and subsequent development of cities. Initially, cities are established in regions with favorable agricultural conditions before expanding into areas with less promising agricultural potential. Following the gradual formation of the Pearl River Delta alluvial plain, a notable shift in urban distribution occurred, establishing a new equilibrium. In recent decades, the rise of megacities has prompted further alterations in the water systems. These dynamic interactions have fostered the co-evolution of urban development and water systems, demonstrating the intricate relationship between environmental factors and urban growth.

References

1. Du, L., Peng, X., Zhao, Y., et al.: From nature-based to engineering-based: the interaction process and turning point of the city-water system relationship in the North China plain. Appl. Geogr. **135** (2021)
2. Zhou, Y.H.: Historical evolution of the water system in the Pearl River delta. Fudan J. Human. Soc. Sci.. Special Issue. **1**, 85–95 (1980)
3. Fang, Y., Jawitz, J.: The evolution of human population distance to water in the USA from 1790 to 2010. Nat. Commun. **10** (2019)

4. Qingzhou, W.: The protection of China's ancient cities from flood damage. Disasters. **13**(3), 193–227 (1989)
5. Skinner, G.W., Ye, G., Chen, Q.: The Late Imperial China's Cities. Zhonghua Book Company, Beijing (2000)
6. Kang, Y.L., Liang, Q.O., Zhang, T., et al.: Study on the changes of the water system and the Spatio-temporal distribution of towns in Zhejiang during the historical period. J. Huzhou Teach. Coll. **41**(5), 59–67 (2019)
7. Sugihara, G., May, R., Ye, H., et al.: Detecting causality in complex ecosystems. Science. **338**(6106), 496–500 (2012)
8. Bodin, Ö., Alexander, S., Baggio, J., et al.: Improving network approaches to the study of complex social-ecological interdependencies. Nat. Sustain. **2**(7), 551–559 (2019)
9. http://srtm.csi.cgiar.org/SELECTION/inputCoord.asp. Accessed 8 March 2023
10. Tan, Q.: Atlas of Historical Maps of China, 1st edn. China Map Publishing House: Distributed by Xinhua Bookstore Beijing Distribution Branch, Beijing (1982)
11. http://www.webmap.cn. Accessed 8 March 2023
12. Yuan, Z.L.: Special Studies on the Population History of Ancient China, 1st edn. Zhongzhou Ancient Books Publishing House, Zhengzhou (1994)
13. He, Y.M., Lu, Y.S.: Spatial distribution and characteristics of cities along the Pearl River during the Qing dynasty. J. Hist. Stud. **5**, 29–47 (2021)
14. Tan, X.M.: Research on the historical evolution of the water environment in the Haihe River basin and its main influencing factors. J. Water Resour. Dev. Res. **12**, 15–20 (2002)
15. Wang, X.H., Dong, W.: An analysis of urban spatial forms in the early historical period of the Ganges River basin in India. J. Archit. His. **01**, 125–134 (2021)
16. Atakara, C., Allahmoradi, M.: Investigating the urban spatial growth by using space syntax and GIS: a case study of Famagusta City. ISPRS Int. J. Geo Inf. **10**(10) (2021)
17. Al-Jabri, M.K.A., Ladik, E.I.: Development stages of planning thought and factors affecting the morphology of modern iraqi cities. IOP Conf. Ser. Earth Environ. Sci. **1129**(1), 012013., IOP Publishing (2023)

Effects of Phosphorus and Cadmium Fractions in Soil on Phosphorus and Cadmium Contents in Bermudagrass in the Water-Level-Fluctuating Zone of the Three Gorges Reservoir Area

Ze Luo[1,2] and Yanxue Jiang[1,2](✉)

[1] Key Laboratory of Ecology and Environment of Three Gorges Reservoir Area, School of Environment and Ecology, Chongqing University, Chongqing 400045, China
944155899@qq.com

[2] College of Environment and Ecology, Chongqing University, Chongqing 400045, China

Abstract. Bermuda grass (*Cynodon dactylon* (L.) pers.) and soil samples were collected in the water-level-fluctuating zone (WLFZ) from Changshou to Fuling in the Three Gorges Reservoir Area (TGR), and the fractions of phosphorus (P) and cadmium (Cd) in the soil and the contents of P and Cd in Bermuda grass were determined. The results show that: The Cd and P content in the soils and the Bermuda grass of the WLFZ from Changshou to Fuling in the TGR increased first and then decreased along the river. The main fractions of Cd in the soil samples were Fe/Mn-Cd and Car-Cd. P fractions at different sites showed a homogeneous trend, which followed the order of HCl-P > Re-P > NaOH-IP > NaOH-OP>NaHCO3-IP > Water-P ≈ $NaHCO_3$-OP. The distribution of Cd content in Bermuda grass was root>stem>leaf, while that in Bermuda grass was stem > leaf > root. The root of Bermuda grass had higher Cd enrichment coefficient and lower migration coefficient The results of Mantal test and Spearman correlation analysis showed that the content of Cd in Bermuda grass was significantly correlated with the content of Fe/Mn-Cd, TCd and $NaHCO_3$-IP; The P content in Bermuda grass was significantly correlated with TCd content, $NaHCO_3$-IP content, Ex-Cd content, Fe/Mn-Cd content, Water-P and NaOH-OP content.

Keywords: Phosphorus · Cadmium · The Water-level-fluctuating Zone · The Three Gorges Reservoir Area · Bermuda grass

1 Introduction

As a transitional zone between terrestrial and aquatic ecosystems, the Water-level-fluctuating Zone (WLFZ) has high spatial and temporal heterogeneity in its habitats, and has unique material exchange, energy transfer and life processes, which shapes the coupled cycle of phosphorus (P) and cadmium (Cd) different from terrestrial and aquatic habitats. Plants in WLFZ are important ecological barriers to intercept P pollution load and accumulate heavy metal Cd in runoff, and play an important role in protecting water

V. Achal et al. (Eds.): EPPC 2024, ESE, pp. 23–32, 2025.
https://doi.org/10.1007/978-3-031-92123-0_2

quality safety in reservoir area [1]. Bermudagrass (*Cynodon dactylon* (L.) pers.) is a perennial herbaceous plant belonging to the genus bermudagrass of Gramineae, which has a good ability to tolerate Cd and accumulate P and Cd, and participates in the important links of material cycle and energy flow of the reservoir ecosystem under the soil background of the WLFZ of the Three Gorges Reservoir Area (TGR) with high P and Cd content can absorb energy from sunlight and convert P obtained from the soil into biological organic matter, thereby reducing the P load into the water body and protecting the water body [2, 3]. Meanwhile, Bermuda grass can reduce the risk of Cd entering the food chain because of its good Cd root enrichment capacity and tolerance [4]. The content of P in soil is an important factor affecting the accumulation and transportation of Cd in Bermuda grass. Chen [2] found that available P was important soil ecological factors to enhance Cd absorption and translocation of bermudagrass. Tan [3] reported that Bermuda grass can schedule the P allocation to adapt to the Cd absorption and enrichment in different organs under different Cd concentrations. However, there is no report on the forms of P and Cd in the WLFZ of the TGR and the absorption mechanism of in-situ Bermuda grass in the TGR. In this paper, the P and Cd forms in soil and the P content in Bermuda grass were determined in the WLFZ from Fuling to Changshou in the TGR, which is rich in P and seriously polluted by Cd, and the accumulation characteristics of Cd in Bermuda grass were predicted, so as to provide a scientific basis for the ecological restoration of Cd and the reduction of water eutrophication by Bermuda grass in the TGR.

2 Materials and Methods

2.1 Sample Collection

The soil and plant samples were collected from 5 sampling sites in the WLFZ from Changshou to Fuling in the TGR in February 2023. The locations of the sampling sites are shown in Fig. 1. All the sampling sites are gentle slopes, and the soil type is purple soil. The soil sample was taken by quartering and placed in a polyethylene self-sealing bag.

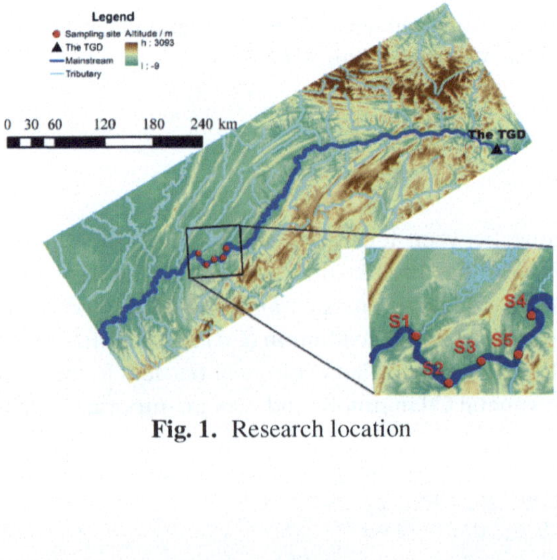

Fig. 1. Research location

After carefully removing plant residues, fine roots, and gravel from the soil, uniformly mix and reduce to about 1 kg. After lyophilization, it was milled through a 100-mesh sieve and stored in a refrigerator at −20 °C. The collected herbaceous plants were taken back to the laboratory and rinsed with tap water, and then rinsed with pure water for 2–3 times, and then de-enzymed at 105 °C for 30 min, and then dried at 65 °C to constant weight. After that, the roots, stems and leaves were separated, crushed by a grinder and a mortar, and milled through a 60-mesh sieve, and then stored at room temperature in a ventilated and dry place.

2.2 Measurements

2.2.1 Fractional Extraction of Soil Phosphorus Forms

The fractions of Cd in soil was determined by Tessier sequential extraction method [5]. In brief, 1.0 g of dry soil was put into a 50 ml centrifuge tube, and different extractants were added continuously: (1) 1 m $MaCl_2$ (exchangeable state, Ex-Cd). (2) 1 M NaAC (carbonate bound state, Car-Cd) (3) 0.04 M HAc (Fe/Mn manganese oxide bound state, Fe/Mn-Cd), (4) 0.02 M HNO_3 + 30% H_2O_2 85 °C water bath treatment (organic matter bound state, Org-Cd), (5) HF and concentrated HNO3 digestion (residue state, Re-Cd)). All the extracts were centrifuged at 4000 rpm for 20 minutes, and then the supernatant was filtered through a 0.45 μm filter and sent to inductively coupled plasma mass spectrometry (ICP-MS, NexION 2000P, Perkinelmer) for Cd determination.

2.2.2 Fractional Extraction of Soil Cadmium Forms

The fractions of P in soil were determined by the modified Hedley chemical sequential extraction method [6]. Briefly, 0.5 g of dry soil was placed in a 50 ml centrifuge tube, and different extractants were continuously added: (1) ultrapure water (Water-P); (2) 0.5 M $NaHCO_3$ (weakly adsorbed total and inorganic phosphorus, $NaHCO_3$-TP&IP); (3) 0.1 M NaOH (total and inorganic P bound to Fe/Al oxides, NaOH-TP&IP); (4) 1 M HCl (Ca-bound phosphorus, HCl-P); (5) 4 mL H_2SO_4 and 1 mL $HClO_4$ digestion (residual phosphorus, Re-P). The first four extractions were centrifuged at 4500 rpm for 15 min, and then the supernatant was passed through a 0.45 μm filter. Potassium persulfate was added to the extract of $NaHCO_3$ and NaOH, and then the extract was digested at 121 °C for the determination of $NaHCO_3$-IP and NaOH-IP. $NaHCO_3$-OP/NaOH-OP is the difference between $NaHCO_3$-TP/NaOH-TP and $NaHCO_3$-IP/NaOH-IP in the corresponding extraction solution.

2.2.3 Determination of Phosphorus and Cadmium Content in Different Tissues of Bermuda Grass

P content in plant: 0.5 g pre-dry put into a digestive tube, 5 mL of H_2SO_4 is added, shaken evenly, H_2O_2 is added twice, 2 mL of H_2O_2 is added each time, shaken evenly, covered, placed on a graphite electric heating plate to be heated and digested at 200 °C after the violent reaction is finished, the solid substance disappears and becomes a solution, when the solution turns brown, stop heating, add 10 drops of H_2O_2 after slightly cooling, continue to heat and digest for 5 min, cool, add H_2O_2 for digestion, continue to heat for

5 min after the solution is colorless or clear, take down and cool, and determine the total P of plants by molybdate spectrophotometry.

Cd content in plant: Take 0.5 g of freeze-dried plant sample, and heat and digest the crushed root, stem and leaf samples with the mixed acid graphite plate of concentrated HNO_3 and HClO4 (3:1 v/v). The supernatant was then filtered through a 0.45 μm filter and then into an ICP-MS for Cd determination.

2.3 Data Analysis

One-way analysis of variance (ANOVA), least significant difference test (LSD test) and Tamheini (T2 method) were used to evaluate the differences of soil and plant indicators among different sample sites, and different letters marked in the chart indicated that there were significant differences among the corresponding groups ($p < 0.05$). Mantel test was used for correlation analysis. The above analysis adopts the statistical analysis of "agricolae," "vegan," "minpack. LM" "and other packages of R language (R-4.2.1), and uses the" ggplot2 "package to visualize the above results.

3 Results and Discussion

3.1 Cadmium Forms

As shown in Table 1, the total amount of Cd in the Changshou to Fuling WLFZ of the TGR soil showed a trend of increasing and then decreasing along the river to the direction, which may be due to that the middle sample point was the Fuling urban sample point and there were many factories nearby, and the human activities were large, resulting in a higher Cd content. There were significant differences in the concentrations of various fractions of Cd among the sampling sites. Compared with the upstream sampling sites S1 and S2, the downstream sampling sites S4 and S5 had more Ex-Cd and Car-Cd ($P < 0.05$); However, there were more Org-Cd and Re-Cd in the upper reaches ($P < 0.05$).It indicating that the ecological risk of Cd in the lower reaches was greater, which might be due to the downward migration of Cd along the river in S3 sample site of Fuling City. Fe/Mn-Cd in S3 and S4 sample points is significantly higher than other sample points, reaching 0.283 ± 0.09 and 0.331 ± 0.006 mg/kg, which may be related to the soil parent material [7].

3.2 Phosphorus Forms

As shown in Table 2, the average value of soil total phosphorus (TP) in the WLFZ from Changshou to Fuling in the TGR increases first and then decreases along the river bank. There is a significant difference ($P < 0.05$) in the TP content of each sample point. Compared with the average TP (600 mg/kg, 452.76 mg/kg and 813.3 mg/kg) in the whole watershed of the TGR [8], the Pengxi River Basin [9] and the Daning River Basin [10], the soil TP content in the WLFZ from Changshou to Fuling in the TGR is at a medium level. There was no significant difference between S3 and S2 and S4 ($p > 0.05$), while the contents of other P fractions in S3 were significantly higher than

Table 1. Distribution of cadmium fractions mass concentration in Water-Level-Fluctuating Zone soil

Cd Forms	S1	S2	S3	S4	S5
Ex-Cd	0.024 ± 0.008c	0.034 ± 0.014bc	0.053 ± 0.018ab	0.064 ± 0.009a	0.074 ± 0.009a
Car-Cd	0.164 ± 0.015b	0.132 ± 0.018b	0.152 ± 0.015b	0.163 ± 0.015b	0.265 ± 0.013a
Fe/Mn-Cd	0.161 ± 0.018c	0.092 ± 0.025d	0.283 ± 0.009b	0.331 ± 0.006a	0.134 ± 0.005c
Org-Cd	0.157 ± 0.015a	0.174 ± 0.015a	0.164 ± 0.004a	0.052 ± 0.005c	0.087 ± 0.001b
Re-Cd	0.032 ± 0.009a	0.116 ± 0.013d	0.062 ± 0.004c	0.073 ± 0.001c	0.088 ± 0.002b
TCd	0.538 ± 0.064c	0.548 ± 0.089c	0.714 ± 0.034a	0.683 ± 0.029ab	0.648 ± 0.024b

Table 2. Distribution of phosphorus fractions mass concentration in Water-Level-Fluctuating Zone soil

P Forms	S1	S2	S3	S4	S5
Water-P	5.502 ± 0.361c	3.83 ± 0.339d	20.665 ± 0.499a	9.658 ± 0.696b	6.138 ± 0.396c
$NaHCO_3$-IP	10.23 ± 0.688e	22.734 ± 0.472d	48.121 ± 1.142a	29.717 ± 0.831c	34.444 ± 0.836b
$NaHCO_3$-OP	4.125 ± 0.856b	7.063 ± 1.245ab	7.906 ± 1.141a	6.202 ± 1.016ab	5.888 ± 1.285ab
NaOH-IP	35.1 ± 1.213e	44.276 ± 2.06d	98.526 ± 0.536a	79.436 ± 1.041b	59.158 ± 1.899c
NaOH-OP	29.423 ± 3.852c	17.7 ± 1.911e	79.622 ± 5.402a	50.721 ± 5.316b	49.333 ± 4.389b
HCl-P	201.758 ± 4.187e	352.163 ± 0.619b	442.499 ± 5.334a	352.608 ± 3.891b	322.065 ± 5.504c
Re-P	176.531 ± 4.187b	202.477 ± 0.619a	199.33 ± 5.334a	199.913 ± 3.891a	171.827 ± 5.504b
TP	462.676 ± 13.149e	809.48 ± 8.252b	896.668 ± 16.069a	728.254 ± 14.739c	648.853 ± 15.225d

other sites ($P < 0.05$), and the contents of other P fractions in S1 were lower than other sites ($P < 0.05$) except that the content of NaOH-OP was higher than that in S2.

The different fractions of P in each point showed a homogeneous trend: HCl-P > Re-P > NaOH-IP > NaOH-OP>$NaHCO_3$-IP > Water-P ≈ $NaHCO_3$-OP. Studies have shown that periodic inundation-outcropping will also reduce the coefficient of variation of each P fraction in the soil of the WLFZ between different sections at the same altitude [6]. The interaction between the soil in the WLFZ and the overlying water is also the main reason for the homogenization of P fraction. The difference of P concentration in the water body of the TGR in time is much greater than the change in space, and after continuous submergence and exposure, the soil in the WLFZ repeatedly reacts with the corresponding overlying water, making the two tend to balance [11]. Zhao [12] also found a homogenization trend of soil TP and other chemical properties at different points after long-term operation of the reservoir in the study on the WLFZ of Manwan Reservoir.

3.3 Characteristics of Phosphorus and Cadmium Content in Bermudagrass

It can be seen from Fig. 2a that the distribution of Cd content in Bermuda grass is root > stem > leaf. Cd contents in roots and stems of Bermuda grass at S3 site were significantly higher than other sites ($P < 0.05$); Cd contents in roots, stems and leaves of Bermuda grass at S5 site were significantly lower than other sites ($P < 0.05$); The above results indicate that Bermuda grass isolates Cd in the root and a small amount of Cd migrates to the aboveground part.

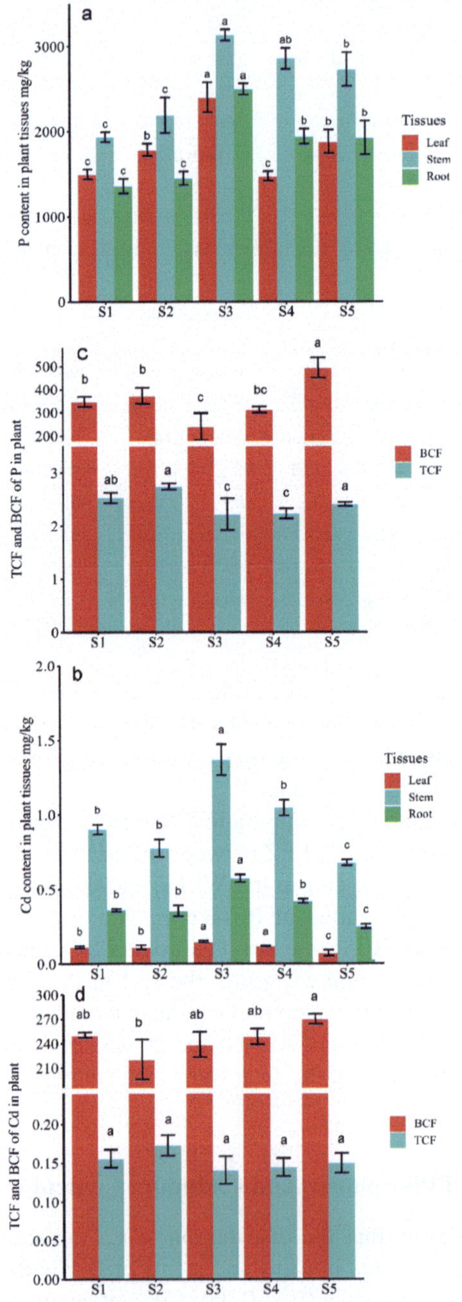

Fig. 2. Distribution, enrichment and transfer coefficient of phosphorus and cadmium in Bermuda grass in Water-Level-Fluctuating Zone soil

It can be seen from Fig. 2b that the distribution of P and Cd in Bermuda grass in different samples is the same, showing a downward trend of first increasing and then decreasing along the river. The P content in the roots and leaves of Bermuda grass in S3 was significantly higher than that in other Bermuda grass samples (P 0.05), and the P content in the stems was significantly higher than that in other Bermuda grass samples except S4 ($P < 0.05$).

It can be seen from Fig. 2c that the P bioconcentration factor (BCF) in the roots of Bermuda grass was very high, exceeding 200, and the BCF and transport coefficient factor (TCF) had no obvious regularity along the river flow direction. The TCF of each sample site was greater than 1, and the TCF of S2 was the largest, reaching 2.7. The BCF of S5 was significantly higher than that of other sample sites ($P < 0.05$); The BCF and enrichment coefficient of S3 sample site were significantly lower than those of other sample sites ($P < 0.05$).

It can be seen from Fig. 2d that the BCF of Cd in the root of Bermuda grass is very high, but the TCF is very low, indicating that Bermuda grass in the TGR can accumulate a large amount of Cd and isolate Cd mainly in the root, and rarely transport Cd to the stem and leaf. There was no significant difference in the BCF of Cd among the sample sites, except that the BCF of S5 was significantly higher than that of S2 ($P < 0.05$); the migration coefficient of Cd in each group was low, and there was no significant difference in the migration coefficient among the sample sites ($P > 0.05$).

3.4 Characteristics of Phosphorus and Cadmium Content in Bermudagrass

As shown in Fig. 3, the results of Mantel analysis showed that the Cd content in Bermuda grass was significantly correlated with the content of Fe/Mn-Cd and TCd ($P < 0.01$), and also significantly correlated with the content of $NaHCO_3$-IP ($P < 0.05$); The P content in Bermuda grass was significantly correlated with TCd content, $NaHCO_3$-IP content and NaOH-OP content in soil ($P < 0.001$), and also significantly correlated with Ex-Cd content, Fe/Mn-Cd and Water-P content ($P < 0.05$). Spearman correlation analysis showed that Ex-Cd was significantly negatively correlated with Org-Cd ($P < 0.05$), and significantly positively correlated with $NaHCO_3$-IP ($P < 0.01$), while Fe/Mn-Cd was significantly positively correlated with TCd ($P < 0.01$); The contents of NaOH-IP and Water-P were significantly positively correlated with the contents of NaOH-OP ($P < 0.001$) and NaOH-IP ($P < 0.05$), and the contents of HCl-P and Re-P were significantly positively correlated ($P < 0.05$); TP content was positively correlated with HCl-P and Re-P ($P < 0.001$), and positively correlated with NaOH-OP ($P < 0.05$).

The fractions of P and Cd in soil affected the content of P in Bermuda grass. Flooding will lead to the mutual transformation of various fractions of P, resulting in more bioavailable fractions of P. In an anaerobic environment, NaOH-IP is converted into Water-P and released into the overlying Water [13]. In addition, NaOH-IP is also considered to be an intermediate form between Water-P and NaOH-OP [14], and the soil in the WLFZ experiences repeated periodic submerging-outcropping, which promotes the mutual transformation between them. Therefore, NaOH-IP was positively correlated with $NaHCO_3$-IP and NaOH-OP. The pH value of the water in the TGR is weak alkaline, and it is almost impossible to release HCl-P in such an environment. Flooding led to the formation of a reducing environment in the WLFZ soil, which was conducive to

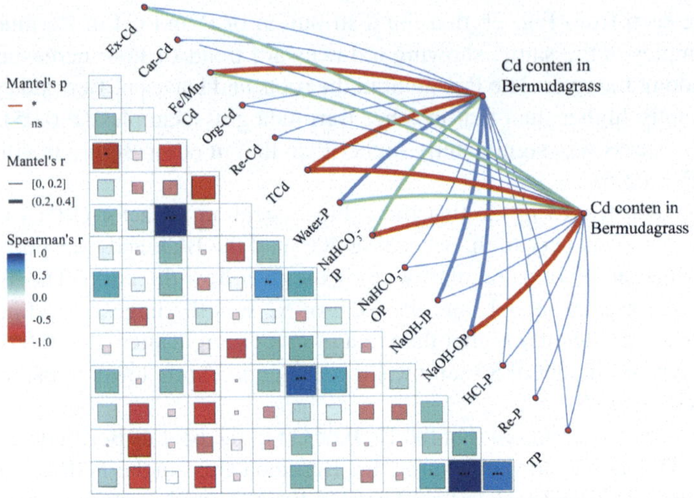

Fig. 3. Correlation analysis of phosphorus and cadmium forms in soils of the Water-Level-Fluctuating Zone and their mantel test with phosphorus and cadmium content of bermudagrass. *, * * *and* * * * *in the figure indicate that the significant levels of the correlation between the two indicators are* $p < 0.05$, $p < 0.01$ *and* $p < 0.001$

the combination of free forms with Ca^{2+} in interstitial water and the transformation into HCl-P [15], resulting in HCl-P in the soil becoming the main P form in the TGR, so there was a significant positive correlation between HCl-P and TP. Studies have shown that Bermuda grass has a low enrichment rate of active P and a high enrichment rate of stable organic P, so there is a significant positive correlation between P content in Bermuda grass and NaOH-OP [16]; Water-P and $NaHCO_3$-IP are typical forms that can be used by plants [17]. Flooding and drying conditions significantly increase the bioavailable P content in the WLFZ, resulting in a significant positive correlation between the P content in Bermuda grass and Water-P and $NaHCO_3$-IP; Studies have shown that low concentration of Cd promotes the P absorption of plants, because Cd stress mobilizes the P absorption and distribution of Bermuda grass to be lower than the toxicity [18], and the phosphorus and cadmium in the soil of the WLFZ are correlated [19], resulting in the correlation between P in Bermuda grass and various forms of Cd.

The Ex-Cd and Car-Cd, which are easy to absorb, are not correlated with the Cd content in Bermuda grass, but interestingly, the Fe/Mn-Cd were significantly positively correlated with the Cd content in Bermuda grass. Fe/Mn-Cd are non-noble metal oxides, which have abundant surface active sites and fast electron transfer ability, and have strong binding energy with Cd. From Sect. 3.1, it can be seen that this is the main existence fraction of most samples. Soil flooding results in the reduction of Fe/Mn-Cd to Ex-Cd or Car-Cd and an increase in soluble Fe and Mn, which are oxidized back to insoluble oxidized components once oxygen is able to diffuse back into the soil as the flood recedes [20]. The sampling period of this experiment was in the drying period, so it was speculated that the Fe/Mn-Cd reduced by flooding might be reduced and transferred to

Bermuda grass during flooding, and then re-fixed in the soil during drying, resulting in a significant correlation between the soil Fe/Mn-Cd and Bermuda grass.

4 Conclusions

1. The Cd content in the soils of the WLFZ from Changshou to Fuling in the TGR increased first and then decreased along the river. The main fractions of Cd in the samples were Fe/Mn-Cd and Car-Cd. Compared with the upstream samples, the downstream samples had more Ex-Cd and Car-Cd, and the upstream samples had more Org-Cd and Re-Cd. It indicates that there is a greater risk of Cd release downstream.
2. The TP content in the soil of the WLFZ from Changshou to Fuling in the TGR also showed a trend of increasing first and then decreasing along the river direction.TP were at the medium level in the WLFZ of the TGR. P fractions at different sites showed a homogeneous trend, which followed the order of HCl-P > Re-P > NaOH-IP > NaOH-OP>$NaHCO_3$-IP > Water-P $\approx$ $NaHCO_3$-OP.
3. The contents of P and Cd in Bermuda grass from Changshou to Fuling in the TGR increased first and then decreased along the river direction. The distribution of Cd content in Bermuda grass was root>stem> eaf, while that in Bermuda grass was stem>leaf>root. The root of Bermuda grass had higher Cd enrichment coefficient and lower migration coefficient, which reduced the toxicity of Cd stress.
4. The results of Mantal test and Spearman correlation analysis showed that the content of Cd in Bermuda grass was significantly correlated with the content of Fe/Mn-Cd, TCd and $NaHCO_3$-IP; The P content in Bermuda grass was significantly correlated with TCd content, $NaHCO_3$-IP content, Ex-Cd content, Fe/Mn-Cd content, Water-P and NaOH-OP content. The correlation results showed that Ex-Cd was significantly negatively correlated with Fe/Mn-Cd, and significantly positively correlated with $NaHCO_3$-IP.

Acknowledgments. This work was supported by the National Natural Science Foundation of China (Grant No. 32001195), and the Fundamental Research Funds for the Central Universities (Grant No. 2019CDQYCH011).

References

1. Bing, H.J., Wu, Y.H., Zhou, J.: Spatial variation of heavy metal contamination in the riparian sediments after two-year flow regulation in the three gorges reservoir, China. J. Sci. Total Environ. **649**, 1004–1016 (2019)
2. Chen, B., Tan, S.D., Zeng, Q.R.: Soil nutrient heterogeneity affects the accumulation and transfer of cadmium in Bermuda grass (Cynodon dactylon (l.) pers.). J. Chemosphere. **221**, 342–348 (2019)
3. Tan, S.S., Liu, Z.S., Zeng, Q.R.: Nutrient allocation might affect the cadmium accumulation of Bermuda grass (Cynodon dactylon). J. Chemosphere. **252**, 274–281 (2020)
4. Yang, Y.T., Tan, S.D., Dong, F.X.: Ecophysiological response of Hainan wild bermudagrass to cadmium stress. J. Southwest Agric. **31**, 2086–2091 (2018)

5. Tessier, A., Campbell, P.G.C., Bisson, M.: Sequential extraction procedure for the speciation of particulate trace-metals. Anal. Chem. **51**, 844–851 (1979)
6. Chao, W.: Phosphorus Forms and Release Risk in the Water-Level-Fluctuating Zone of Tributaries in the Three Gorges Reservoir Area. D. Chongqing University (2022)
7. Zhong, Z., Bing, H.J., Wu, Y.H.: Distribution characteristics of cadmium and its speciation in soils at different elevations in the water-level-fluctuating zone of Fengdu-Zhongxian section of the Three Gorges Reservoir. Lake Sci. **31**, 1601–1611 (2019)
8. Li, S., Chen, M., Wang, Y.C.: Spatial and temporal distribution characteristics of soil nitrogen and phosphorus the water-level-fluctuating zone of the Three Gorges Reservoir in recent 10 years. Environ. Sci. **33**, 2448–2457 (2020)
9. Feng, L.: Phosphorus Forms and Adsorption Characteristics of Soil in the Water-Level-Fluctuating Zone of the Pengxi River Basin in the Three Gorges Reservoir Area. D. Chongqing University (2017)
10. Xu, D.X., Qin, Y.W., Zhang, L.: Phosphorus forms and their distribution characteristics in sediments and soils of water-level-fluctuating zone in the backwater area of the Daning river, the inflow river of the Three Gorges Reservoir. Environ Sci. **30**, 1337–1344 (2009)
11. Tang, J.L., Wang, T., Zhu, B.: Tempo-spatial analysis of water quality in tributary bays of the Three Gorges Reservoir region (China). J. Environ. Sci. Pollut. Res. **22**(21), 16709–16720 (2015)
12. Zhao, Q., Liu, S., Deng, L.: Soil degradation associated with water-level fluctuations in the Manwan Reservoir, Lancang River Basin. J. Catena. **113**, 226–235 (2014)
13. Jalali, M., Matin, N.H.: Soil phosphorus forms and their variations in selected paddy soils of Iran. J. Environ. Monit. Assess. **185**, 8557–8565 (2013)
14. Zhang, T.: Chemical Behavior of Phosphorus over Time in Fertilized Soils. ProQuest Dissertations Publishing, Canada (1997)
15. Hupfer, M., Zak, D., Rossberg, R.: Evaluation of a well-established sequential phosphorus fractionation technique for use in calcite-rich lake sediments: identification and prevention of artifacts due to apatite formation. J. Limnol. Oceanogr. Methods. **7**, 399–410 (2009)
16. Wang, X.F., Liu, T.T., Gong, X.J.: Characteristics of phosphorus forms in rhizosphere soil of typical plants in hydro-fluctuation belt of Three Gorges Reservoir area. J. Acta Ecol. Sin. **40**, 1342–1356 (2020)
17. Li, X.T., Ye, D.H., Zhang, X.Z.: Research progress on response characteristics of plants to different forms of phosphorus. J. Plant Nutrit. Fertil. **23**, 1536–1546 (2017)
18. Li, H.Y., Su, Y.H.: Effects of cadmium on tolerance physiology and nutrient absorption and accumulation of grain amaranth. J. Ecol. Environ. **21**, 308–313 (2012)
19. Tang, J.: Migration, Enrichment and Transformation of Cadmium and Other Heavy Metal Elements in the Three Gorges Reservoir Area. D. Chengdu University of Technology (2006)
20. Wang, M., Wang, L.F., Zhao, S.W.: Manganese facilitates cadmium stabilization through physicochemical dynamics and amino acid accumulation in rice rhizosphere under flood-associated low pe plus pH. J. Hazard. Mater. **416**, 304–318 (2021)

Research Hot Spots and Evolutionary Trends of Ecological Compensation Transfer Payments-Analysis Based on CiteSpace Knowledge Graph

Qianqian Du, Yuhong Tang(✉), and Yuanyuan Li

Tibet University, Lhasa 850000, China
duqianqian2020@163.com

Abstract. This paper uses more than 4500 pieces of ecological compensation transfer payment literature from the WOS and CNKI databases as the foundation for understanding the hot spots and emerging trends in the field. CiteSpace software is used to precisely draw keyword highlighting, keyword clustering, and other related knowledge maps, as well as to analyze the annual publication of literature, research authors, research institutions, country distribution, and research on ecological compensation transfer payments. The annual publication of literature, research authors, research institutions, national distribution, research hot spots, frontier concerns, etc. were all subjected to a scientific bibliometric study. ① The findings indicate that the annual output of literature has basically progressed through three stages: beginning, developing, and maturing. ② The United States, China, England, Australia, and other nations make up the majority of the writers and institutions cited in the literature, which serves as the foundation for ecological compensatory transfer payments. ③ Research hot zones include the value of environmental services, fiscal policy, transfer payments, and ecological compensation. The primary areas of frontier research are payment for environmental services, ecological compensation methods, and important ecological function regions.

Keywords: ecological compensation · transfer payments · CiteSpace

1 Introduction

The ecological basis for sustainable social and economic growth has been significantly threatened in recent years by global environmental degradation, ecological destruction, and resource scarcity. Every nation must decide whether to pursue economic development or ecological environmental protection in order to solve the "double-choice" question and create a better world. With the advancement of ecological civilization, ecological compensation has drawn the interest of many academics as an economic tool to support ecological environmental protection. As an economic means of promoting ecological environmental protection, ecological compensation has attracted the attention of many scholars at home and abroad as the construction of ecological civilisation advances. The exploration of ecological compensation began with the British

V. Achal et al. (Eds.): EPPC 2024, ESE, pp. 33–43, 2025.
https://doi.org/10.1007/978-3-031-92123-0_3

ecologist Tansley, who gave an accurate definition of ecosystem: in a specific space, by virtue of the role of material circulation, coupled with the flow of energy, biological and non-biological components affect each other, and depend on each other to form a unique ecological functional unit. Based on the effective compensation of solar energy, ecosystems are basically able to maintain stability through self-regulation, and realise the development process from simple to complex with the flow of energy and material circulation. There are some differences in the definition of ecological compensation given by different researchers, some scholars call ecological compensation ecosystem service compensation, some scholars call ecological compensation environmental service compensation, and some scholars call it payment for environmental services and so on. The study focuses on the definition of eco-compensation, the subject and object of compensation, compensation methods, eco-compensation policies, eco-compensation mechanisms, eco-compensation effects, and the relationship between eco-compensation and transfer payments (Turner and Van Den Bergh et al., 2000 [1]; Persson and Alpízar, 2013 [2]; Hatton MacDonald and Bark et al., 2014 [3]; Selva and Pauli et al., 2019 [4]).

To gain a deeper understanding of the focal areas and emerging trends in research on ecological compensation transfer payments, this study employs CiteSpace software to visualize and analyze the authors, institutions, country distribution, hotspots, and frontiers of both domestic and international studies. Furthermore, it explores future research directions that warrant attention to contribute to global investigations on ecological compensation transfer payments as well as advancements in ecological environmental protection, development, and resource utilization.

2 Research Methodology and Data Sources

CiteSpace is bibliometric software built on the Java programming language that assesses the state of the field's research frontiers by producing a number of graphic maps that highlight recently developed patterns and advancements. The authors, contents, and hot spots of both national and international ecological compensation transfer research are visualized and analyzed in this work using CiteSpace 5.7.R5 software's keyword co-occurrence and emergence, time zone perspective mapping, as well as additional modules. The CNKI was used to search and collect domestic literature materials, and 409 records were obtained after "ecological compensation" and "transfer payment" were selected as keywords in the search and the WOS database was used to search and collect literature by selecting "payment for environmental services," "compensation for ecosystem services," "compensation for environmental services," "payment for ecosystem services," "compensation and reward for environmental services" and "financial transfer payment." The keywords "transfer payment from the exchequer" were filtered and the language was set to English, and 4419 records were obtained after filtering. The search period was from 2004 to 2023.

3 Results and Analysis

3.1 Annual Distribution of Literature

The number of topics on ecological compensation transfer payments released each year shows an overall increasing trend (Fig. 1). From 2004 to 2008, the total number of articles published per year was no more than 50. From 2008 to 2018, the number of published literatures increased year by year. Since 2018, more than 300 papers have been published every year, and research on ecological compensation transfer payments has gradually matured.

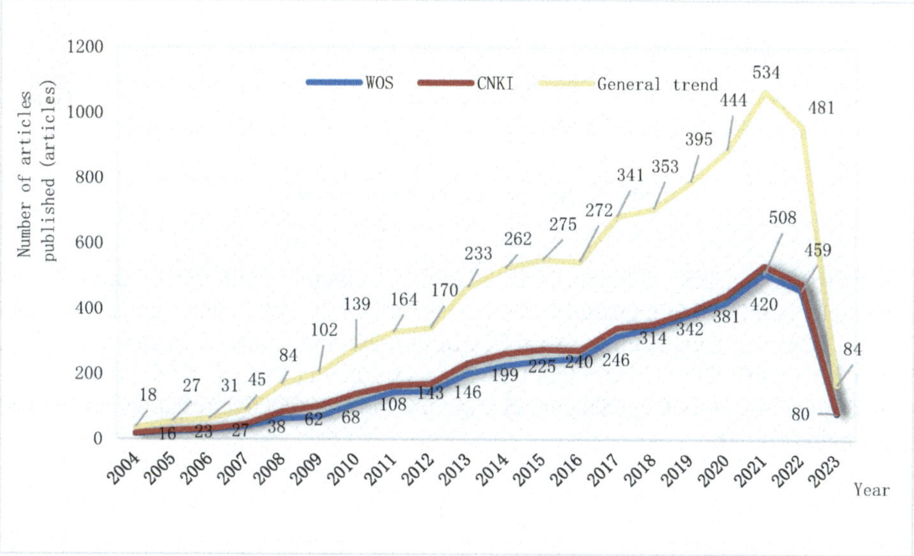

Fig. 1. Annual distribution of ecological compensation transfer payment documents

3.2 Distribution of Study Authors, Institutions, and Countries

Researchers play an irreplaceable role as the core force of discipline construction and development. In CiteSpace, we set the node type as "Author" and the time slice as 1 year respectively. In the CNKI database, analysed from the perspective of high-profile authors, Ren Junlin, Kuang Yang, Zhang Yi, Liu Guihuan, Li Guoping, Zhang Wenbin, Zhang Dongmei, Dong Zhanfeng, Zhang Anlu, and Wang Jinnan are the high-producing authors of this field. In the WOS database, analysed in terms of high frequency authors, SVEN WUNDER, BETTINA MATZDORF and JULIA P G JONES are the most prolific authors in the field (Fig. 2).

In CiteSpace, set the node type as "Institution" and the time slice as 1 year respectively. In the CNKI database search, a total of 308 institutions were involved, with the China Academy of Fiscal Science, Wuhan Textile University Accounting School, and

Fig. 2. Author co-occurrence mapping

Yangtze River Academy of Sciences as the research institutes with rich research results in this field. A total of 596 institutions were involved in the WOS database search, with Chinese Acad Sci, Univ Chinese Acad Sci, Beijing Normal Univ, Michigan State Univ, Wageningen Univ, Univ Cambridge, Univ Nacl Autonoma Mexico, Univ Oxford, Colorado State Univ, Univ Copenhagen, etc. are the most productive research institutions in this field (Fig. 3).

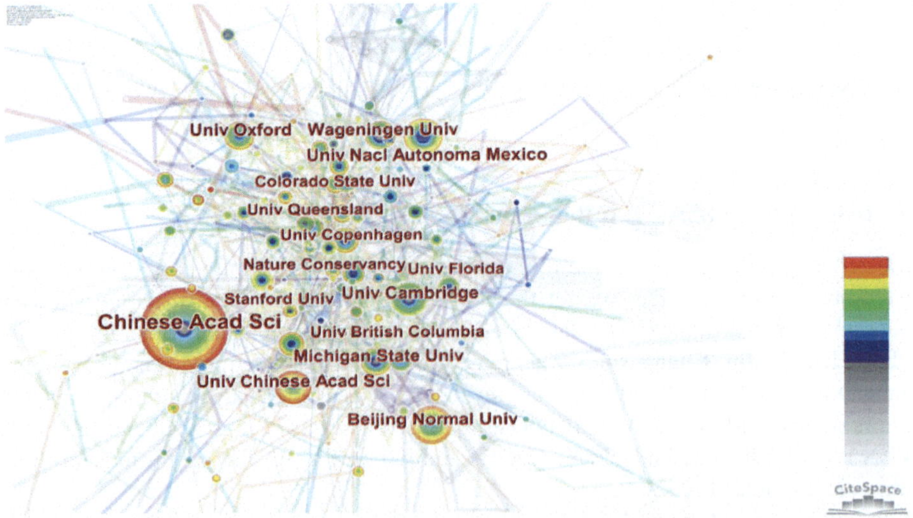

Fig. 3. Institutional co-existence diagram

Many countries in the world are committed to ecological compensation transfer research, as seen in Fig. 4, analyzed from the perspective of high-frequency countries, dominated by the United States, China, Germany, England, Australia, Brazil, and Canada.

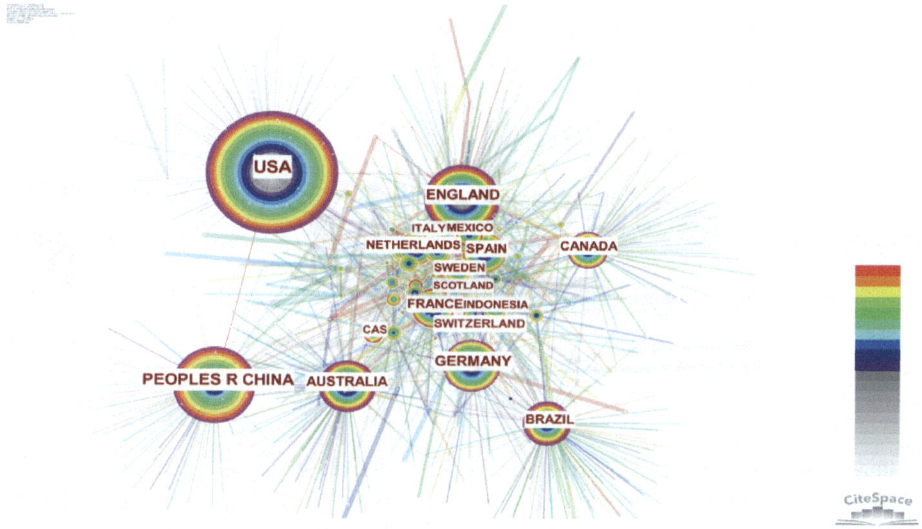

Fig. 4. Country co-existence map

3.3 Research Hot Spots

When searching the CNKI database, the key words used frequently in the research of ecological compensation transfer payment include ecological compensation, main functional areas, transfer payment, ethnic minority areas, ecological compensation mechanism, national key ecological functional areas, etc. The frequency of ecological compensation was 172 times, the frequency of transfer payment was 62 times, the frequency of ecological compensation mechanism was 46 times, the frequency of main functional areas was 29 times, the frequency of ethnic areas was 23 times, the frequency of national key ecological functional areas was 22 times, and the frequency of river basin ecological compensation was 21 times. When searching the WOS database, there are 699 keywords involved (N = 699), and there are 3516 line segments (E = 3516), and the line segments represent the linkage between keywords, which can be understood as the linkage between keywords in the same literature with a density of 0.0144 (Density = 0.0144).The high-frequency terms of ecological compensation transfer research are ecosystem service, payment, conservation, environmental service, management, biodiversity, impact, policy, deforestation, land use, etc., with frequencies of 1259, 964, 813, 752, 537, 488, 395, 330, 257, 247, respectively(Fig. 5).

In CNKI database retrieval, the keyword co-occurrence network is clustered into 12 categories, and in WOS database retrieval, the keyword co-occurrence network is

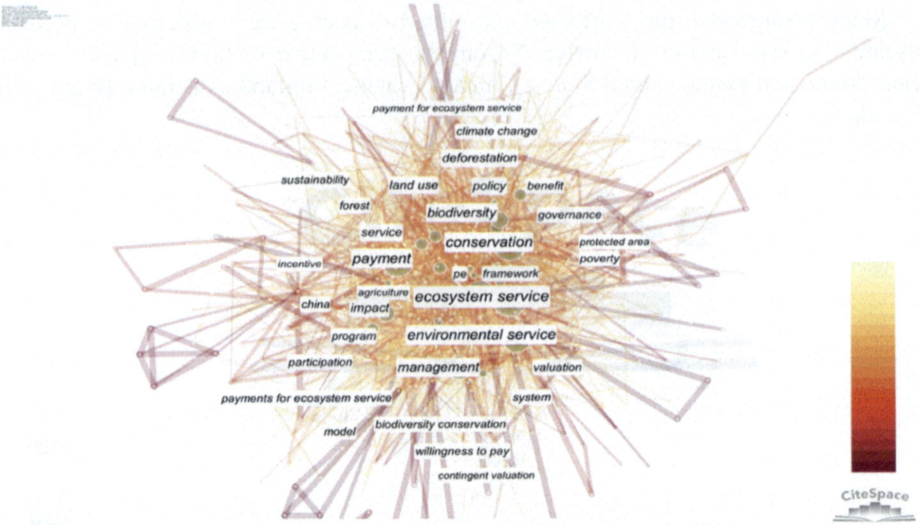

Fig. 5. Co-occurrence of keywords in the literature

clustered into 9 categories. As shown in the keyword clustering diagram in Fig. 6, the keyword co-occurrence network clustered into #0redd, #1willingness to pay, #2ecological compensation, #3land use, #4agri-environment scheme, #5neoliberalism, #6soil erosion, #7blockchain, #8payments for environmental services 9 categories.

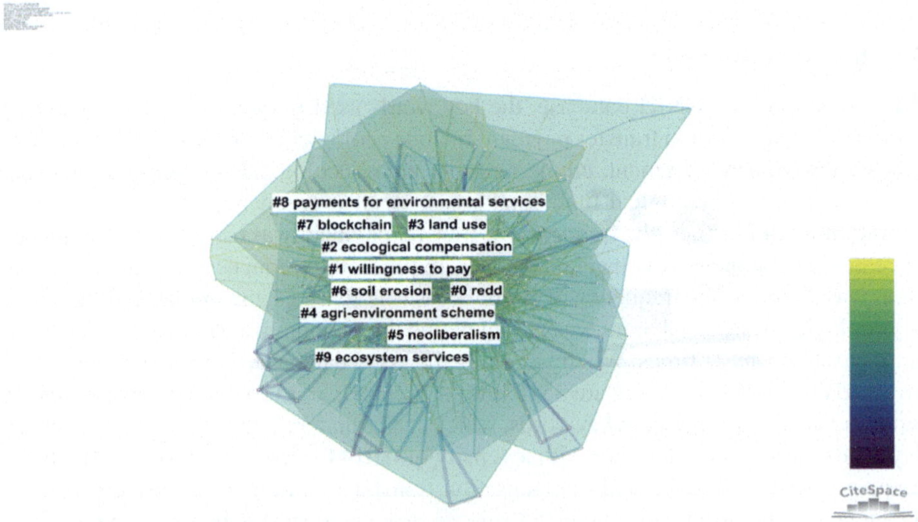

Fig. 6. Foreign keyword clustering chart

The keyword co-occurrence network collection constitutes a very irregular region, the region and the label corresponds to each other, the smaller the value means that the number of keywords covered by the clusters the more keywords, multiple keywords form a cluster. At this point, how to analyse and evaluate the clustering effect? We need to consider two values, the first is off the clustering module value (Q value), the second is the average profile value of the clusters (S value), if the Q value is larger than 0.3, which fully demonstrates that the clustering structure is unusually significant, and the S is higher than 0.5, which indicates that the clustering is very reasonable. The data from the figure shows that this clustering is reliable.

The research focuses on the following aspects: According to their individual research viewpoints, different academics have interpreted the study on the connotation of ecological compensation differently. Since a very long time ago(Wunder and Engel et al., 2008) [5], a precise description of ecological compensation has been offered, describing it as voluntary transactions where suppliers and buyers of environmental services bargain and engage in exchange for the provision of those services. With the strong contradiction between economy and ecology gradually coming to the fore, the study of ecological compensation has received more and more attention (Chen and Mirza et al., 2022 [6]; Zhao and Chau et al., 2022 [7]). The use of economic methods to adjust the relationship between different stakeholders in ecological protection and internalise the external costs of the ecological environment is conducive to the promotion of regional sustainable development (Grilli and Fratini et al., 2020 [8]; Liu and Zhang et al., 2022 [9]). Natural resource use regulations have been widely implemented in recent decades to protect ecosystems and address climate change. In forested areas, for example, ecological compensation policies are prevalent regulations that provide monetary subsidies and restrictions on the harvesting of timber or non-timber products by local forest owners (Blicharska and Hedblom et al., 2022) [10], usually in areas characterised by forest degradation or incorporation into nature reserves (Leberger and Rosa et al., 2020 [11]; Liu and Feng et al., 2022 [12]; Liao and Ye et al., 2023 [13]). In addition, the value of ecosystem services reflects the value provided by the ecosystems of a country or region and can measure the level of coordination between environmental quality and the economy (Hu and Wang et al., 2020) [14].

3.4 Research Frontiers

The keyword co-occurrence time zone map of ecological compensation transfer research was generated by selecting "Timezone" in CiteSpace software (Fig. 7). The "Business" function was used to draw the keyword emergence map (Fig. 8), and the keyword co-occurrence time zone map was used to focus on the latest keyword emergence and explore the frontier topics.

In the CNKI database, 7 keywords appeared in the study of ecological compensation transfer payment with "ecological compensation mechanism" having the highest intensity of emergence, followed by "key ecological functional zone," and "National key ecological functional zone" having the longest duration since 2014. In the WOS database, 25 keywords emerged in the study of ecological compensation transfer payments (Fig. 8). Since 2006, the main focus was on "redd," "payments for environmental service," "ecological compensation", "issue", "poverty," and ecological compensation,"

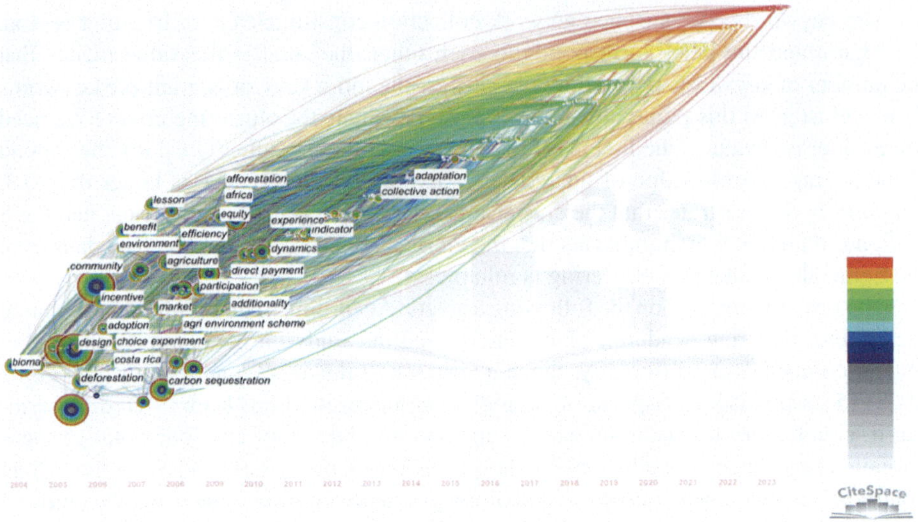

Fig. 7. Keyword co-occurrence time zone map for the study of ecological compensation transfer payments

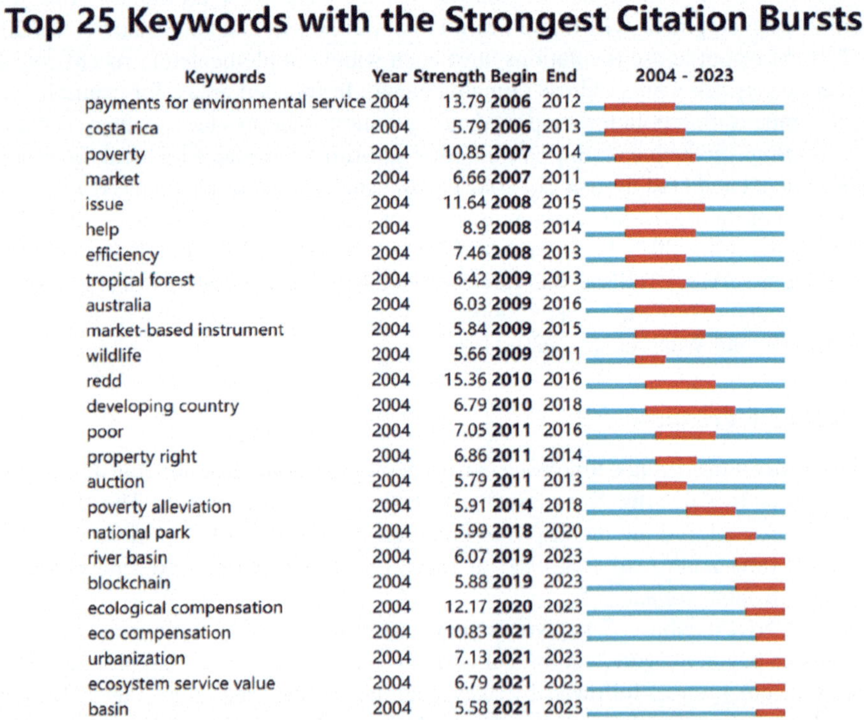

Keywords	Year	Strength	Begin	End
payments for environmental service	2004	13.79	**2006**	2012
costa rica	2004	5.79	**2006**	2013
poverty	2004	10.85	**2007**	2014
market	2004	6.66	**2007**	2011
issue	2004	11.64	**2008**	2015
help	2004	8.9	**2008**	2014
efficiency	2004	7.46	**2008**	2013
tropical forest	2004	6.42	**2009**	2013
australia	2004	6.03	**2009**	2016
market-based instrument	2004	5.84	**2009**	2015
wildlife	2004	5.66	**2009**	2011
redd	2004	15.36	**2010**	2016
developing country	2004	6.79	**2010**	2018
poor	2004	7.05	**2011**	2016
property right	2004	6.86	**2011**	2014
auction	2004	5.79	**2011**	2013
poverty alleviation	2004	5.91	**2014**	2018
national park	2004	5.99	**2018**	2020
river basin	2004	6.07	**2019**	2023
blockchain	2004	5.88	**2019**	2023
ecological compensation	2004	12.17	**2020**	2023
eco compensation	2004	10.83	**2021**	2023
urbanization	2004	7.13	**2021**	2023
ecosystem service value	2004	6.79	**2021**	2023
basin	2004	5.58	**2021**	2023

Fig. 8. Keyword emergent mapping of ecological compensation transfer studies

"issue," "poverty," "eco compensation," among which "eco compensation", "urbanization," "ecosystem service value," and "ecological compensation" are the key words, The keywords "eco compensation," "urbanization," "ecosystem service value," "basin," etc. only started to appear in 2021, which is a late appearance. From the abrupt changeability keywords in the ecological compensation transfer research literature, most of the research hot spots erupted mostly in 3–4 years, and the eruption phase is more concentrated. before 2018, the keywords abrupt change mainly around environmental service charges, poverty, Costa Rica, Australia, and tropical forests, which are closely related to the implementation of national ecological and environment-related policies. After 2018, with the implementation of international acts and national guidelines and policies, new keywords began to appear in the study of ecological compensation transfer payments, amoIn the CNKI database, 7 keywords appeared in the study of ecological compensation transfer payment with "ecological compensation mechanism" having the highest intensity of emergence, followed by "key ecological functional zone," and "National key ecological functional zone" having the longest duration since 2014.In the WOS database, 25 keywords emerged in the study of ecological compensation transfer payments (Fig. 8). Since 2006, the main focus was on "redd," "payments for environmental service," "ecological compensation," "issue," "poverty," and ecological compensation," "issue," "poverty," "eco compensation," among which "eco compensation," "urbanization," "ecosystem service value," and "ecological compensation" are the key words, The keywords "eco compensation," "urbanization," "ecosystem service value," "basin," etc. only started to appear in 2021, which is a late appearance. From the abrupt changeability keywords in the ecological compensation transfer research literature, most of the research hot spots erupted mostly in 3–4 years, and the eruption phase is more concentrated. Before 2018, the keywords abrupt change mainly around environmental service charges, poverty, Costa Rica, Australia, and tropical forests, which are closely related to the implementation of national ecological and environment-related policies. After 2018, with the implementation of international acts and national guidelines and policies, new keywords began to appear in the study of ecological compensation transfer payments, among which keywords such as national parks, ecological compensation mechanisms, and ecosystem service values emerged with greater intensity.

4 Conclusion and Outlook

4.1 Conclusion

From the annual distribution of literature, the literature related to ecological compensation transfer payment goes through slow growth, rapid growth and steady increase stages, but the overall trend of steady increase; In terms of research authors, Sven Wunder, Bettina Matzdorf, Julia P G Jones, Junlin Ren, and Yang Kuang constitute the high-producing authors of ecological compensation transfer research, and each high-producing author has a high average frequency of publication, but there is a lack of certain cooperation among authors; in terms of research institutions, Chinese Academy of Sciences, Beijing Normal University, Michigan State University, and In terms of research institutions, Chinese Academy of Sciences, Beijing Normal University, Michigan State University, and Wageningen University are the main research institutions with high average frequencies, and they are the backbone of the field "ecological compensation," "issue," "poverty," etc. The keywords "payments for environmental service," "ecological compensation," "issue," "poverty" and so on have become hot research directions. Meanwhile, according to the keyword emergence results, the research frontiers point to "payments for environmental service," "ecological compensation" and other fields.

4.2 Research Outlook

By analyzing the relevant literature in the WOS database, it is easy to find that the research field of ecological compensation transfer payments is broadening, and the innovation from basic theory to technical methods has laid a solid foundation for future expansionary research. However, the international level should also focus on the following points in terms of basic theory, research depth, and integration with regional policies: Strengthening cooperation among researchers and promoting cooperation among research units across regions. Increase research efforts on ecological compensation transfer payments. Accelerate the innovation of ecological compensation mechanisms. Ecological protection is a global issue, and each of our countries has the responsibility and obligation to contribute to ecological protection and development. Through global conferences, the international community can discuss the frontier issues of ecological compensation and share and learn from successful cases between regions.

Acknowledgments. Tibet University Graduate High-level Talent Cultivation Programme Project "Financial Transfer for Ecological Compensation in Tibet" (Project No. 2020-GSP-B035).

References

1. Blicharska, M., Hedblom, M., et al.: Operationalisation of ecological compensation—obstacles and ways forward. J. Environ. Manage. **304**, 114277 (2022)
2. Chen, Z., Mirza, N., et al.: Green banking—can financial institutions support green recovery? Econ. Anal. Policy. **75**, 389–395 (2022)
3. Grilli, G., Fratini, R., et al.: A spatial-based tool for the analysis of payments for forest ecosystem services related to hydrogeological protection. Forest Policy Econ. **111**, 102039 (2020)
4. Hatton MacDonald, D., Bark, R.H., et al.: Is ecosystem service research used by decision-makers? A case study of the Murray-Darling Basin, Australia. Landsc. Ecol. **29**, 1447–1460 (2014)
5. Hu, M., Wang, Y., et al.: How to balance ecosystem services and economic benefits? A case study in the Pearl River Delta, China. J. Environ. Manage. **271**, 110917 (2020)
6. Leberger, R., Rosa, I.M.D., et al.: Global patterns of forest loss across IUCN categories of protected areas. Biol. Conserv. **241**, 108299 (2020)
7. Liao, W.M., Ye, D.Y., et al.: Financial compensation for natural forest logging ban: standard calculation based on willingness to accept. Sci. Prog. **106**(1) (2023)
8. Liu, F., Feng, C., et al.: Effectiveness of functional zones in National Nature Reserves for the protection of forest ecosystems in China. J. Environ. Manage. **308**, 114593 (2022)
9. Liu, M.B., Zhang, A.L., et al.: Research on the game mechanism of cultivated land ecological compensation standards determination: based on the empirical analysis of the Yangtze River Economic Belt, China. LAND. **11**(9), 1583 (2022)

10. Persson, U.M., Alpízar, F.: Conditional cash transfers and payments for environmental services—a conceptual framework for explaining and judging differences in outcomes. World Dev. **43**, 124–137 (2013)
11. Selva, G.V., Pauli, N., et al.: Can environmental compensation contribute to socially equitable conservation? The case of an ecological fiscal transfer in the Brazilian Atlantic forest. Local Environ. **24**(10), 931–948 (2019)
12. Turner, R.K., Van Den Bergh, J.C., et al.: Ecological-economic analysis of wetlands: scientific integration for management and policy. Ecol. Econ. **35**(1), 7–23 (2000)
13. Wunder, S., Engel, S., et al.: Taking stock: a comparative analysis of payments for environmental services programs in developed and developing countries. Ecol. Econ. **65**(4), 834–852 (2008)
14. Zhao, L., Chau, K.Y., et al.: Enhancing green economic recovery through green bonds financing and energy efficiency investments. Econ. Anal. Policy. **76**, 488–501 (2022)

Exploration of Design for a Small-Scale Water Purification Plant in Mountainous Rural Areas

Fan Xiang[1,2](✉), Zongkai Ma[1,2], Ming Li[1], Li He[1,2], Binbin Qiu[1,2], and Youli Xiao[1]

[1] Powerchina Chengdu Engineering Co. Ltd., Chengdu 610000, Sichuan, China
xiangfanxpy@163.com

[2] Sichuan Province Engineering Technology Research Center of Municipal Water Environment Treatment, Chengdu 610000, Sichuan, China

Abstract. Water sources of many mountainous rural areas are usually mountain streams in which the turbidity increases significantly during the rainy season due to the influx of sediment and debris carried by the rainwater. This paper takes the A village water purification plant project as an example, analyzes the existing problems of the current water purification plant, and proposes the adoption of ultrafiltration technology to effectively address these issues. At the same time, a technical and economic comparison between conventional treatment process and ultrafiltration technology is made, highlighting the advantages of ultrafiltration technology in both technical and economic aspects. The practical application of ultrafiltration technology in the design of water purification plant in mountainous rural areas is discussed, and the practical operational performance is revealed that ultrafiltration process has superior effluent quality and remains highly stable under conditions of unattended operation and maintenance.

Keywords: Rural areas · Water purification · Ultrafiltration

1 Introduction

In recent years, China has been advancing the comprehensive revitalization of its rural areas, proposing implementation plans to address shortcomings in rural infrastructure, with a focus on improving the rural water supply engineering system [1]. The strategy involves promoting integrated urban-rural water supply and scaling up centralized water supply where conditions permit, and strengthening the standardized construction and renovation of small-scale water supply projects where integration is not feasible [2].

Numerous scattered villages are in a dispersed distribution amidst the mountains in mountainous cities, often located far from urban centers with significant elevation differences. Applying large-scale centralized water supply projects to address drinking water issues in these rural areas would lead to long pipeline extensions, high energy consumption for pressurization, uncertain water supply safety, and high construction costs. Therefore, from a technical and economic perspective, constructing decentralized small village-level water purification plants appears to be a reasonable and feasible solution for addressing drinking water issues in mountainous rural areas [2].

V. Achal et al. (Eds.): EPPC 2024, ESE, pp. 44–51, 2025.
https://doi.org/10.1007/978-3-031-92123-0_4

However, due to the unique geographical and cultural factors, the construction and operation of small-scale water purification plants in mountainous rural areas face several difficulties and challenges compared to urban water treatment plants construction: (1) The mountain streams, as primary water source, have significant fluctuations in turbidity all over the year—ranging from 5–10 NTU during non-rainy seasons to nearly 1000 NTU during rainy seasons; (2) The plants are distributed across numerous remote locations, resulting in extremely difficult operations and maintenance management; (3) Maintenance relies on local villagers who may lack the necessary skills to operate conventional water treatment facilities effectively, leading to inadequate management and reduced processing capabilities of the purification equipment. Guided by these issues, this paper takes the rural water supply project of a mountainous city in southern China as an example to explore the key design points for small-scale water purification plants in mountainous areas.

2 Project Overview and Current Situation Analysis

2.1 Project Overview

A Village is situated in a mountainous city in the northwest of Fujian Province. The city features towering ridges, undulating hills, and staggered valleys and small basins, with mountainous areas accounting for 80% of its total area. The city has an subtropical marine monsoon climate, characterized by mild temperatures and abundant rainfall, with an annual precipitation ranging from 1600 to 1800 mm. The rainy and dry seasons are distinct, with frequent flooding during the rainy season. Most locals rely on surface water as their drinking water source. In accordance with the local integrated urban-rural water supply planning, after completing the large-scale centralized water supply projects, the city still needs to construct 176 village-level small water purification plants to address the drinking water issues in remote rural areas. A Village's new small water purification plant, with a capacity of 100 m^3/day, is one of these 176 plants, serving the domestic drinking water needs of A Village.

2.2 Current Situation Analysis

A Village currently operates a small water purification plant with a capacity of 100 m^3/day, taking mountain streams as water source. The plant consists of an integrated water purifier and an underground clearwell, utilizing a water treatment process of coagulation reaction, inclined tube sedimentation, and valve-less filtration. Chlorine tablets are used for disinfection, with the addition occurring through a gravity-fed system. Powdered PAC is used as a coagulant, with a dissolution tank and metering pump in place. According to water quality testing data, the quality of the water source in this plant, except for excessive turbidity and coliform bacteria, meets the "Standards for drinking water quality (GB5749-2022)" After investigation, following main issues of the water plant have been identified.

1. The effluent water quality does not meet the standards. Specifically, during the rainy season, both turbidity and coliform bacteria exceed standards, and occasionally, aluminum ions do as well. Analysis shows that heavy rains wash the surface, causing

the raw water turbidity to increase sharply, reaching nearly 1000 NTU. As the water purification plant lacks automatic dosing facilities and 24-hour staff presence, it cannot promptly increase the dosage of coagulants and disinfectants, leading to suspended solids penetrating the filter and resulting in excessive turbidity and coliform bacteria in the effluent. Occasional aluminum ion excess is due to the lack of professional knowledge among maintenance staff, who over-dose coagulants based on experience.
2. Professional maintenance staff is absent. A Village is located in a relatively remote mountainous area, far from the city center and with inconvenient transportation, so only villagers living in the village can be selected as maintenance staff. Due to the lack of professional water treatment knowledge, maintenance staff usually determine the dosage of coagulants based on experience, which leads to either too much or too little coagulant being added when the raw water turbidity changes. Additionally, there is no duty room at the small water purification plant, and maintenance staff are unwilling or unable to stay at the plant for 24 h, preventing them from promptly detecting changes in raw water turbidity. Even if they notice such changes, they cannot quickly adjust the coagulant dosage at the plant.
3. Villagers struggle to adapt to the taste of disinfectant in the water. As the raw water contains a high level of coliform bacteria, disinfectants must be added during the water treatment process. However, villagers accustomed to natural water find the taste of the disinfectant unacceptable, leading to low satisfaction with the current water purification plant in the village.

In light of these issues, it is crucial for the small-scale water purification plant design to select water treatment processes that offer higher quality effluent, more stable treatment effects, simpler operation and maintenance, and even the possibility of unattended operation.

3 Treatment Process Discussion

Small-scale water purification plants typically adopt integrated water purifier to minimize land use and construction difficulties. The current mainstream technologies for removing turbidity and coliform bacteria from raw water are conventional treatment processes (consisting of coagulation, sedimentation, sand filtration, and disinfection) and ultrafiltration processes (including disinfection).

3.1 Conventional Treatment Analysis

The existing water purification equipment at Village A employs a conventional treatment process. However, due to its early construction date, it lacks automated dosing facilities constrained by the technology available at the time. If a new water purification plant were to continue using conventional processes, modern technological advancements would necessitate certain optimizations. Firstly, the addition of an automatic dosing system is essential to adjust the coagulant dosage based on influent turbidity and flow rate, and the disinfectant dosage according to flow rate and residual chlorine in the effluent. Additionally, considering the significant increase in turbidity of raw water during the rainy season, fully automatic adjustment of appropriate dosages under large turbidity variations may not be feasible without manual intervention. Therefore, when employing conventional processes, it is necessary to establish a duty room with two operators working in shifts to ensure 24-hour duty.

3.2 Ultrafiltration Technology Analysis

Membrane separation technology uses membranes with extremely fine pores as the separation medium. When pressure is applied on one side, various components in the raw water selectively pass through the membrane, achieving separation or purification purposes. With pore sizes ranging from 0.03 to 0.1 μm, ultrafiltration membranes lie between microfiltration and nanofiltration, primarily used for solid-liquid phase separation [3]. They effectively remove suspended particles, colloidal proteins, bacteria, and viruses from solutions. In recent years, ultrafiltration membrane technology has been increasingly utilized in drinking water treatment due to its high-precision filtration capabilities, which provide high-quality potable water and ensure both biological and chemical safety [4]. The advantages of ultrafiltration membrane technology in small-scale water purification plants include:

1) Adhering to the "Technical specification for water supply projects in towns and villages (SL310-2019)", which recommends non-chemical ultrafiltration treatment methods for small-scale plants, eliminating the need for coagulants and thus avoiding issues related to inaccurate dosing during the rainy season.
2) Effectively removing most bacteria, requiring only periodic replenishment of residual chlorine to meet standards, significantly reducing the odor of disinfectants in the water, making it more acceptable to villagers [5].
3) Ultrafiltration treatment needs simplified operation and maintenance, with fully automated water treatment processes requiring no human intervention. The backwashing process for ultrafiltration membranes can also operate automatically, better suited for rural water purification plants with limited maintenance resources in mountainous areas.

Therefore, ultrafiltration technology effectively addresses the challenges faced by small-scale rural water purification plants.

3.3 Economic Analysis

In the past, the higher initial investment costs and additional membrane replacement expenses associated with ultrafiltration technology deterred market adoption. However, with advances in materials science in China, the production cost of ultrafiltration membranes has decreased annually, providing economic competitiveness. A comparative analysis of the procurement and operational costs for integrated water purifier using conventional treatment process and ultrafiltration technology was conducted.

Market research indicates that a 100 m^3/d integrated water purifier with conventional treatment process costs ¥347,000, while a similar capacity ultrafiltration system costs ¥430,000. In terms of procurement costs, conventional treatment process is less expensive than ultrafiltration technology.

The operating costs for conventional treatment process include PAC procurement, disinfectant procurement, electricity fees, and operator salaries. Calculations are as follows:

PAC procurement cost: Annual average daily PAC dosage of 5 mg/L (as Al_2O_3) and using 10% PAC solution priced at ¥900 per ton result in an annual PAC procurement cost of ¥1642.5.

Disinfectant procurement cost: Annual average daily dosage of 1 mg/L (as active chlorine) and 10% sodium hypochlorite solution priced at ¥1700 per ton result in an annual disinfectant procurement cost of ¥620.5;

Electricity fees: Based on engineering data, the electricity consumption of the plant averages 4.2 kWh/day, with electricity priced at ¥0.6/kWh, resulting in annual electricity fees of ¥932.92;

Operator salaries: As mentioned earlier, two operators are required for conventional treatment process, with each village operator's salary set at ¥20,000/year. Consequently, the total annual operator salaries amount to ¥40,000;

Total: The annual operating cost for conventional treatment process plant is ¥43,195.94.

Ultrafiltration technology does not require coagulant addition but necessitates annual chemical cleaning and membrane replacement every 10 years. Its operating costs include disinfectant procurement, ultrafiltration membrane cleaning, electricity fees, operator salaries, and membrane replacement costs. Calculations are as follows:

Disinfectant procurement cost: Considering only maintaining residual chlorine in the distribution pipeline, an annual average daily disinfectant dosage of 0.5 mg/L (as active chlorine) and using 10% sodium hypochlorite solution priced at ¥1700 per ton, result in an annual disinfectant procurement cost of ¥310.25;

Annual cleaning expenses: Annual ultrafiltration membrane cleaning is typically performed by professionals sent by manufacturer with chemicals needed to the site. According to manufacturers, total labor and chemical costs for each cleaning amount to approximately ¥3000;

Electricity fees: Based on engineering data, the electricity consumption of the plant averages 3.6 kWh/day, with electricity priced at ¥0.6/kWh, resulting in annual electricity fees of ¥788.40;

Operator salaries: Due to simplified maintenance, ultrafiltration technology allows for unattended operation, with one operator responsible for inspection at two water purification plants. Village operator salaries are set at ¥20,000/year, resulting in annual operator salaries of ¥10,000 for one plant;

Membrane replacement Cost: Membranes are replaced every 10 years, with a 100 m^3/d membrane system costing ¥50,000 for each replacement, amortized over 10 years, resulting in an annual membrane replacement cost of ¥5000;

Total: The annual operating cost for ultrafiltration technology plant is ¥19,098.65.

Based on the above analysis, although the procurement cost of ultrafiltration technology is higher than that of conventional treatment process, its annual operating cost is lower, mainly due to the reduced need for operators. Considering both procurement and operating costs, the payback period for the incremental investment of ultrafiltration technology over conventional treatment process is 3–4 years, while typical operating periods range from 25 to 30 years. Thus, ultrafiltration technology offers significant economic advantages.

3.4 Water Treatment Process Selection

Based on the technical and economic analysis presented above, Table 1 can be derived.

Table 1. A table for comparison of water treatment processes

Comparison metrics	Conventional treatment process	Ultrafiltration technology
Effluent quality	Turbidity <1 NTU with noticeable disinfectant odor	Turbidity <1 NTU with noticeable disinfectant odor
Impact load tolerance	Strong when there is appropriate dosage of coagulants and professional staff	Strong and achieving fully automated operation
Land use	Comparatively larger	Comparatively smaller
Operational and maintenance	More intricate, necessitating adjustments to coagulant and disinfectant dosage according to turbidity in raw water	Simpler management, involving routine checks on equipment status
Required maintenance staff	Two skilled operators on 24-hour duty	Fully automated operation with one operator conducting regular inspection
Procurement cost	¥347,000	¥430,000
Annual operating cost	¥43,195.94	¥19,098.65
25-year life cycle cost	¥1,424,500	¥907,500

As revealed in Table 1, under the application conditions of rural areas, ultrafiltration technology demonstrates superior effluent quality, simplified fully automated operation and maintenance, capability for unattended operation, low operational costs, and lower 25-year life cycle cost. In terms of both technical and economic aspects, ultrafiltration technology holds significant advantages over conventional treatment process. Consequently, the water purification plant in the present case is recommended to adopt integrated water purifier employing ultrafiltration technology.

4 Water Treatment Process Design

Following the aforementioned analysis, the water purification plant in this case adopts a process combining "ultrafiltration membrane + sodium hypochlorite disinfection." The plant consists of an integrated water purifier employing ultrafiltration technology and a clearwell. Raw water is directly fed into the ultrafiltration membrane system for filtration, and then the filtrated water is disinfected before getting into the clearwell.

The integrated ultrafiltration treatment purifier utilizes submerged ultrafiltration membrane process, with a processing capacity of 100 m^3/d. The purifier encompasses main modules such as the ultrafiltration membrane filtration system, backwash aeration system, disinfectant dosing system, and electrical automation control system.

The ultrafiltration membrane components employ modified PVDF with an average pore size <0.1 μm. Parameters of the purifier are specified as follows: a design flux of 17 L/(m^2·h), a water backwash intensity of 60 L/(m^2·h), an air scrubbing intensity of 70 m^3/(m^2·h), a maximum transmembrane pressure $\leq$0.03 MPa, and a water production rate $\geq$ 90%. The entire set of equipment meets the functions of automatic water production, automatic backwashing, automatic wastewater discharge, automatic maintenance cleaning and automatic disinfection.

Considering the small scale of the water purification plant in this case, poor power supply reliability, and the potential increase in the number of membrane system backwashes during the rainy season due to high turbidity of raw water, which may reduce the water production rate, it is recommended to take a higher effective volume for the clearwell to ensure water supply, meeting the needs during periods of raw water with high turbidity, peak water usage, and power outages. According to the "Technical specification for water supply projects in towns and villages (SL310-2019)", the effective volume of the clearwell is set at 60% of the design scale of the plant. Therefore, one clearwell is designed in this plant, with an effective volume of 60 m^3.

5 Operational Performance

Since its completion and operation, the A village water purification plant has maintained a water production rate of up to 95% during non-rainy seasons and stabilized at 90% during rainy seasons over the course of 1 year. The raw water turbidity fluctuated between 3 NTU-800 NTU, with the effluent quality consistently meeting national standards, where the turbidity of effluent remained stable <0.2 NTU.

Throughout the operation of the A village water purification plant, only one villager appointed by the village committee as a maintenance staff to conduct regular patrols of the plant, ensuring the normal operation of treatment process. This demonstrates that the ultrafiltration process remains highly stable under conditions of unattended operation and maintenance.

6 Conclusions

1. The main problems faced by water purification plants in mountainous rural areas include scattered locations, difficulty in controlling coagulant dosing, lack of professional operation and maintenance management staff, and unstable outlet water quality.
2. The procurement cost of small-scale integrated ultrafiltration purifier is higher than that of conventional treatment process purifier, but the operating cost is lower, and the 25-year life cycle cost is also lower. The main reason for the low operating cost of the ultrafiltration process is the stable outlet water quality and simplized operation and maintenance, which greatly reduces the need for operation and maintenance staff, achieving unattended operation.

3. The actual case of the A village water purification plant indicates that when applied to water purification in mountainous rural areas, the ultrafiltration technology can ensure stable outlet turbidity below 0.2 NTU despite large fluctuations in raw water turbidity, and the water production rate can remain stable at 90%. Additionally, the ultrafiltration process can achieve the real unattended operation.

References

1. Fan, X.: Investigation of building urban and rural water supply integrated system in mountain city. Water & Wastewater. Engineering. **58**(SI), 51–56 (2022)
2. Huiyun, L., Ning, Z.: Technical points and case analysis of urban and rural water supply integrated project. China Water Wastewater. **36**(22), 72–76 (2020)
3. A L F, Jenifer V, Tomas T, Alejandro R, A M C.: Advances in drinking water treatment through piloting with UF membranes. Water. **15**(6), 1031 (2023)
4. Hagen, K.: Removal of particles, bacteria and parasites with ultrafiltration for drinking water treatment. Water Supply. **17**(1), 269–363 (1999)
5. Qidong, W., Chen, C., Yongli, Z., Peng, T., Xiaoyu, R., Jingyu, S., Xinyu, L., Xin, C., Alberto, T., Baicang, L.: Safe purification of rural drinking water by biological aerated filter coupled with ultrafiltration. Sci. Total Environ. **868**, 161632 (2023)

Study on Environmental Impact of Water Quality and Sediment in Qionghai Tanmen Sea Area

Yingying Chen[1(✉)], Qiaoling Xie[2], Yanxia Ni[1(✉)], and Dan Ling[1]

[1] CCTEG Hangzhou Research Institute Co., Ltd., Hangzhou 310000, Zhejiang, China
{chenyingying,Niyanxia}@ccteghz.com

[2] Hangzhou Tianjin Environmental Technology Consulting Development Co., Ltd., Hangzhou 310000, Zhejiang, China

Abstract. In this paper, the navigation survey data of Tanmen Sea area in January 2019, March 2021 and March 2022 were collected to study the pollution characteristics of seawater quality and sediment. The results show that COD, DIN (inorganic nitrogen) and active phosphate in seawater water quality exceed the standard, the water quality in Tanmen Harbor presents a state of severe eutrophication. In March 2021, petroleum exceeded the standard, copper, lead and zinc in heavy metals exceeded the standard slightly, and other factors could meet the requirements of the class of seawater water quality standards. In 2019 and 2021, the petroleum in the sediment can not meet the quality standards of the third class sediment, the organic carbon in 2021 appeared in the phenomenon of exceeding the quality standards of the second class sediment, sulfide, copper and cadmium in 2019 slightly exceeded the standard, and other factors can meet the requirements of the first class sediment standard.

Keywords: Tanmen Sea · Pollution characteristics · Seawater · Sediment

1 Introduction

Nutrients and heavy metals are Marine chemicals closely related to plankton reproduction and play an important role in aquatic ecosystems [1]. With the development of economy, a large number of pollutants such as nutrient salts, organic substances, nutrient salts, petroleum and heavy metals are discharged into the sea, causing serious impacts on the coastal waters [2]. A large number of land-based pollutants produced by human activities gather in the nearshore sea area. Under the influence of hydrological dynamics, if the pollutants cannot be dispersed well, a high-pollution area will be formed in the nearshore sea area [3]. After entering the coastal waters, land-based pollutants are transferred to sediments through adsorption, flocculation, bioaccumulation and other processes, thus exerting a continuous impact on the ecological environment [4]. Some studies have also pointed out that eutrophication will affect the absorption of pollutants by plankton, thus affecting the migration and transformation of Marine ecosystems, and then affecting the content of pollutants in sediments and organisms in them [5]. The

V. Achal et al. (Eds.): EPPC 2024, ESE, pp. 52–59, 2025.
https://doi.org/10.1007/978-3-031-92123-0_5

absorption of nitrogen and phosphorus by different algae varies, but their absorption of nitrogen and phosphorus is carried out in a certain proportion. Therefore, the nitrogen and phosphorus ratio between nutrient salts can better explain the type of water eutrophication [6].

Qionghai Tanmen is located in the east coast of Hainan Province, rich in natural resources and fishery resources. In particular, Qionghai Tanmen Town has become a beautiful name card of Tanmen by means of Marine culture tourism and other customs town [7]. However, the development of local traditional fisheries has seriously damaged the Marine ecological environment [8]. The purpose of this study is to study the seawater quality and Marine sediment environment of Qionghai Tanmen, and to provide reference for the sustainable and healthy development of Tanmen fishery.

2 Research Method

2.1 Sample Collection and Quality Control

This study collected the Marine environment status survey data of three spring voyages around Tanmen Sea area in March 2022, March 2021 and January 2019 for impact analysis. In order to analyze and plan the change trend of Marine environmental quality in a more targeted way, the monitoring points selected in this study are shown in Table 1 and Fig. 1.

Table 1. Monitoring station table

Monitoring time	Water quality monitoring station	Sediment monitoring station
January 2019	1–8	2,3,4,6,8
March 2021	14–20,6	6,15,18,19,20
March 2022	5,20,25–29	5,27,28

2.2 Eutrophication Index Method

Eutrophication index (EI) was adopted for evaluation in this study, and its formula is as follows:

$$\mathrm{EI} = C_{COD} \times C_{DIN} \times D_{DIP} \div 4500 \times 10^6$$

Where, EI is the eutrophication index, C_{COD}, C_{DIN} and C_{DIP} are respectively the concentrations of COD, DIN and active phosphate, mg/L.

An EI value less than 1 indicates non-eutrophication, an EI value between 1 and 3 indicates mild eutrophication, an EI value between 3 and 6 indicates moderate eutrophication, and an EI value greater than 9 indicates severe eutrophication.

In order to calculate the contribution rate of each nutrient salt to the eutrophication index, the following formula was obtained after the eutrophication index formula was deformed in this study [9].

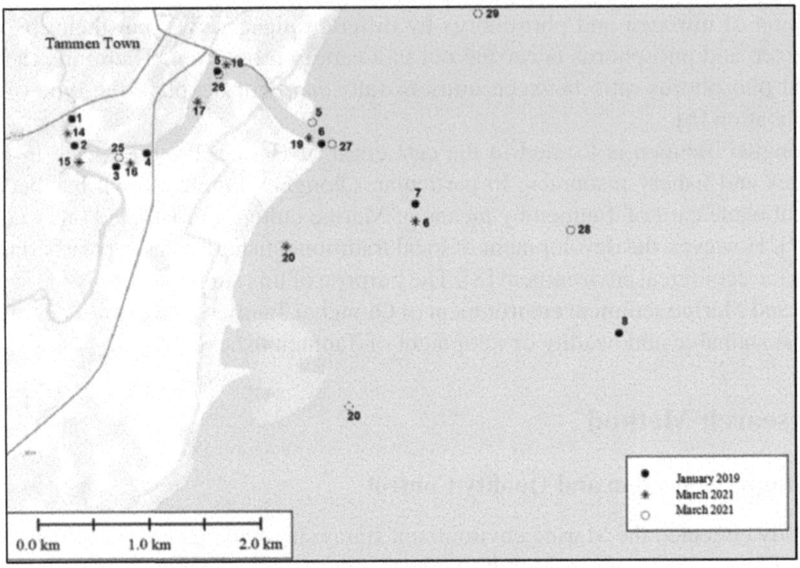

Fig. 1. Bitmap of monitoring points

2.3 Single Factor Pollution Coefficient

The single-factor pollution index (P) has been identified as an effective tool for monitoring pollution over a period of time [10]. In the formula $Pi = \frac{Ci}{Si}$, Pi is the pollution index of i, Ci is the measured value of heavy metal i, and Si is the standard value of pollutant I .

3 Results and Discussion

3.1 Pollution Level

Based on the survey results of seawater quality status in spring 2019, 2021 and 2022, the change trends of average and extreme values of major water quality factors are listed in Table 2. As can be seen from Table 2, the phenomenon of excess of active phosphate is the most serious from 2019 to 2021, and there are monitoring points that fail to meet the four categories of seawater quality standards in all the years, and there are inorganic nitrogen points that exceed the four categories of seawater quality standards in January 2019 and January 2021, and inorganic nitrogen can meet the four categories of seawater quality standards in March 2022. But can not meet the three water quality standards. COD also exceeded the standard in 2019 and 2021, among which some stations in 2019 could not meet the four types of water quality standards, and by March 2022, the COD of all monitoring stations could meet the first class of seawater water quality standards. This phenomenon is consistent with the statistical results in the Marine ecological environment status Bulletin of Hainan Province. According to the bulletin, the water quality of Qionghai Tanmen fishing Port cannot meet the second class water quality standards, and the main indicators exceeding the standard are active phosphate, chemical oxygen demand and inorganic nitrogen. From the perspective of interannual variation trend, COD, inorganic nitrogen and active phosphate showed a decreasing trend year by year.

Table 2. Comparison table of seawater quality standard rate (%)

Index	Time	Minimum value	Maximum value	Mean value	Over standard rate (%)			
					First class	Secondary	Three-class	Fourth
COD (mg/L)	2019.1	0.87	5.85	2.73	62.5	37.5	25	12.5
	2021.3	0.11	4.53	1.91	37.5	25	25	0
	2022.3	0.23	0.86	0.54	0	0	0	0
DIN (mg/L)	2019.1	0.0615	1.7722	0.9468	75	75	75	75
	2021.3	0.0858	0.7416	0.4105	75	62.5	62.5	37.5
	2022.3	0.026	0.476	0.1509	28.57	28.57	14.29	0
Active phosphate (mg/L)	2019.1	0.0055	0.1596	0.0964	75	75	75	75
	2021.3	0.0009	0.2343	0.1014	75	75	75	75
	2022.3	0.0026	0.0663	0.0213	28.57	28.57	28.57	28.57
Petroleum (mg/L)	2019.1	0.004	0.023	0.0128	0	0	0	0
	2021.3	0.008	0.166	0.0808	37.5	37.5	0	0
	2022.3	0.0102	0.0204	0.0145	0	0	0	0
Cu (μg/L)	2019.1	0.34	1.95	1.28	0	0	0	0
	2021.3	0.7	1.89	1.37	0	0	0	0
	2022.3	0.4	1.4	0.80	0	0	0	0
Pb (μg/L)	2019.1	0.81	2.3	1.43	75	0	0	0
	2021.3	0.05	2.46	0.77	25	0	0	0
	2022.3	0.1	0.18	0.15	0	0	0	0
Zn (μg/L)	2019.1	10.48	26.19	14.82	12.5	0	0	0
	2021.3	6.33	16.33	11.81	0	0	0	0
	2022.3	10.5	15.8	12.36	0	0	0	0
Cd (μg/L)	2019.1	0.011	0.018	0.014	0	0	0	0
	2021.3	0.01	0.88	0.268	0	0	0	0
	2022.3	0.03	0.1	0.056	0	0	0	0
Hg (μg/L)	2019.1	0.01	0.04	0.026	0	0	0	0
	2021.3	0.012	0.031	0.021	0	0	0	0
	2022.3	0.013	0.033	0.02	0	0	0	0
As (μg/L)	2019.1	0.51	0.79	0.60	0	0	0	0
	2021.3	0.51	0.62	0.55	0	0	0	0
	2022.3	0.7	0.7	0.70	0	0	0	0

As can be seen from Table 2, only in 2021, some oil stations exceeded Class I and Class II seawater quality standards, and the overall excessive phenomenon was not serious. Heavy metals (copper, lead, zinc, cadmium, arsenic, mercury) in the three survey voyages, only lead and zinc appear to exceed the phenomenon of Class I seawater water quality standards, but can also meet the Class II seawater water quality standards, copper, cadmium, arsenic, mercury can meet the Class II seawater water quality standards. This research result is also consistent with the conclusion of the Marine ecological Environment status Bulletin of Hainan Province in 2019, 2021 and 2022. The water quality of Tanmen Port is affected by inorganic nitrogen, active phosphate, lead and zinc, and some stations have exceeded the standard.

Based on the survey results of spring status in 2019, 2021 and 2022, the change trends of average and extreme values of major factors in sediments are listed in Table 3. As can be seen from Table 3, the phenomenon of exceeding the standard of oil in the surveyed sea area is relatively serious, and the requirements of the three types of sediment quality standards cannot be met in January 2019 and March 2021, while the

Table 3. Comparison table of excess rate (%) of sediment survey factors for each voyage

Index	Time	Minimum value	Maximum value	Mean value	Over standard rate (%)		
					First class	Secondary	Three-class
Organic carbon ($\times 10^{-2}$)	2019.1	0.81	1.93	1.45	0	0	0
	2021.3	0.21	3.27	1.34	40	20	0
	2022.3	0.35	0.68	0.5	0	0	0
Petroleum ($\times 10^{-6}$)	2019.1	4.8	9626.5	4709.5	60	60	60
	2021.3	9.93	3514.79	1173.92	40	40	40
	2022.3	48.8	69.6	57.7	0	0	0
Sulphide ($\times 10^{-6}$)	2019.1	189.6	988.8	573.8	60	40	0
	2021.3	1.02	237	71.97	0	0	0
	2022.3	1.6	1.8	1.7	0	0	0
Cu ($\times 10^{-6}$)	2019.1	3.49	36.84	23.46	20	0	0
	2021.3	8.08	12.83	10.67	0	0	0
	2022.3	2.6	10.8	6.6	0	0	0
Pb ($\times 10^{-6}$)	2019.1	25.67	138.35	94.98	0	0	0
	2021.3	59.91	145.31	84.2	0	0	0
	2022.3	28.7	32.4	30.5	0	0	0
Zn ($\times 10^{-6}$)	2019.1	0.33	8.36	4.85	0	0	0
	2021.3	1.19	31.3	13.02	0	0	0
	2022.3	5.1	8.5	7	0	0	0
Cd ($\times 10^{-6}$)	2019.1	0.05	0.79	0.33	40	0	0
	2021.3	0.17	0.31	0.23	0	0	0
	2022.3	0.05	0.31	0.2	0	0	0
Hg ($\times 10^{-6}$)	2019.1	0.012	0.11	0.064	0	0	0
	2021.3	0.005	0.2	0.076	0	0	0
	2022.3	0.025	0.054	0.035	0	0	0
As ($\times 10^{-6}$)	2019.1	2.57	7.68	5.28	0	0	0
	2021.3	4.29	11.32	8.18	0	0	0
	2022.3	2.99	11.9	6.3	0	0	0

three types of sediment quality standards can be met only in March 2022. Organic carbon can not meet the requirements of the second type of sediment quality standard in 2021; Sulfide, copper, cadmium in 2019 can not meet the requirements of a class of sediment quality standards. Other factors (zinc, lead, mercury, arsenic) can meet the requirements of a class of sediment quality standards. For the petroleum with more serious excessive phenomenon, its concentration shows a downward trend, and by March 2022, the concentration of petroleum in the sediment can meet the quality standard of first-class sediment; Organic carbon in March 2021 slightly exceeded the standard, but in 2019 and 2022 can meet the first class sediment quality standards. The concentrations of sulfide, copper and cadmium show a downward trend, and can meet the first class sediment quality standards in 2021 and 2022.

3.2 Eutrophication Index and Its Influencing Factors

In order to further study the eutrophication degree of Tanmen Sea area, the eutrophication index of the studied sea area is listed in Table 4. The results show that the EI index in January 2019 is very large, and the maximum value is 367.70 at station 1. The EI index of stations 1 to 6 in Tanmen Port gradually decreases, but they are all heavier than the severe eutrophication state. Stations 14 to 18 in March 2021 are in a state of severe eutrophication, stations 19 are in a state of moderate eutrophication, and stations 25 and 26 in Tanmen Gangnei in March 2022 are in a state of severe eutrophication. The results of three adjustments show that the seawater in Tanmen Port is not easily dispersed due to the discharge of land-based pollutants, and the sea area in Tanmen Port is in a state of severe eutrophication. At the mouth gate, with the improvement of hydrodynamic diffusion conditions, the eutrophication state of water quality has been greatly improved, and the water body is in a state of non-eutrophication. According to monitoring data, the average N/P ratio in Tanmen Harbor is 9.78 in January 2019, 3.89 in March 2021, and 6.54 in March 2022, all of which are less than 10. According to existing studies, N:P < 10 indicates that water body eutrophication becomes N-limiting type [11, 12]. The results show that the water body in Tanmen Harbor is eutrophication into nitrogen limiting type. According to the research results of Zhang Jibiao et al., in the offshore seawater of Hainan Island, the average value of N/P is (46.8 ± 78.6) on the whole, and the overall N is relatively excess and P is relatively insufficient, which is shown as a P-limited feature [13]. In this study, the N/P in January 2019 ranged from 6.89 to 11.18, the N/P in March 2021 ranged from 3.05 to 124.89, and the N/P in March 2022 ranged from 5.90 to 22.76. The monitoring stations with N/P less than 10 were located in Tanmen Bay. The monitoring station with N/P greater than 10 is located in Tanmen Bay, which is also consistent with the conclusions of Zhang Jibiao et al., and also indicates that the phosphorus emission of local pollution sources in Tanmen Bay is high.

The research results of ECOD, EDIN and EDIP show that the average ECOD in January 2019, March 2021 and March 2022 is 0.43, 0.41 and 0.51, respectively. The mean values of EDIN in January 2019, March 2021 and March 2022 were 0.36, 0.40 and 0.34 respectively, and the mean values of EDIP in January 2019, March 2021 and March 2022 were 0.26, 0.38 and 0.26 respectively, which contributed to water eutrophication. In general, COD contributed the most, DIN followed, and DIP the least. According to the correlation analysis, in January 2019, DIP showed a significant positive correlation

Table 4. Eutrophication index

2019.1					2021.3					2022.3				
Position	EI	E_{COD}	E_{DIN}	E_{DIP}	Position	EI	E_{COD}	E_{DIN}	E_{DIP}	Position	EI	E_{COD}	E_{DIN}	E_{DIP}
1	367.70	0.41	0.42	0.24	6	0.002	0.50	0.32	0.17	5	0.01	0.52	0.32	0.16
2	126.23	0.42	0.40	0.23	14	42.81	0.44	0.42	0.32	20	0.01	0.54	0.33	0.13
3	156.03	0.40	0.41	0.24	15	22.38	0.46	0.38	0.31	25	2.64	0.38	0.36	0.26
4	85.29	0.40	0.40	0.24	16	86.88	0.39	0.46	0.38	26	5.47	0.39	0.36	0.25
5	51.62	0.40	0.39	0.25	17	20.16	0.37	0.46	0.34	27	0.03	0.56	0.35	0.09
6	23.70	0.37	0.40	0.26	18	20.93	0.39	0.44	0.34	28	0.01	0.61	0.29	0.10
7	0.21	0.49	0.26	0.18	19	3.75	0.36	0.42	0.32	29	0.005	0.54	0.36	0.10
8	0.07	0.54	0.23	0.14	20	0.03	0.37	0.29	0.34					
Min	0.07	0.37	0.23	0.14		0.002	0.36	0.29	0.17		0.005	0.38	0.29	0.09
Max	367.7	0.54	0.42	0.26		86.88	0.5	0.46	0.38		5.47	0.61	0.36	0.26
Mean	101.36	0.43	0.36	0.22		24.62	0.41	0.40	0.32		1.17	0.51	0.34	0.16

with DIN with a correlation of 0.97, DIP showed a significant positive correlation with salinity with a correlation of −0.86, and in March 2021, DIP showed a significant positive correlation with DIN with a correlation of 0.96, DIP showed a positive correlation with salinity with a correlation of −0.26. In March 2022, DIP showed a significant positive correlation with DIN with a correlation of 0.99, and DIP showed a positive correlation with salinity with a correlation of −0.99, which indicates that DIN and DIP have homology, while DIP and salinity are both negatively correlated, which also indicates that DIP in seawater is mainly influenced by terrigenous input salt [2].

4 Conclusion

The phenomenon of COD, inorganic nitrogen and active phosphate exceeding the standard of seawater water quality is more serious, petroleum exceeded the standard in March 2021, copper, lead and zinc in heavy metals exceeded the standard slightly, and other factors can meet the requirements of a class of seawater water quality standards. The petroleum in the sediments in 2019 and 2021 cannot meet the quality standards of the third class of sediments, the organic carbon in 2021 appears to exceed the quality standards of the second class of sediments, sulfide, copper and cadmium in 2019 slightly exceeded the standard, and other factors can meet the requirements of the first class of sediment standards. Affected by the hydrodynamic process of the input of terrigenous pollutants, the water quality in Tanmen Harbor presents a state of severe eutrophication, and the water quality environment outside Tanmen Harbor is good. According to the study of N/P ratio, the N/P ratio in Tanmen Harbor is less than 10, and the eutrophication becomes nitrogen limiting type. The research results of ECOD, EDIN and EDIP show that the contribution of eutrophication of water body is the largest, followed by DIN, and DIP is the smallest. According to correlation studies, DIP and DIN are significantly positively correlated, while DIP and salinity are both negatively correlated, which also indicates that DIP in seawater is mainly influenced by terrigenous input salt.

References

1. Shou, Y.P., Zhao, J.J., Zhu, Y.X., et al.: Heavy metals pollution characteristics and risk assessment in sediments and waters: the case of Tianjin, China [J]. Environ. Res. **212**, 1–100 (2022)
2. Hou, Q.H., Li, W.L., Ye, Y.Y., Zhang, Y.B.: Research progress of heavy metal pollution in Zhanjiang Bay and Adjacent Sea Areas [J]. Guangdong Chem. Ind. **9**(50), 105–109 (2023)
3. Zhang, Z.X., Li, Z.L., Ye, H., et al.: Spatial characteristics and sources of water pollution in nearshore area: a case study on the nearshore area of Ruian in southern Zhejiang [J]. J. 0Zhejiang Univ. (Sci. Ed.). **50**(3), 332–345 (2023)
4. Chen, R., Mubashir, M., Khoo, K., et al.: Perovskite oxide for emerging photo(electro)catalysis in energy and environment [J]. Environ. Res. **205**, 112544 (2022)
5. Chen, M., Ding, S., Gao, S., et al.: Efficacy of dredging engineering as a means to remove trace metals from lake sediments[J]. Sci. Total Environ. **665**, 181–190 (2019)
6. Justic, D., Rabalais, N.N., Turner, R.E., et al.: Changes in nutrient structure of river-dominated coastal waters: stoichiometric nutrient balance and its consequences [J]. Estcoastshelf. **40**(3), 339–356 (1995)
7. Wu, R., Wang, D.R.: Analysis on marine ecology and environmental protection in Hainan province [J]. Mar. Dev. Manage. **30**(9), 61–65 (2013)
8. Zeng, W.T., Yang, Y.P., Zhang, D.Q., et al.: Sources, distribution of main controlling factors, and potential ecological risk assessment for heavy metals in the surface sediment of Hainan Island north bay, South China [J]. Environ. Sci. **39**(3), 1085–1094 (2018)
9. Zhang, P., Lai, J.Y., Dai, P.D., et al.: Spatiotemporal variation of COD and its contribution to the eutrophication in the Hainan Island Adjacent Seawater [J]. J. Guangdong Ocean Univ. **40**(2), 64–70 (2020)
10. Cui, M.K., Xu, S.L., Song, W.Q., Ye, H.B., Huang, J.L., Liu, B.H., Dong, B., Xu, Z.X.: Trace metals, polycyclic aromatic hydrocarbons and polychlorinated biphenyls in the surface sediments from Sanya River, China: Distribution, sources and ecological risk [J]. Environ. Pollut. **294**, 118614 (2022)
11. Redfield, A.C.: The influence of organisms on the composition of seawater [J]. Sea. **2**, 26–77 (1963)
12. Cao, H.S., Deng, Y.W., Chen, F.J., PU X Q.: Analysis of trace element characteristics and ecological risk assessment in the surface of sediments in the Zhanjiang Bay [J]. J. Ocean Technol. **3**(2), 71–77 (2020)
13. Zhang, J.B., Zhang, P., Dai, P.D., et al.: Spatiotemporal distributions of DIP and the eutrophication in Hainan Island adjacent coastal water [J]. China Environ. Sci. **39**(6), 2541–2548 (2019)

Composite Approaches for Sustainable Water Quality Improvement: A Case Study of the Artificial Lake at Guangzhou Foreign Language School Campus

Zhang Qin(✉)

Shanghai Yueli Education Technology Co., Ltd., Shanghai 200000, China
frankqin70320@gmail.com

Abstract. This study presents a comprehensive sustainable water quality improvement model and an indicator evaluation model specifically applied to the artificial lake at the Guangzhou Foreign Language School Campus. Situated adjacent to an active construction site, the lake's deteriorating water quality posed significant risks to both the local ecosystem and the health of students and staff. The results show that we observed a significant improvement in water quality over a treatment period of approximately 730 days. The WQI_{UL} increased from a bad level of 7.35 to a good level of 60.58. Key indicators such as dissolved oxygen and the permanganate index approached their target values, demonstrating substantial progress in the lake's ecological health. Furthermore, total phosphorus, ammonia nitrogen, nitrate, and suspended solids showed marked improvements, although they have not fully reached the desired targets. This case study underscores the effectiveness of composite water treatment strategies in rehabilitating small water bodies in educational environments. The findings highlight the potential for similar approaches to be implemented in other areas facing comparable environmental challenges, contributing to the broader field of sustainable water management.

Keywords: Sustainable water quality improvement model · Indicator evaluation model · Environmental Management

1 Introduction

The quality of water in small water bodies, particularly those in educational and residential areas, is an urgent and critical environmental concern that demands immediate attention. This issue is of utmost importance to environmental researchers, policymakers, and professionals in water management. In Guangzhou, a rapidly growing metropolis, these water bodies often face pollution challenges due to urbanization, increased human activity, and inadequate waste management. For example, Nansha District, located at the southern tip of Guangzhou, has undergone significant development in recent years, partly due to land reclamation projects. These projects have been essential for expanding available land for industrial, commercial, and residential use. However, land reclamation activities have considerable environmental implications, particularly on water quality.

V. Achal et al. (Eds.): EPPC 2024, ESE, pp. 60–71, 2025.
https://doi.org/10.1007/978-3-031-92123-0_6

Land reclamation, a process that typically involves extensive dredging and soil deposition, significantly impacts water quality [1]. It increases sediment load in nearby water bodies, reducing water clarity, affecting light penetration, and disrupting the aquatic ecosystem. Suspended solids can clog fish gills, smother benthic habitats, and alter the physical properties of the water. The materials used in land reclamation, including dredged sediments and construction materials, can introduce various pollutants into the water. These may include heavy metals, hydrocarbons, and other organic contaminants. The Permanganate Index values can indicate the presence of such pollutants, posing risks to aquatic life and human health. Meanwhile, runoff from newly created land, often laden with fertilizers and other organic materials, can increase concentrations of nutrients like Total Phosphorus and Ammonia Nitrogen. This nutrient enrichment can lead to eutrophication, characterized by excessive algal blooms and subsequent oxygen depletion. Besides, it can significantly alter the natural hydrodynamics of the coastal area. Changes in water flow patterns can affect the distribution and dilution of pollutants, leading to localized areas of poor water quality. These alterations can also impact tidal flushing, potentially leading to stagnant water and hypoxia.

Addressing the water quality in small water bodies is a complex issue that requires a comprehensive approach. This approach must encompass physical, chemical, biological, and bio-ecological restoration methods [2]. These diverse methods collectively enhance water quality. Physical methods focus on directly removing pollutants and improving water conditions. The first step, intercepting pollution sources and removing sludge, involves identifying and blocking the entry of contaminants while dredging and disposing of accumulated sludge. Water regulation equipment helps maintain optimal water flow and distribution, ensuring balanced water conditions. Aeration and oxygenation are essential for enhancing dissolved oxygen levels, promoting the natural degradation of organic pollutants through aerobic microbial activity [3]. Chemical methods address specific contaminants through various reactions. Coagulation and precipitation technology involves adding coagulants to aggregate and settle suspended particles for easier removal. Heavy metal fixation technology stabilizes and immobilizes heavy metals in the sediment, preventing their spread in the water. Robust oxidation technology uses oxidizing agents to break down complex organic pollutants and pathogens, significantly improving water quality. Biological methods utilize natural processes to restore water quality. Microbial enhancement techniques introduce beneficial microorganisms to accelerate the breakdown of organic matter and pollutants. Aquatic plant restoration involves using water plants to absorb and assimilate nutrients and contaminants. Sediment biological oxidation promotes microbial activity within the sediment, enhancing the degradation of organic pollutants [4]. Bio-ecological methods combine biological and ecological principles for long-term water quality improvement. Aeration ecological purification integrates mechanical aeration with ecological systems to boost natural purification processes. Stabilization ponds (oxidation ponds) are shallow, engineered ponds that treat wastewater through natural biological processes, promoting algae and bacteria growth to degrade pollutants. Biological barrier restoration technology establishes biological barriers, such as constructed wetlands or riparian buffers, to intercept and treat pollutants before they enter the water body [5].

Therefore, this study has chosen the artificial lake at the Guangzhou Foreign Language School Campus in Nansha District as its case study. This lake is adjacent to an active construction site and has become a focal point of concern due to its deteriorating water quality. The pond, which accumulates runoff and wastewater from nearby construction activities, poses potential risks to both the local ecosystem and the health of students and staff.

Maintaining high standards of water quality is essential, especially in educational settings where the safety and well-being of individuals are paramount. Poor water quality in this context can lead to contamination of local water sources, health hazards, and a diminished learning environment. Addressing these issues through effective remediation strategies is crucial for protecting the immediate school community and setting a precedent in environmental stewardship.

This paper aims to assess the wastewater pond's current state, identify contamination sources, and propose viable remediation strategies. The study seeks to enhance water quality and ensure a safer environment for the school by employing a combination of water sampling, analysis, and treatment techniques. The findings of this research will contribute to a broader understanding of environmental management in educational settings and offer insights into similar cases elsewhere.

2 Methodology

2.1 Details of the Research Area

This research is centered on an artificial lake situated outside the premises of Guangzhou Foreign Language School, located in the Nansha District of Guangzhou. The region experiences a South Asian subtropical maritime monsoon climate, with an average temperature of 24.2 °C, a maximum temperature of 39.3 °C, and a minimum temperature of 6.3 °C. The average wind speed is 1.8 m/s, and the annual average precipitation is 2053.8 mm, with the peak rainy season falling between July and September, reaching 950.2 mm. The study area encompasses 14,190 m^2, consisting of three segments: the first water body covers 6000 square meters, the second water body covers 4000 m^2, and the third water body covers 4190 m^2. The average water depth is around 1.2 m, with the deepest point reaching 2.0 m (see Fig. 1).

Fig. 1. Research area.

2.2 Sampling Method

This study employed equal proportional composite sampling to assess water quality in the research area. This approach evaluates overall water quality before sewage treatment and the improvements in water quality post-treatment. The water body was divided into three regions, with water quality monitoring points (SS1-SS7) distributed, as shown in Fig. 2. SS1 and SS2 are located at the 1/3 and 2/3 points along the central axis of region 1. SS3 and SS4 are positioned at the corresponding locations in region 2. In region 3, SS5 and SS7 are located in the central area, with SS6 at the center of a fan-shaped area, forming a right angle with SS5, SS6, and SS7 [6].

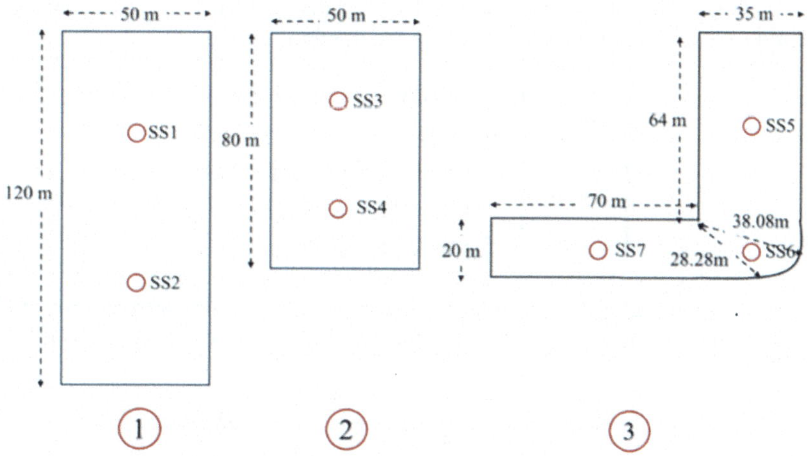

Fig. 2. Sampling site for three research areas.

2.3 Monitor Indicators

After collecting water samples, SS1 and SS2 are mixed in equal proportions, SS3 and SS4 are combined in equal proportions, and SS5, SS6, and SS7 also make the same progress. The water quality indicators were measured, as shown in Fig. 3. This water area is designated for landscape purposes and is predicted to be classified as worse than Class V water. The primary monitor includes five indicators (shown as a blue circle in Fig. 3), demanding to satisfy the minimum requirements of contact water area. Functional indicators contain eight parts (shown as the middle circle in Fig. 3), designed according to environmental quality standards for surface water. Considering the long-term presence of a subway construction team near the water area, the potential of heavy metal pollution and the biochemical treatability should be quantified [7]. Therefore, additional indicators are added to the work (shown as the right circle in Fig. 3).

2.4 Indicator Evaluation Modelling

$$WQI_{UL} = \prod_{i=1}^{8} q_i^{w_i} \tag{1}$$

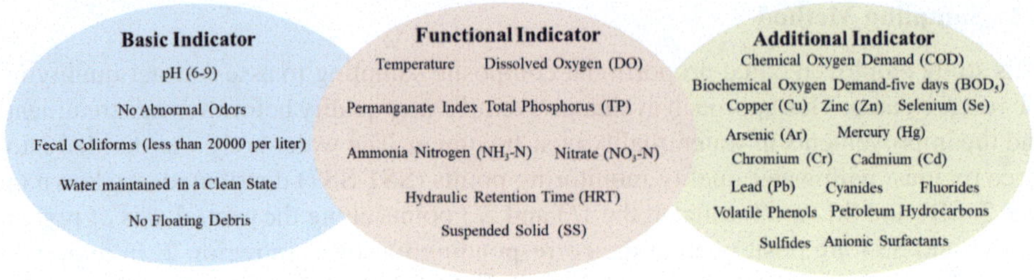

Fig. 3. Monitor indicators for the research area.

Where q_i is the quality of factor i ($i = 1$ to 8), the range is from zero to one hundred. w_i is the weight value of each factor i ($i = 1$ to 8), the total of w_i is 1.

$$q_i = \begin{cases} q_T = \begin{cases} 100 \quad , X_T \leq 12\,^{\circ}\mathrm{C} \\ 100 \times [1 - F(X_T)], X_T > 12\,^{\circ}\mathrm{C} \end{cases} \\ q_{DO} = \begin{cases} 100 \times 2[1 - F(X_{DO})], \quad X_{DO} > 1 \\ 100 \times \left[2F(X_{DO}), X_{DO} \leq 1 \right. \end{cases} \\ 100 \times [1 - F(X_i)] \end{cases} \tag{2}$$

Where $F(X_T)$ is the lognormal cumulative distribution function of temperature, $F(X_{DO})$ is for Dissolved Oxygen, $F(X_i)$ is for other factors, $F(X_i)$ can be calculated as following:

$$F(X_i) = \int_0^{X_i} \frac{1}{\sigma X_i \sqrt{2\pi}} e^{-\frac{(\ln X_i - \mu_i)^2}{2\sigma_i^2}} dX_i \tag{3}$$

$$X_i = \frac{a_i}{a_{i0}} \tag{4}$$

μ_i and δ_i are sample mean and bias, respectively. X_i is the test value to theoretical value ratio. a_i is the test value, and a_{i0} is the theoretical value. The test value will be taken through the method in Sects. 2.2 and 2.3. In this modelling, all parameters can be found in Table 1.

According to the water quality classification standard, the research area's WQI_{UL} is 7.35, which belongs to bad classification. Compared with the target water quality, which the WQI_{UL} value is 95.145, the improvement of permanganate index, total phosphorus, ammonia nitrogen, nitrate and suspended solid is much more critical [8]. And the water quality classification is shown in Table 2.

2.5 Water Quality Improvement Modelling

For the current water quality, in the absence of point source pollution in the research area, the water restoration, improvement, and ecosystem stabilization of ecological management technologies can be from the control of surface source pollution, contaminated substrate control and resource utilization, and water ecosystem restoration of the three key points of entry. The modelling route can be found in Fig. 4.

Table 1. Parameter list.

Indicator	μ_i	δ_i	w_i	a_{i0}	a_i
Temperature	0.256	0.123	0.058	20 °C	26 °C
Dissolved oxygen	−0.016	0.185	0.030	6 mg/L	1.8 mg/L
Permanganate index	0.793	0.15	0.105	4 mg/L	16 mg/L
Total phosphorus	1.129	0.402	0.153	0.1 mg/L	0.5 mg/L
Ammonia nitrogen	0.505	0.374	0.123	0.5 mg/L	2.5 mg/L
Nitrate	0.774	0.612	0.125	0.5 mg/L	2.5 mg/L
Suspended solid	1.250	0.418	0.261	10 mg/L	15 mg/L
Hydraulic retention time	1.333	0.346	0.145	3 days	10 days

Table 2. Water quality classification.

WQI_{UL} value	<30	30–60	60–80	>80
Classification	Bad	Average	Good	Excellent

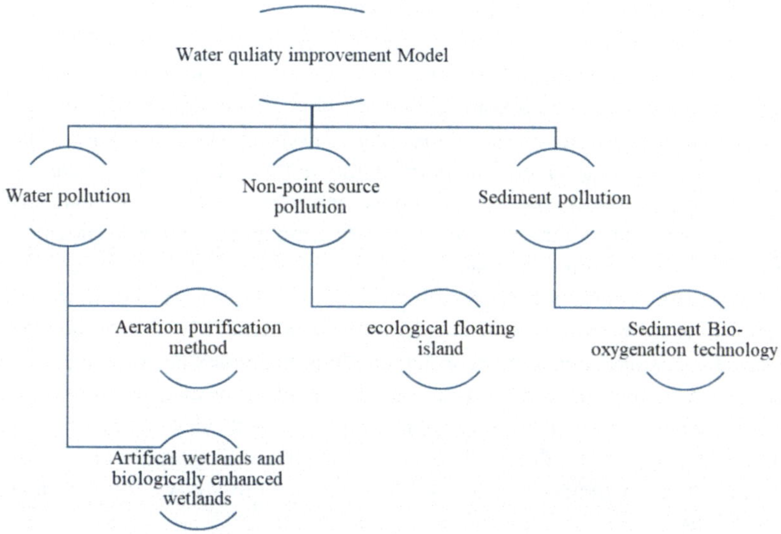

Fig. 4. Water quality improvement model

Firstly, the aeration purification method decomposes nutrients in the water body undergoing restoration, thereby increasing its oxygen content [9]. At the same time, it has uniformly suspended solid particles within the water, which enhances the contact surface area between microorganisms and the polluted water. This ensures that the microorganisms receive adequate dissolved oxygen to break down the pollutants in the water. Therefore, it is evident that the key technical factor in the aeration purification method for restoring water pollution is the management of the dissolved oxygen source and suspended solid.

This method can beautify the ecosystem with artificial wetlands and biologically enhanced wetlands. Biologically enhanced wetland enables the timely treatment of water sources through the cleaning capacity of microorganisms. In addition, through aeration recovery and treatment, as well as the diversity of species within oxidation ponds, the cleaning efficiency of oxidation ponds can be further enhanced.

Since the research area is adjacent to the campus, aesthetics and sustainability in the treatment process are particularly important. However, more than the area is needed to construct a large-scale ecological wetland park. Therefore, in this study, it is more appropriate to establish a miniature artificial wetland that combines purification and landscape effects.

Considering filler cost, effectiveness, availability, and engineering applicability, volcanic rock, vermiculite, and zeolite were selected as the main components of wetland fillers. The wetland functional plants were selected based on their tolerance to waterlogging and ammonia nitrogen in the water body [10]. Evaluation criteria included plant survival rate and leaf senescence rate. Consequently, eight plant species were identified as capable of average growth and water purification in aquatic environments: Alpinia oxyphylla, Alpinia zerumbet, Alpinia purpurata, Alpinia galanga, Alpinia speciosa, Alpinia mutica, Curcuma longa, and Curcuma zedoaria.

To address non-point source pollution and eutrophication, ecological floating islands, also known as ecological floating beds, are employed in constructing wastewater ecological restoration systems to reduce water pollutants [11]. The restoration system primarily includes relevant aquatic and terrestrial organisms, floating island structures, plant floating beds, and positioning devices. The purification method of ecological floating islands predominantly relies on aquatic plants, whose extensive root systems and adsorption capabilities enable comprehensive nutrient purification and pollutant removal.

Sediment Bio-oxidation of the substrate is an effective way to manage eutrophication in the water environment by promoting micro-populations growth in the substrate, thereby inhibiting the release of nutrients from the sediment [12]. The influence of pharmaceuticals on chemical substances in the substrate promotes the rate of air oxidation and degradation of the substances, thus promoting the rate of air oxidation of black odorous substances in the substrate, suppressing the black odor condition to a certain extent, and improving the quality of the water environment to a significant extent. Based on the interdisciplinary methodologies, Table 3 contains this study's input indicators and parameter values.

Table 3. List of parameters in modelling.

Parameter	Unit	Value	Definition
External ammonia input	mg/L/day	0.1	The amount of ammonia nitrogen entering the water body from external sources
External nitrate input	mg/L/day	0.05	The amount of nitrate nitrogen entering the water body from external sources
External total phosphorus input	mg/L/day	0.02	The amount of total phosphorus entering the water body from external sources
External organic matter input	mg/L/day	0.2	The amount of organic matter entering the water body from external sources
External suspended solids input	mg/L/day	0.1	The amount of suspended solids entering the water body from an external source
Nitrification rate constant	1/day	0.1	The rate at which ammonia is converted to nitrate by bacteria in the water
Denitrification rate constant	1/day	0.05	The rate at which nitrate is converted to nitrogen gas by bacteria in the water
Plant uptake rate constant	1/day	0.05	The rate at which plants absorb nutrients from the water
Sediment release rate is constant	1/day	0.02	The rate at which nutrients and other substances are released from the sediment into the water
Organic matter decomposition rate constant	1/day	0.05	The rate at which aquatic plants and algae produce oxygen through photosynthesis
Respiration rate constant	1/day	0.02	The rate at which organisms consume oxygen in the water through respiration
Photosynthesis rate constant	1/day	0.03	The rate at which aquatic plants and algae produce oxygen through photosynthesis
Aeration rate	1/day	1	The rate at which oxygen is introduced into the water through aeration

(*continued*)

Table 3. (*continued*)

Parameter	Unit	Value	Definition
Sedimentation rate constant	1/day	0.01	The rate at which suspended solids settle out of the water column and into the sediment
Cooling rate constant	1/day	0.01	The rate at which the temperature of the water is reduced
Ambient temperature	°C	26	The temperature of the surrounding environment

3 Results and Discussion

Since temperature and hydraulic residence time had little effect on the results and geographic conditions influence these two factors, there is no way to analyze them singularly in terms of quantitative metrics. In this section, this paper discusses the treatment of the other six influencing factors. The water quality treatment time was based on a 2-year (730 days) cycle, considering plant growth and the environmental impacts of the four seasons.

The concentration of dissolved oxygen, permanganate index, total phosphorus, ammonia nitrogen, nitrate and suspended solid changes over 730 days can be found in Fig. 5. The results show that the dissolved oxygen and permanganate index results are close to the target. Besides, the total phosphorus, ammonia nitrogen, nitrate and suspended solids are much better than before but don't achieve the target. Because the leading technology is aeration purification, DO levels can be relatively easily increased through aeration techniques, introducing atmospheric oxygen into the water. Meanwhile, Aquatic plants and algae produce oxygen during photosynthesis, contributing to the increase in DO, especially in well-lit environments. The Permanganate Index measures the amount of organic matter that can be oxidized in the water. The combined Bio-oxidation method effectively breaks down organic matter, reducing the permanganate index.

Phosphorus often comes from agricultural runoff (fertilizers), detergents, and sewage. It binds strongly to soil particles and sediments, making removing it difficult. Phosphorus can be recycled back into the water column from sediments, especially under anoxic conditions, making it harder to control in the long term [13]. Nitrates are highly soluble and can easily leach into groundwater, making them persistent in the environment. Denitrification requires anaerobic conditions and a carbon source, which can be challenging to maintain consistently. Additionally, high nitrate levels in inflow water make the process more demanding. Significantly, the observed pattern of nitrate concentration initially increasing and then decreasing during water pollution treatment is due to the sequence of nitrification and denitrification processes. Nitrification converts ammonia to nitrate under aerobic conditions, causing the initial rise in nitrate levels. Later, under anaerobic conditions, denitrification reduces nitrate to nitrogen gas, leading to a decline in nitrate concentration [14]. Given the conditions near the specific research

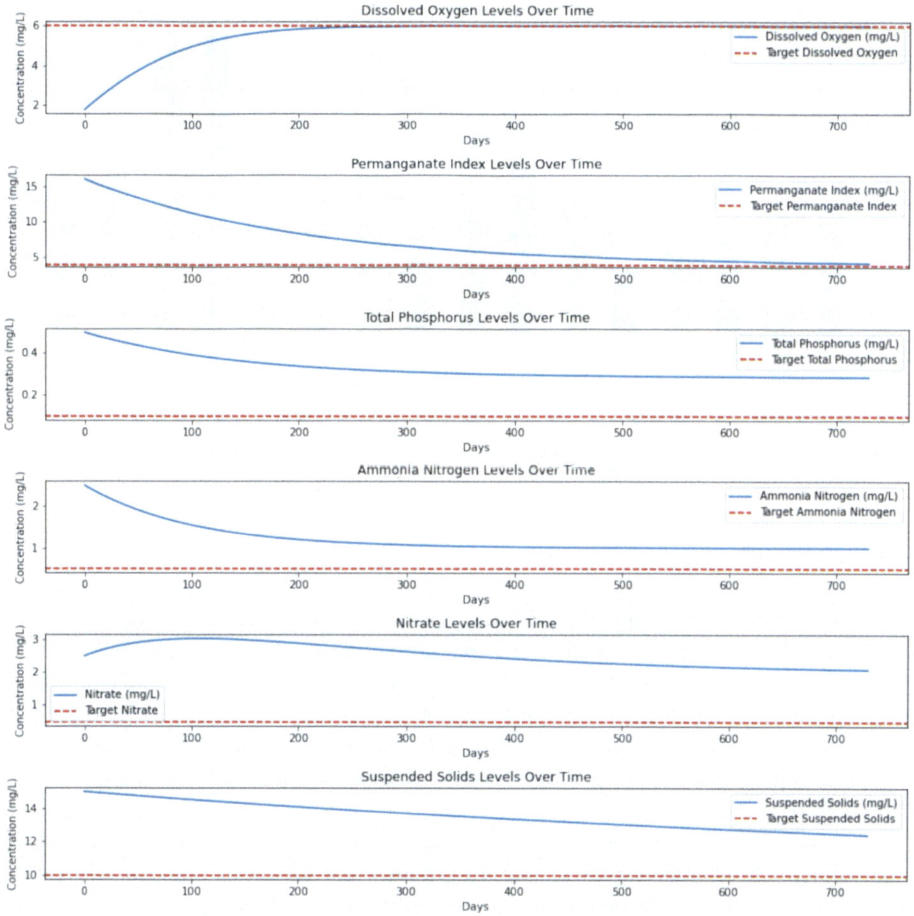

Fig. 5. Six indicators concentration change over 730 days

area, where long-term construction and frequent rainfall in Guangzhou are factors, managing suspended solids becomes particularly challenging. Construction activities disturb the soil, generating large amounts of dust and debris. This particulate matter can quickly enter nearby water bodies, increasing the concentration of suspended solids.

$\boldsymbol{a_i}$ and $\boldsymbol{a_{i0}}$ present the current and target water quality. The water quality after 730 days of treatment is shown in Fig. 5. Six indicators concentration change over 730 days. According to these results, WQI_{UL} value for each indicator and WQI_{UL} both can be calculated (as shown in Fig. 6). The WQI_{UL} value of 730-day treatment is 60.58, much better than 7.35. The classification improves from 'Bad' to 'Good' level, indicating that the treatment solution and modelling are effective.

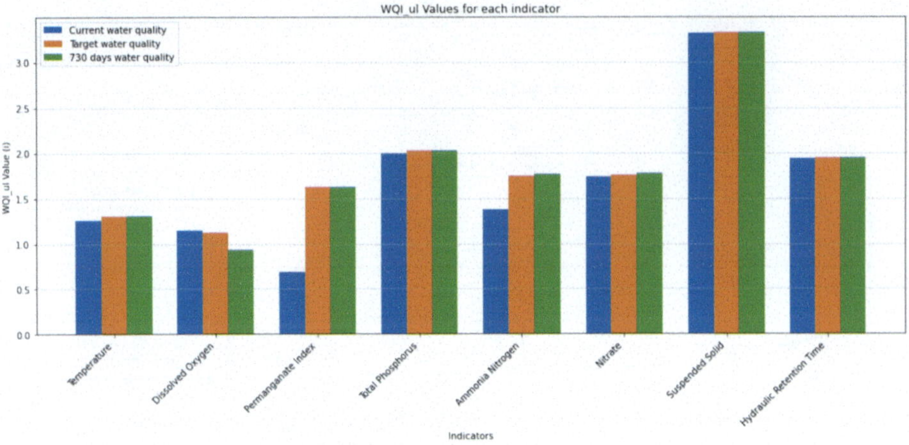

Fig. 6. WQI_{UL} values for each indicator.

4 Conclusion

This study proposes a composite treatment method for small-scale school water bodies. The environmental modification and overall budget are limited compared to large-scale water pollution treatment projects. Therefore, this paper must make the most beneficial plan under limited economic and engineering conditions. By comprehensively considering various environmental factors and optimizing resource allocation, this research provides a feasible and efficient solution for water quality management in school environments.

This paper divides the entire study area into three sections during the modelling establishment. It establishes seven sampling points and identifies eight highly significant indicators, such as Temperature, Dissolved Oxygen, Permanganate Index, Total Phosphorus, Ammonia Nitrogen, Nitrate, Suspended Solids, and Hydraulic Retention Time. This paper proposes a composite method for water quality improvement and a method for water quality evaluation.

The results indicate that the modelling has significantly improved water quality. Initially categorized as 'bad' level, the water quality improved to the 'good' level after 2 years (730 days). Although the indicators for Total Phosphorus, Ammonia Nitrogen, Nitrate, and Suspended Solids have yet to reach their optimal values, this paper identifies effective treatment methods. By giving more time, these parameters will also return to usual standards.

The primary contribution of this research is the integration of various environmental factors into a comprehensive predictive model, which other educational institutions can utilize to manage and improve water quality in similar settings. By providing a robust model and treatment plan, this paper offers a valuable tool for schools to ensure healthier and more sustainable water resources, which is essential for the well-being of students and the surrounding community. However, this study has certain limitations. Regional environmental variations and specific site conditions could influence the model's accuracy. Future research should consider these factors and expand the model to accommodate a broader range of scenarios. Additionally, future studies could examine the long-term impact of implementing such treatment solutions on the overall ecosystem health and student engagement in environmental stewardship.

References

1. Yan, F., Wang, X., Huang, C., Zhang, J., Fenzhen, S., Zhao, Y., Lyne, V.: Sea reclamation in mainland China: process, pattern, and management. Land Use Policy. **127**, 106555 (2023)
2. Gadipelly, C., et al.: Pharmaceutical industry wastewater: review of the technologies for water treatment and reuse. Ind. Eng. Chem. Res. **53**(29), 11571–11592 (2014)
3. Ahmed, S.F., Mofijur, M., Nuzhat, S., Chowdhury, A.T., Nazifa Rafa, M., Uddin, A., Inayat, A., et al.: Recent developments in physical, biological, chemical, and hybrid treatment techniques for removing emerging contaminants from wastewater. J. Hazard. Mater. **416**, 125912 (2021)
4. Arabzadeh, M., Eslamidoost, Z., Rajabi, S., Hashemi, H., Aboulfotoh, A., Rosti, F., Nazari, F., Borj, B.P., Hajivand, M.: Wastewater quality index (WWQI) as an indicator for the assessment of sanitary effluents from the oil and gas industries for reliable and sustainable water reuse. Groundwater Sustain. Dev. **23**, 101015 (2023)
5. Abbasi, H.N., Feng, X., Xiwu, L.: A modified bio-ecological process for rural wastewater treatment. Appl. Sci. **7**(1), 66 (2017)
6. Arun, S., Xin, L., Omkar Gaonkar, B., Neppolian, G.Z., Chakraborty, P.: Antibiotics in sewage treatment plants, receiving water bodies and groundwater of Chennai city and the suburb, South India: occurrence, removal efficiencies, and risk assessment. Sci. Total Environ. **851**, 158195 (2022)
7. de Paul Obade, V., Moore, R.: Synthesizing water quality indicators from standardized geospatial information to remedy water security challenges: a review. Environ. Int. **119**, 220–231 (2018)
8. Maansi, R.J., Wats, M.: Evaluation of surface water quality using water quality indices (WQIs) in Lake Sukhna, Chandigarh, India. applied water. Science. **12**, 1–14 (2022)
9. Kader, S., Raimi, M.O., Spalevic, V., Iyingiala, A.-A., Bukola, R.W., Jaufer, L., Butt, T.E.: A concise study on essential parameters for the sustainability of lagoon waters in terms of scientific literature. Turk. J. Agric. For. **47**(3), 288–307 (2023)
10. Yang, J.: et al. The impact of urbanization on wetland conservation: on the conservation of Nansha Binhai Wetland in Guangzhou City. (2023)
11. Zheng, Q., Yun, H., Lin, H.: Ecological designed experiment method based on pragmatism: a case study of Haizhu wetland restoration project in Guangzhou, China. Landsc. Archit. Front. **12**(1), 66–87 (2024)
12. Ni, S., et al.: Demonstration research project of a new three-stage bio-oxidation pond for purifying black smelly water bodies. J. Water Process. Eng. **47**, 102695 (2022)
13. Li, Y., et al.: Advances in the treatment of phosphorus-containing wastewater. IOP Conf. Ser. Earth Environ. Sci. **647**(1) IOP Publishing (2021)
14. Thakur, I.S., Medhi, K.: Nitrification and denitrification processes for mitigation of nitrous oxide from waste water treatment plants for biovalorization: Challenges and opportunities. Bioresour. Technol. **282**, 502–513 (2019)

Study on Water Conservation Services of the Jinghe River Basin Ecosystem in the Context of Climate Change

Shuyan Cai, Xiaoyu Zhang, and Zixiang Zhou(✉)

Xi'an University of Science and Technology, Xi'an 710699, Shaanxi, China
zhouzixiang@xust.edu.cn

Abstract. Over the past few years, amidst the dual challenges of climatic change and anthropogenic impacts, comprehensively investigating the effects of shifting climates and vegetation alterations on hydrographic cycles has emerged as pivotal for the Loess Plateau's sustainable future. This paper examines the land utilization and cover changes across different times and locations in the Jinghe River Basin over the past two decades. Furthermore, the study predicts prospective patterns of land utilization in the basin and examines the characteristics of the basin's ecosystem conservation of water across different times and locations. Additionally, the study forecasts the characteristics of ecosystem conservation services of water under disparate climate change and land utilization scenarios in the medium and long term. The study revealed that the conservation services of water in the basin across different times and locations from 2000 to 2020 exhibited a cyclical pattern. The protection of forests and other high-nutrient-rich land types, along with the restriction of land utilization expansion in the scenario of ecological protection, are identified as key factors that enhanced the ecosystem's conservation services of water.

Keywords: water conservation services · land utilization · NDVI · SWAT · LSTM · PLUS

1 Introduction

The functions that ecosystems perform, such as regulating river runoff, trapping precipitation, purifying water, improving regional hydrological conditions and maintaining ecological security, are collectively referred to as ecosystem water conservation services [1]. The conservation of water plays a crucial role in upholding ecological well-being and fostering societal progress, and is a key factor in ensuring human-water harmony. In recent times, a multitude of human activities have resulted in the irrational utilization of water resources [2, 3], leading to the overexploitation of vital ecological zones and a decline in hydrological regulation. Consequently, the quantitative assessment and in-depth investigation of ecosystem water conservation services have emerged as a prominent area of research [4–6]. Serving as the heartland of China's ecological security barrier, the study of the water conservation of the Loess Plateau has an irreplaceable strategic value.

V. Achal et al. (Eds.): EPPC 2024, ESE, pp. 72–81, 2025.
https://doi.org/10.1007/978-3-031-92123-0_7

Particularly amidst the scenarios of climate variability and vegetative fluctuations, the effect of these elements on hydrological processes has turned into a central concern for securing the sustainable progression of the Loess Plateau, and therefore in-depth research is needed to comprehensively analyze their complex effects.

2 Data and Methods

2.1 Summary of the Research Zone

Nestled in the heart of China's Loess Plateau, lies the basin of the Jinghe River, and it constitutes a secondary stream of the Yellow River. The chief river flows for an aggregate distance of 455.1 km, while the geographic coordinates of the basin lie between 106°14′–108°42′ E longitude and 34°46′–37°19′ N latitude. The total area covered by the basin is approximately 45,421 km^2, exhibiting a mean yearly temperature around 8 °C, an average yearly rainfall spectrum of 400–800 mm, and water resources that are relatively scarce.

The distribution of land utilization within the basin is markedly heterogeneous, featuring extensive stretches of farmland, grassland and woodland. Cultivated land and grassland are distributed in the central part of the basin, while forest land is concentrated at the western and eastern rims in the basin. Primarily, construction land is situated in the downstream area of the Jinghe River, with a small amount in the central part of the basin.

2.2 List of Data

① Digital Elevation Model (spatial resolution 30 m); ② Land use data (spatial resolution 30 m); ③ Soil data (resolution 1000 m); ④ Climatological information (Daily weather elements, 1997–2020); ⑤ MODIS data (MODIS 13Q1 data products);⑥ Geographic information data (Water systems, Railways, motorways, Important roads and primary to tertiary roads); ⑦ Socio-economic data (GDP data 1000 m, Population density data 1000 m); ⑧ Measured hydrological data (Daily runoff 2000–2020 measured data); ⑨ Future climate data (NEX-GDDP Data Products).

2.3 Research Methodology

1. Multi-model fusion

   i. Patch-generating Land Use Simulation (PLUS) model: to predict land utilization [7]; ii. Long Short-Term Memory (LSTM) network model: to model and forecast the future time trend of normalized vegetation index (NDVI) [8]; iii. Soil and Water Assessment Tool (SWAT) model: to simulate and predict the ecosystem conservation services of water [9]. Considering the influence of vegetation change factors, this paper introduced NDVI data into the SWAT model [10].
2. Technical route

   This paper employs the PLUS model to predict land utilization of the basin in the context of the three presetting situations of land utilization in the future. The LSTM network deep learning model was used to forecast the future NDVI in the context of the three climate scenarios: SSP126, SSP245, and SSP585. Furthermore, the basin was predicted for the years 2030 and 2060. Furthermore, this paper forecasts the ecosystem conservation services of water within the basin for the years 2030 and 2060 under nine integrated climate-land utilization scenarios.

3. Multi-scenario demand forecasting for land utilization
   i. Natural development: The natural development scenario assumes that the future trend of conversion of various land utilization types follows the development pattern from 2000 to 2020.
   ii. Ecological protection: In the ecological protection scenario, emphasis is placed on the protection of forest land, grassland and water resources, and moderate restrictions are placed on the rapid growth of urban construction land, thereby decreasing the 20% chance of transforming forest land, grassland, and water bodies into building land.
   iii. Economic growth: In the economic growth scenario, there's a 30% decrease in the odds of building land transitioning to different land utilization, juxtaposed with a 20% surge in the probability of other land types turning into building land.
4. Scenarios for ecosystem water services

   Based on the meteorological data of three climate scenarios (SSP126, SSP245, SSP585) and three land utilization/vegetation cover scenarios in the Beijing Climate Center—Climate System Model Version 2-Medium Resolution (BCC-CSM2-MR) applicable to China, nine integrated climate-land utilization scenarios have been constructed (Refer to Table 1).

Table 1. Future 9 scenarios set for the Jinghe River Basin

Future scenarios	SSP126	SSP245	SSP585
Natural Development (ND)	ND-SSP126	ND-SSP245	ND-SSP585
Ecological Protection (EP)	EP-SSP126	EP-SSP245	EP-SSP585
Economic Growth (EG)	EG-SSP126	EG-SSP245	EG-SSP585

3 Results

3.1 Spatial Variation Characteristics of Land Utilization

Figure 1 illustrates the proportion of land utilization types within the basin in 2000, 2010 and 2020. Over the 21-year period, the basin has undergone a significant process of urbanization, accompanied by the rapid expansion of land designated for urban construction, a reduction in the area of watersheds and a drastic decline in the area of arable land. This reflects the spatial disparities in regional economic growth and the diversification of land utilization, which can be attributed to the guiding effects of industrial structure adjustment, agricultural modernization and the initiative of reverting farmland to meadows and forests.

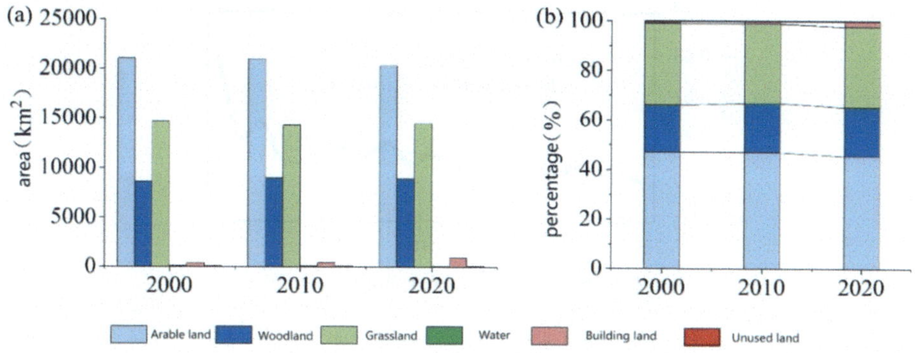

Fig. 1. Characteristics of land use structure, 2000–2020

3.2 Time-Related Fluctuation Traits of Land Utilization

This paper employs the land utilization transfer matrix to elaborate the evolution of land utilization within the basin over a period exceeding 20 years. Tables 2 and 3 illustrate the area transfer matrix of land utilization types in the basin from 2000 to 2010 and from 2010 to 2020, respectively.

Table 2. Matrix of land use area transfers, 2000–2010 (km^2)

2000	2010					
	Arable land	Woodland	Grassland	Water	Building land	Unused land
Arable land	22778.272	173.982	207.997	16.855	132.659	0.002
Woodland	182.282	9000.826	308.887	2.436	0.603	0
Grassland	221.166	718.269	15328.15	4.956	3.725	0
Water	42.265	2.73	5.759	24.036	0.119	0.05
Building land	55.926	0.466	2.993	0.085	310.956	0
Unused land	0.001	0	0	0	0	90.439

Table 3. Matrix of land use area transfers, 2010–2020 (km^2)

2010	2020					
	Arable land	Woodland	Grassland	Water	Building land	Unused land
Arable land	20681.054	504.594	1430.082	11.105	651.155	1.922
Woodland	308.728	8486.651	1088.212	3.928	8.719	0.035
Grassland	1429.194	861.138	13508.492	2.914	36.004	16.044
Water	28.485	2.089	4.345	11.8	1.649	0
Building land	86.798	2.466	4.067	0.273	354.458	0
Unused land	2.963	0.038	15.279	0	0.109	72.102

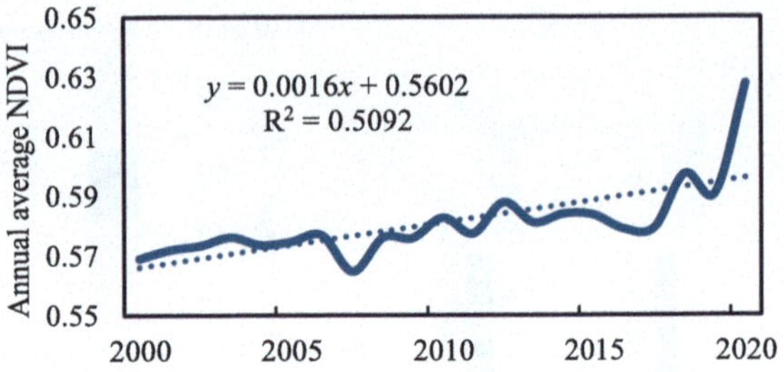

Fig. 2. Trends of NDVI, 2000–2020

Over the 21-year period, the inter-conversion of land utilization types was dominated by the transfer of farmland and grassland. The majority of farmland was converted to forest land and grassland, while the largest amount of transfer was observed for grassland. The decrease in the area of cultivable land and the substantial increase in building land can be attributed to the effective implementation of measures to return farmland to forest and grassland. The rapid economic growth, urbanization and population growth resulting from the fertility policy have also accelerated the expansion of construction land.

3.3 Characteristics of Temporal Changes in Vegetation Cover Factors

This paper employs the MODIS-NDVI product to analyze the chronological evolution of NDVI within the basin over the past two decades, as illustrated in Fig. 2. As illustrated in the figure, the annual average NDVI exhibited a discernible upward trajectory, indicative of a sustained enhancement in vegetation conditions. This can be attributed to the effective execution of the initiative of reverting farmland to meadows and forests.

In 2007, the NDVI in the basin exhibited a notable decline, reaching its lowest value over the past two decades. The occurrence of a severe drought in the Loess Plateau region during this same year [11] is posited to have contributed to the observed deterioration in vegetation condition.

3.4 Characteristics of Spatial Variation in Vegetation Cover Factors

From a spatial perspective, the vegetation condition of the watershed displays an encouraging improvement from north to south, notably, in the East and West border regions in the drainage basin. These regions exhibit exemplary vegetation coverage, with the Liupanshan mountainous region and the Ziwuling forested area standing out as notable examples of excellent natural conditions. The Slop River sub-basin, Puhe sub-basin, Heihe sub-basin, Yuanchengchuan sub-basin and Honghe sub-basin, along with their riparian periphery, are characterized by relatively abundant water resources and favorable soil conditions, which are reflected in relatively good vegetation conditions.

3.5 Characteristics of Temporal Changes in Ecosystem Water Conservation Services Simulated by the SWAT Model

1. Analysis of inter-annual variability: Fig. 3 illustrates the trends of precipitation, evapotranspiration and ecosystem conservation services of water over the course of the study period. As evidenced by the figure, there are notable inter-annual differences in ecosystem conservation services of water within the basin, with fluctuations closely correlated with the trends of precipitation and evapotranspiration during the same period.
2. Analysis of monthly changes: Fig. 4 illustrates the month-by-month changes in ecosystem conservation services of water over the past two decades. It demonstrates considerable variation on a monthly basis, with notable fluctuations in both positive and negative values. Negative values indicate a deficiency in precipitation within the basin, and that the ecosystem conservation services of water in the basin are in a state of deficit during that particular month.

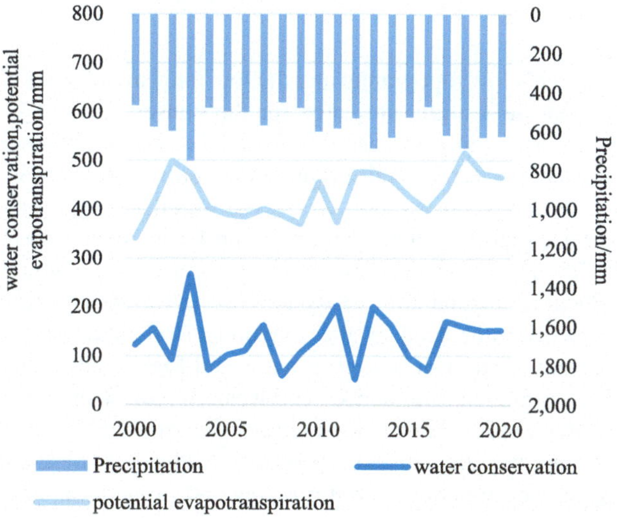

Fig. 3. Trend map of changes in water conservation services

3.6 Characteristics of Spatial Changes in Ecosystem Water Conservation Services Simulated by the SWAT Model

The north-western part of the basin exhibits significantly higher precipitation levels than the south-eastern part. The Hedaochuan sub-basin records the highest precipitation, followed by the Dongchuan, Wonsungchuan, Puhe and Ruhe sub-basins. Additionally, evapotranspiration is relatively high in each of these sub-basins, which results in the ecosystem conservation services of water remaining low. Despite the low precipitation levels observed in the southern part of the basin, evapotranspiration remains low. This is

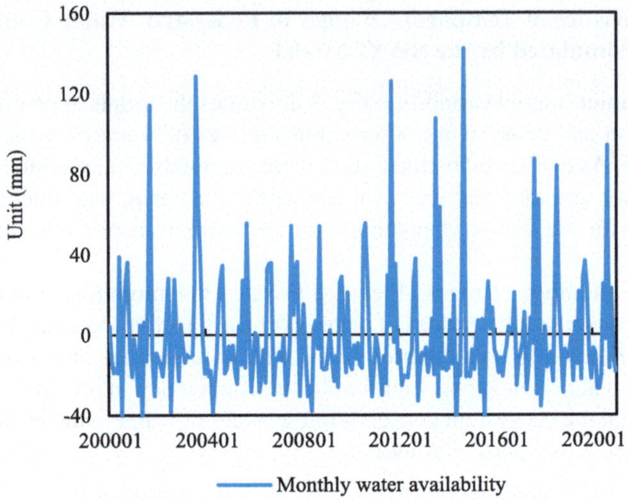

Fig. 4. Monthly changes in conservation services of water

attributed to the region's low topography and the low intensity of solar radiation, which collectively contribute to a high value area for the conservation of water within the basin's ecosystem. Furthermore, surface runoff exerts a certain influence on ecosystem conservation services of water. Effective reduction of surface runoff can also lead to an improvement in ecosystem conservation services of water, and vice versa. For instance, in areas where surface runoff is relatively high in sub-basins such as the Ru River, Hong River and Jinghe River, ecosystem conservation services of water are relatively low.

The geographical arrangement of ecosystem conservation services of water in the basin is predominantly influenced by the zonation of precipitation and temperature, showing a distribution pattern of low north-west and high south-east [12]. Secondly, topography exerts a significant influence. The western part of the basin exhibits relatively high runoff due to the relatively high terrain, as evidenced by the basins of Ruhe, Honghe, Jinghe (Kongtong-Jingchuan section), and Ruihezi. In these areas, the ecosystem conservation service function of water is low and prone to natural disasters such as landslides and soil erosion. Furthermore, the land utilization mode exerts a significant influence. The southernmost Slop River sub-basin in the downstream segment of the basin is situated in the Guanzhong Plain, characterized by low terrain, flat topography, extensive dry-crop farmland, and vegetation dominated by high-water-consuming crops. Consequently, the ecosystem conservation services of water within this basin are relatively low.

3.7 Projections of Land Utilization Change Under Three Scenarios

Relying on the forecast outcomes, the transformation in the expanse of land usage categories and the degree of fluctuation within the basin under the three scenarios were statistically calculated as shown in Fig. 5.

i. By 2030, under the natural development scenario, the areas of farmland, forested land, and unused land will have shrunk to varying degrees. The most conspicuous decline will be noted in the expanse of farmland, succeeded by a considerably lesser reduction in the domain of wooded areas, and culminating in the least diminishment affecting the portion of unutilized land. In contrast, the coverage of grasslands, water bodies, and building lands will have grown; In the situation of ecological protection, the area of building land exhibits the smallest increase in comparison to the other two scenarios; The economic growth scenario exhibits a significantly greater extension of building land than the other two situations.

ii. By 2060, the alterations in the expanse of farmland, forest land, and unused land under the natural development scenario are not significantly different from those in 2030. The area of all land categories except cropland and unused land increases; In the scenario of the ecological protection, the increase in building land remains relatively modest in comparison to the other two scenarios; In contrast, the economic growth scenario exhibits a notable rise in building land, which is considerably larger than in the other two scenarios.

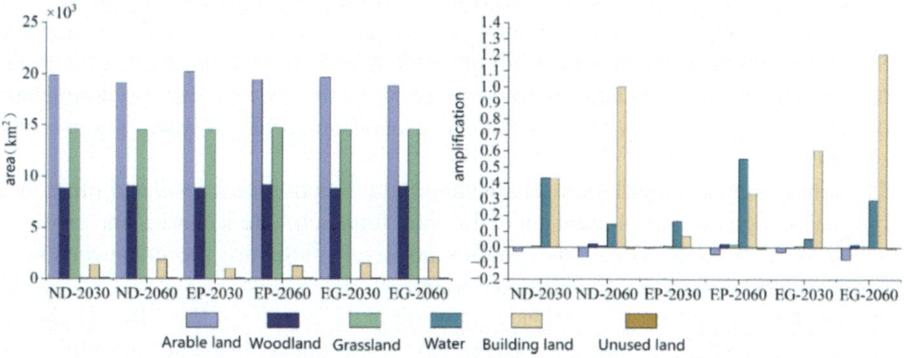

Fig. 5. Changes in the area of land use types under each scenario

3.8 Future Spatial Changes in Ecosystem Conservation Services of Water

Figure 6 illustrates the projected conservation services of water under nine composite scenarios for the years 2030 and 2060. The figure demonstrates that precipitation is the primary driver of the ecosystem's water services, influencing the amount of water that can be absorbed, stored, and purified by the ecosystem. Furthermore, the figure demonstrates that ecological protection strategies enhance vegetation cover and, consequently, the ecosystem's water services.

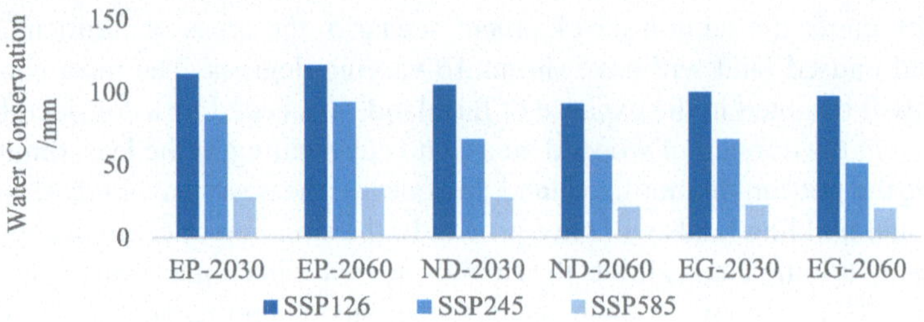

Fig. 6. Comparison of ecosystem conservation services of water for the nine composite scenarios for the future 2030 and 2060

4 Conclusions

1. The ecosystem conservation services of water changes across different times and locations over the past two decades exhibited a cyclical pattern.

   From 2000 until 2011, the peak value of water conservation services occurred in a 4-year cycle, with the lowest value observed in the subsequent year. From 2012 to 2020, a continuous downward trend was observed every 4 years, while the average annual change in ecosystem conservation services of water exhibited stability. Significant fluctuations in ecosystem of water were noted within the year, with the highest values concentrated in June–September and the lowest values predominantly occurring in spring and winter, when the conservation service of water was in a state of deficit.
2. The diminishment of carbon discharge and the protection of forests and other land types with high nutrient content under different future climate scenarios and medium- and long-term land utilization scenarios positively influence the enhancement of ecosystem conservation capabilities of water.

   The level of conservation services of water in the context of the different climate scenarios is SSP126 > SSP245 > SSP585, while under different land utilization scenarios, it is ecological protection > natural development > economic growth. The 'EP-SSP126' scenario is associated with the highest level of ecosystem conservation services of water in 2060. This is attributable to the protection of highly nutritious land types, such as forests, the imposition of restrictions on land expansion under the ecological protection scenario, and the highest total annual rainfall and relatively good vegetation cover.

References

1. Qing, L.: Research on the Water Conservation Service Function of Ecosystems in Liaoning Province Based on the InVEST Model. Liaoning Normal University (2022). https://doi.org/10.27212/d.cnki.glnsu.2022.000314
2. Junxia, Q.: A review of research on sustainable use of water resources in the Yellow River Basin. Water Safety. **7**, 94–97 (2023). https://kns.cnki.net/kcms2/article/abstract?v=bb_EGB5rVPLYbFEaAGNghzVaj9ldFUzKfwcC2nR-mDI_lYb4Wd8g_ltCUPO14VKeCNPqZm5dCwbBRAQOdjTtzrw4_6nzVVf1_C3STxiCFLa4mxgQPkvkL90qyw

FDpWcGfmJLIisM4q1pTTNLy3gfihqCaCHf82QG8b5_R5B9C6fvF46LhMqvnQ= = &uniplatform=NZKPT&language=CHS

3. Yu, J., Zhou, J., Zhao, J., et al.: Agroecological risk assessment based on coupling of water and land resources-a case of Heihe River Basin. Land. **12**(4), 794 (2023). https://doi.org/10.3390/land12040794
4. Wu, S., Zhou, W., Yan, K., et al.: Response of the water conservation function to vegetation dynamics in the Qinghai-Tibetan Plateau based on MODIS products. IEEE J. Sel. Top. Appl. Earth Observ. Remote Sens. **13**, 1675–1686 (2020). https://doi.org/10.1109/jstars.2020.2984830
5. Dongdong, W., Yangwen, J., Cunjian, N., et al.: Temporal and spatial changes of different functions of water conservation in the source area of the Yellow River in the past 60 years and their influencing factors. J. Ecol., 1–14 (2024-10-05) http://kns.cnki.net/kcms/detail/21.1148.Q.20240102.1011.002.html
6. Yan, X., Cao, G., Cao, S., et al.: Spatiotemporal variations of water conservation and its influencing factors in the Qinghai Plateau, China. Ecol. Indic. **155**, 111047 (2023). https://doi.org/10.1016/j.ecolind.2023.111047
7. Guofu, L., Shengyan, D.: Study on the impacts of climate and land use change on runoff change—A case study of the upper Yi River area in the Yilo River Basin. Geoscience **32**(5), 635–640 (2012). https://doi.org/10.13249/j.cnki.sgs.2012.05.018
8. Liang, X., Guan, Q., Clarke, K.C., et al.: Understanding the drivers of sustainable land expansion using a patch-generating land use simulation (PLUS) model: a case study in Wuhan, China. Comput. Environ. Urban. Syst. **85**, 101569 (2021). https://doi.org/10.1016/j.compenvurbsys.2020.101569
9. Wu, T., Fu, H., Feng, F., et al.: A new approach to predict normalized difference vegetation index using time-delay neural network in the arid and semi-arid grassland. Int. J. Remote Sens. **40**, 1–14 (2019). https://doi.org/10.1080/01431161.2019.1624870
10. Bai, J., Zhou, Z., Zou, Y., et al.: Watershed drought and ecosystem services: spatiotemporal characteristics and gray relational analysis. ISPRS Int. J. Geo Inf. **10**(2), 43 (2021). https://doi.org/10.3390/ijgi10020043
11. Li, C., Zheng, F.L., Liu, W.Z.: Analysis of spatial and temporal variations of extreme precipitation events on the loess plateau from 1961 to 2007. J. Nat. Resour. **25**(2), 291–299 (2010). https://doi.org/10.11849/zrzyxb.2010.02.013
12. Pan Tao, W., Shaohong, D., et al.: Spatial and temporal changes of water supply services in the ecosystems of the Sanjiangyuan area based on the InVEST model. J. Appl. Ecol. **24**(1), 183–189 (2013). https://doi.org/10.13287/j.1001-9332.2013.0140

Assessment of China's Water Ecological Security and Diagnosis of Obstacles Based on the DPSIR Model

Aishan Ye[1,2], Jiayi Cai[2], Xiaohua Li[2](✉), and Yangyang Deng[2]

[1] Faculty of Humanities and Social Sciences, Macao Polytechnic University, Macao 999078, China

[2] College of Yonyou Digital and Intelligence, Nantong Institute of Technology, Nantong 226002, China

1484043391@qq.com

Abstract. Water ecological security is a crucial element of national ecological civilization construction. Based on the DPSIR model, this study constructed an evaluation index system of China's water ecological security and applied the comprehensive index method and obstacle degree model to evaluate water ecological security and diagnose obstacle factors in China's 31 provinces from 2012 to 2022. The results show that: (1) China's water ecological security situation is severe, with an overall fluctuating downward trend. The performance of each subsystem of water ecological security varies. (2) The level of water ecological security in China's eight economic regions differs significantly, showing a spatial distribution pattern of stepwise decline from east to west. The eastern coast performs better than the national average, while the situation in the southwestern, northeastern, and northwestern regions is critical. (3) The main obstacles to China's water ecological security lie in the impact, status, and response subsystems, reflected in the poor condition of water environment, intensified impact of water problems, and insufficient governance response. (4) Water scarcity, groundwater over-exploitation, inadequate water pollution control, and insufficient water supply capacity are common issues across China's eight economic regions, but the focus varies regionally. Accordingly, recommendations are made to promote the steady improvement of China's water ecological security.

Keywords: water ecological security · DPSIR model · comprehensive index method · obstacle degree model

1 Introduction

Water is the lifeblood of economic and social development and a key element in the construction of ecological civilization. With the rapid development of China's economy and urbanization, problems such as water scarcity and deterioration of the water environment have become increasingly prominent, becoming bottlenecks restricting the sustainable development of the region [1, 2]. In 2023, per capita comprehensive water consumption

V. Achal et al. (Eds.): EPPC 2024, ESE, pp. 82–95, 2025.
https://doi.org/10.1007/978-3-031-92123-0_8

is 419 m^3, and water consumption per 10,000 yuan of gross domestic product (GDP) is 46.9 m^3. Compared to 2022, water consumption per 10,000 yuan of GDP and water consumption per 10,000 yuan of industrial added value decreased by 6.4% and 3.9%, respectively. However, there is still a significant gap compared with the world advanced level, with China's water consumption being approximately twice that of advanced countries. To ensure the steady progress of water ecological civilization construction, water ecological security has become a crucial frontier issue for green development [3]. China, a vast country with unbalanced regional development, generally faces common problems such as uneven spatial and temporal distribution of water resources, prominent contradiction between supply and demand, and degradation of water ecology. Different economic regions have distinct differences in water resources development and utilization, industrial structure, and pollution discharge, making the regional water ecological security situation more complex and variable [4]. Given the significant differences in resource and environmental endowments, development stages, and problems faced by different regions in China, it is necessary to strengthen the comprehensive assessment and diagnosis of obstacles to water ecological security in various economic regions. This will not only help reveal the spatial and temporal patterns of water ecological security and obstacles in each region but also provide an important reference for formulating water resources management policies and ecological civilization construction measures tailored to local conditions.

2 Study Design

2.1 Construction of China's Water Ecological Security Evaluation Index System

The DPSIR model consists of five subsystems: driving, pressure, state, impact, and response [5]. The driving is generally understood to be the primary factor that induces problems in water ecological security and is mainly composed of population and economic development indicators for each province. The pressure refers to the direct factors that contribute to the degradation of the water environment, mainly consisting of water consumption data and water quality data, with water consumption indicators often having a significant impact on the water environment. The state represents the current condition and status of the water ecosystem and is composed of water consumption and water resource indicators. The impact can be understood as the changes triggered or brought about by the current state of water ecology, caused by both human and non-human factors, and is mainly composed of wastewater treatment and urban infrastructure indicators. The response is usually regarded as the measures taken by human activities to enhance and improve water ecological security, which is primarily composed of water conservancy and water supply indicators. A more comprehensive set of evaluation indicators allows for better categorization and a holistic consideration of the integrated effect of each element on the system. The water ecological security evaluation index system constructed in this study is shown in Table 1.

2.2 Research Methodology

This study divided 30 indicators from 31 provinces in China into DPSIR model based on their time and stage of impact on water ecology. To ensure objective and accurate

Table 1. China's water ecological security evaluation index system

Subsystem	Indicator	Unit	Attribute
Driving (D)	Urbanization rate (D1)	%	–
	Per capita GDP (D2)	yuan/person	–
	Industrial added value (D3)	100 million yuan	–
	Value added of agriculture, forestry, animal husbandry and fishery (D4)	100 million yuan	–
	Per capita disposable income of rural residents (D5)	yuan	–
	Per capita disposable income of urban residents (D6)	yuan	–
Pressure (P)	Total agricultural water consumption (P1)	100 million cubic meters	–
	Total industrial water consumption (P2)	100 million cubic meters	–
	Total domestic water consumption (P3)	100 million cubic meters	–
	Total ecological water consumption (P4)	100 million cubic meters	–
	Chemical oxygen demand emissions (P5)	10,000 tons	–
	Ammonia emissions (P6)	10,000 tons	–
State (S)	Per capita water consumption (S1)	cubic meter/person	–
	Water consumption per 10,000 yuan of GDP (S2)	cubic meters/10,000 yuan	–
	Total water resources (S3)	100 million cubic meters	+
	Effective irrigated area (S4)	1000 hectares	+
	Surface water resources (S5)	100 million cubic meters	+
	Groundwater resources (S6)	100 million cubic meters	+
Impact (I)	Completed investment in industrial pollution control (I1)	10,000 yuan	+
	Completed Investment in wastewater treatment projects (I2)	10,000 yuan	+
	Integrated production capacity for water supply (I3)	10,000 cubic meters/day	+

(*continued*)

Table 1. (*continued*)

Subsystem	Indicator	Unit	Attribute
	Length of water pipeline (I4)	kilometers	+
	Length of urban drains (I5)	10,000 kilometers	+
	Green coverage of built-up areas (I6)	%	+
Response (R)	Urban water penetration rate (R1)	%	+
	Local finance expenditure on agriculture, forestry and water affairs (R2)	100 million yuan	+
	Total reservoir capacity (R3)	100 million cubic meters	+
	Total surface water supply (R4)	100 million cubic meters	+
	Total groundwater supply (R5)	100 million cubic meters	+
	Total other water supply (R6)	100 million cubic meters	+

Note: "+" indicates a positive indicator, "–" indicates a negative indicator.

evaluation of water ecological security in various Chinese provinces, the entropy weight method is employed to calculate the weight of indicators, followed by the application of the comprehensive index method for measurement. Simultaneously, each subsystem of the DPSIR model is evaluated and analyzed to understand the degree of influence of each subsystem. The water ecological security level is then determined according to the evaluation criteria, and the temporal changes and spatial distribution are comprehensively characterized and analyzed. Furthermore, the obstacle degree model is used to calculate the indicator obstacle degree, subsystem level obstacle degree, and main obstacle factors for the 31 provinces [6].

2.3 Evaluation Levels and Criteria

To better understand and manage water ecological security, following the usual practices in academia, the water ecological security status is classified into five levels [7]. The water ecological safety classification criteria are shown in Table 2.

2.4 Data Sources and Regional Division

This study comprehensively considers the impacts of human activities, economic and social environments, and selects 30 representative indicators related to economy, water resources, water quality, water use, greening, and water conservancy. The data for this study were obtained from the statistical yearbooks of each province and the website of the National Bureau of Statistics, with missing data supplemented by interpolation. The sample includes 31 provinces in China, excluding Hong Kong, Macao, and Taiwan due

Table 2. Criteria for classification of water ecological safety level

Evaluation results	Level	Description of water ecological security
(0–0.2)	Severe Warning (I)	Near collapse of water ecosystem services
[0.2–0.4)	Medium warning (II)	Significant damage to water ecology
[0.4–0.6)	Early warning (III)	Water ecology is prone to deterioration after disturbance
[0.6–0.8)	Relatively Safe (IV)	Water ecology issues are not significant
[0.8–1)	Safe (V)	Strong regenerative capacity of water ecosystems

to the lack of relevant data. The study period is selected from 2012 to 2022 due to the lag in the release of relevant data.

To present the differences in water ecological security levels among regions in China, the eight major economic regions are used as the main research objects. The basis for this division is derived from the research report "Strategies and Policies for Coordinated Regional Development" by the Development Center of the State Council. Accordingly, China is divided into the Northeast, Northern Coast, Eastern Coast, Southern Coast, Middle reaches of the Yellow River, Middle reaches of the Yangtze River, Southwest, and Northwest regions.

3 Results and Analyses of China's Water Ecological Security Assessment

3.1 Results of Water Ecological Security Assessment

By using the above method, the comprehensive evaluation results of water ecological security for 31 provinces in China from 2012 to 2022 can be obtained. Table 3 presents the water ecological security evaluation values of these provinces for selected years. The overall fluctuating and decreasing trend of China's regional water ecological security level from 2012 to 2022 reflects the persistent severity of the water ecological security situation in recent years. Specifically, during the period 2012–2015, the average value of water ecological security remained at a high level of about 0.336, which is at the lower limit of the early warning level. During this period, China's economy maintained relatively rapid growth, but environmental governance efforts were also constantly increasing, the overall water ecological security situation remained generally stable. From 2016 to 2018, the average value of water ecological security showed a continuous decline, from 0.329 in 2016 to 0.320 in 2018, approaching the upper limit of the medium warning level. This may be related to increased economic downward pressure during this period, relaxation of environmental regulations in some areas, and insufficient investment in water environment management. In 2019, the average value of water ecological safety rebounded slightly to 0.333, indicating that various regions had increased their efforts to prevent and control water pollution, leading to an improvement in water environment quality. However, in 2020, affected by the major infectious disease outbreak, the average value of water ecological safety declined again to 0.318. From 2021 to 2022, the average value of water ecological safety fluctuated slightly, at 0.337 and 0.321 respectively, generally remaining at the lower limit of the early warning level.

At the provincial level, Shandong, Jiangsu, Guangdong, and Zhejiang maintained relatively high water ecological safety evaluation results, generally between 0.4 and 0.6 over the years, corresponding to the early warning level. Despite being economically developed and investing heavily in governance, these provinces still face certain pressures on water ecological safety due to their dense populations and high development intensity. The evaluation results for Henan, Hubei, Sichuan, and other provinces mostly ranged between 0.35 and 0.45, generally transitioning from medium warning to early warning. Qinghai, Gansu, Shaanxi, Ningxia, and other provinces with relatively scarce water resources due to natural conditions generally had low water ecological security ratings, between 0.2 and 0.3, in the medium warning level. The water ecological safety status of Heilongjiang, Jilin, and Liaoning varied greatly, with Heilongjiang slightly better than Jilin and Liaoning. Most provinces showed a downward trend in their water ecological safety level, reflecting the increased pressure of economic and social development on the water ecological environment in recent years. However, a few provinces, such as Nei Mongol (from 0.334 in 2012 to 0.364 in 2022), Xinjiang (from 0.406 to 0.434), and Zhejiang (from 0.373 to 0.478), exhibited relatively stable or slightly increasing water ecological safety grades, indicating that water environment management in these areas has achieved certain results.

3.2 Overview of System Layers of Water Ecosystem Security

To comprehensively analyze the internal situation of water ecological security, Table 4 demonstrates the changes in the mean values of each system layer of water ecological security in China from 2012 to 2022. The data in Table 4 reveal that the driving subsystem layer has exhibited relatively stable values with small fluctuations over the past decade. It declined from 0.056 in 2012 to 0.053 in 2018, then increased slightly thereafter, reaching 0.056 in 2021 before declining to 0.055 in 2022. This indicates that the external factors driving water ecological security have remained stable to a certain extent, but it also reflects the gradual impacts of factors such as policies or economic development on water ecological security. The pressure subsystem layer has not changed significantly during this period, showing a "V" development trend. It gradually decreased from 0.044 in 2012 to 0.039 in 2019, indicating a short-term increase in external pressure on the water ecosystem. However, in 2020, the pressure subsystem layer suddenly rose to 0.047, which may be closely related to the increasing popularity of pollution control equipment. The state subsystem layer shows some fluctuation, dropping from 0.076 in 2012 to 0.065 in 2018, then rebounding to 0.063 in 2021 and rising slightly to 0.067 in 2022. This reflects a gradual stabilization of the water ecosystem's health after experiencing some degree of deterioration. However, the overall level remains low, indicating the need to address certain water ecology problems. The impact subsystem layer exhibits an "N" pattern of development during the study period, remaining between 0.071 and 0.083 from 2012 to 2019, then slightly decreasing in 2020 and returning to 0.074 in 2022. This suggests that the impacts on the water ecosystem have not significantly improved in

Table 3. Comprehensive evaluation results of water ecological security in 31 provinces in China

Province	2012	2014	2016	2018	2020	2022
Beijing	0.295	0.301	0.292	0.286	0.267	0.210
Tianjin	0.205	0.206	0.198	0.196	0.217	0.194
Hebei	0.405	0.407	0.378	0.408	0.354	0.344
Shanxi	0.286	0.275	0.272	0.306	0.297	0.251
Nei Mongol	0.334	0.336	0.337	0.373	0.348	0.364
Liaoning	0.348	0.318	0.314	0.297	0.305	0.306
Jilin	0.270	0.251	0.257	0.249	0.272	0.265
Heilongjiang	0.382	0.385	0.400	0.376	0.393	0.376
Shanghai	0.201	0.207	0.226	0.181	0.217	0.230
Jiangsu	0.492	0.489	0.535	0.529	0.558	0.396
Zhejiang	0.373	0.395	0.386	0.390	0.401	0.478
Anhui	0.340	0.350	0.395	0.358	0.374	0.332
Fujian	0.323	0.316	0.321	0.245	0.272	0.287
Jiangxi	0.365	0.336	0.342	0.328	0.331	0.323
Shandong	0.548	0.492	0.522	0.530	0.500	0.483
Henan	0.381	0.383	0.400	0.403	0.399	0.363
Hubei	0.365	0.359	0.385	0.346	0.373	0.401
Hunan	0.368	0.352	0.362	0.321	0.331	0.329
Guangdong	0.483	0.459	0.488	0.503	0.457	0.487
Guangxi	0.361	0.349	0.345	0.321	0.333	0.340
Hainan	0.206	0.192	0.187	0.180	0.199	0.201
Chongqing	0.226	0.221	0.217	0.211	0.252	0.228
Sichuan	0.459	0.409	0.395	0.403	0.420	0.379
Guizhou	0.292	0.289	0.264	0.251	0.288	0.300
Yunnan	0.359	0.341	0.356	0.353	0.360	0.374
Tibet	0.354	0.364	0.337	0.340	0.346	0.349
Shaanxi	0.287	0.255	0.249	0.243	0.277	0.247
Gansu	0.264	0.247	0.255	0.244	0.261	0.259
Qinghai	0.238	0.228	0.205	0.221	0.245	0.228
Ningxia	0.187	0.191	0.191	0.172	0.206	0.193
Xinjiang	0.406	0.404	0.395	0.365	0.010	0.434

recent years, or that people's awareness and attention to water ecology are insufficient, necessitating more attention to ongoing environmental impacts. The response subsystem layer shows a fluctuating downward trend during the study period, reaching a maximum value of 0.093 in 2021 before dropping slightly to 0.081 in 2022. This indicates a gradual strengthening of the response to water ecosystem protection, along with the enhancement of public environmental awareness and the implementation of relevant policies, reaching a historical high in 2021, which suggests that the state is actively responding to the issue of water ecosystem safety. Overall, China's water ecological security shows a degree of volatility during the 2012–2022 period, with an overall trend that is not optimistic across various indicators. Despite some alleviation at the pressure and impact subsystem levels, the state of water ecology remains worrying due to the combined effects of the driving, state, and response subsystem levels. In particular, the smoothness of the impact subsystem layer indicates that concern for water ecology issues at the public and policy levels needs to be further deepened.

Table 4. Mean values of water ecological security at various subsystem levels in China from 2012 to 2022

Year	Driving (D)	Pressure (P)	Status (S)	Impact (I)	Response (R)
2012	0.056	0.044	0.076	0.072	0.086
2013	0.056	0.042	0.072	0.074	0.096
2014	0.055	0.042	0.072	0.071	0.087
2015	0.054	0.042	0.078	0.074	0.088
2016	0.055	0.039	0.073	0.076	0.085
2017	0.055	0.039	0.069	0.078	0.086
2018	0.053	0.039	0.065	0.083	0.079
2019	0.055	0.039	0.071	0.079	0.088
2020	0.053	0.047	0.063	0.074	0.080
2021	0.056	0.056	0.063	0.071	0.093
2022	0.055	0.045	0.067	0.074	0.081

It is also necessary to analyze the weights of each subsystem layer of China's water ecological security to clarify the importance of each layer. As shown in Fig. 1, the weight of the response subsystem layer is the highest, fluctuating between 27.92% and 34.18%, generally maintaining a high level. This indicates that Chinese governments at all levels and various sectors are paying increasing attention to water ecological security, and governance and protection efforts have been continuously strengthened. The weights of the impact subsystem layer and the state subsystem layer are second only to the response subsystem layer, but much higher than those of the driving and pressure subsystem layers. The weight of the impact subsystem layer is generally rising, from 28.85% in 2012 to 34.11% in 2022, indicating the increasing impact of water ecological safety issues on sustainable economic and social development and people's quality of life. The weight of the state subsystem layer is maintained at a high level of more than 24% from 2012 to 2019, indicating that the water environment quality and water ecosystem health are generally good during this period. However, the weight decreases significantly to 21.33% after 2020, urgently requiring attention to strengthen the stability of the state subsystem layer. The weights of the driving and pressure subsystem layers are relatively low. The weight of the driving subsystem layer does not change much, indicating that factors affecting water ecological security, such as population and economic development, are relatively stable. The weight of the pressure subsystem layer shows a decreasing trend from 2012 to 2019, from 6.3% to 5.01%, reflecting an easing of pressure on water resource exploitation and use during this period. However, the pressure value rebounds sharply to 8.79% after 2020.

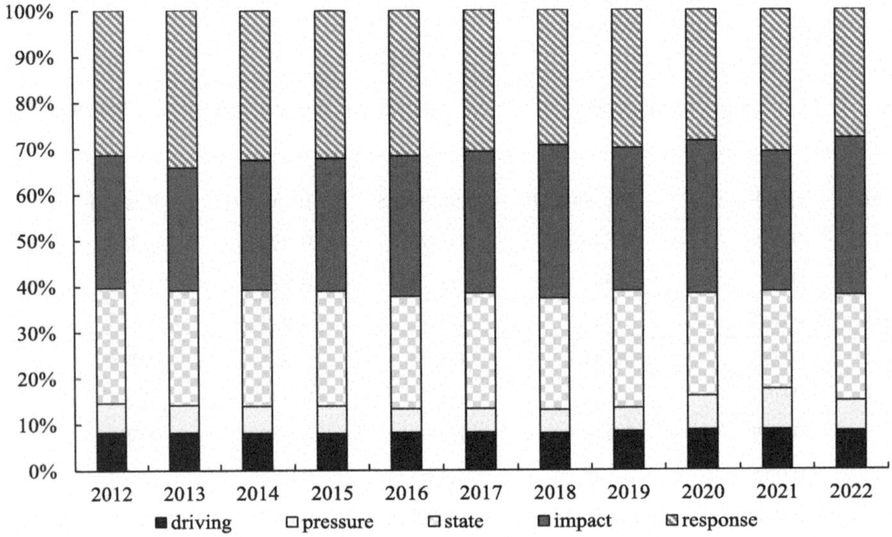

Fig. 1. Weighting values of each subsystem layer of water ecological security in China

3.3 Overview of Water Ecological Safety in China's Eight Economic Regions

Considering the differences in natural environment and social development among China's eight economic regions, a detailed analysis of each region is necessary. Figure 2 shows the development dynamics of water ecological security levels in these regions from 2012 to 2022. There are significant differences in water ecological security levels among the eight regions, with the Eastern Coast ranking first and the Northwest region last. The water ecological security levels exhibit a stepwise distribution: Eastern Coast > Middle Reaches of Yangtze River > Northern Coast > Middle Reaches of Yellow River > Southern Coast > Southwest Region > Northeast Region > Northwest Region. Specifically, the water ecological safety value of the Eastern Coast has been maintained between 0.35 and 0.39 for many years, generally better than the national average and in the middle to upper level of the early warning range. This is due to the region's strong economic strength, significant investment in governance, and more complete water environment infrastructure. However, a significant decline in 2021 and a rebound in 2022 indicate that the regional water ecological security situation remains unstable. The water ecological security value of the Southern Coast is generally lower than that of the Eastern Coast, fluctuating between 0.30 and 0.35, close to the national average. It dropped to 0.309 for two consecutive years in 2017 and 2018, which is at the upper-middle level of the medium warning range. This may be related to factors such as regional industrial restructuring and increased governance efforts. The Southern Coast's water ecological safety has improved since 2019, but progress has been slow. The trend of the water ecological safety value in the Middle Reaches of the Yangtze River is basically consistent with that of the whole country, showing an overall declining trend. It rebounded after falling to a low of 0.338 in 2018 and reached 0.346 in 2022, higher than the national average. This shows that the Yangtze River Economic Belt has achieved positive results in ecological environmental protection, but there is still a gap compared to the Eastern Coast.

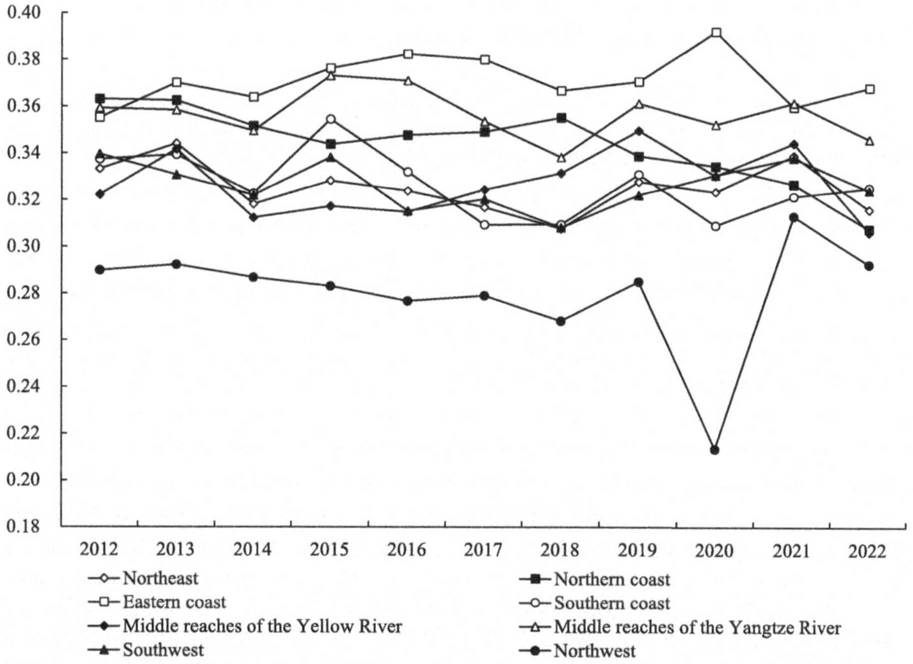

Fig. 2. Development Dynamics of Water Ecological Security Level in China's Eight Major Economic Regions from 2012 to 2022

The water ecological safety value of the Southwest Region is generally lower than the national average, fluctuating between 0.30 and 0.34. In 2018, it fell to a nearly 10-year low of 0.308 and has since rebounded, but in 2022, it fell again to 0.324. This reflects the grim situation of water ecology and environment in the Southwest Region, the constraints of natural conditions, and the weak foundation of governance, urgently requiring greater coordination and policy support. The water ecological safety values of the Northeast Region, the Middle Reaches of the Yellow River, and the Northern Coast show similar trends, generally exhibiting a "V" development pattern of initial decline followed by a rise. However, in the past 2 years, there has been a significant decline. In 2022, the regional water ecological safety values of the Northeast Region, the Middle Reaches of the Yellow River, and the Northern Coast were 0.316, 0.306, and 0.308, respectively, all in the middle to lower levels of the medium warning range. This is mainly attributed to the fact that although the Northeast Region has relatively abundant water resources, problems such as population outflow and industrial decline constrain the improvement of water environment quality. The Middle Reaches of the Yellow River face outstanding water shortage problems, high degrees of water resource exploitation, and a fragile ecological environment. The Northern Coast, on the other hand, experiences a decline in water ecological carrying capacity due to high population density and development intensity.

4 Analysis of Water Ecosystem Security Obstacles and Identification of Obstacle Factors

This section analyzes China's water ecological security from the perspectives of obstacle degree of each subsystem and individual indicators. Based on the obstacle degree model, the obstacle degree of each system layer of China's water ecological security from 2012 to 2022 was calculated. Table 5 shows the obstacle degree of each subsystem during the study period, revealing that the main barriers to water ecological security are the impact, state, and response subsystems. The driver, pressure, and impact subsystem layers showed an increasing trend in the degree of obstacles to China's regional water ecological security. The impact subsystem layer has the largest mean value and the largest increase, reflecting the continued increase in the negative impacts of water environment problems on economic and social development and people's well-being, making it the biggest obstacle to water ecological security. Although the obstacle degree of the response subsystem has slightly decreased in recent years, it remains a main obstacle subsystem layer, indicating that the effectiveness of response measures such as water resources control and water pollution control has improved, enhancing their role in guaranteeing water ecological security. The state subsystem layer ranks third, much higher than the driving and pressure subsystem, suggesting that water environment quality, water ecological health, and other state elements are also major obstacles affecting water ecological security. Overall, the obstacles to water ecological security in China are mainly reflected in the poor state of the water environment, the aggravation of the impacts of water problems, and the lack of governance response. The obstacle degree of driving and pressure is relatively low but has rebounded in recent years and should not be ignored.

Due to differences among China's eight economic regions, the obstacles to water ecosystems in each region also vary. Table 6 presents the main obstacles to water ecosystem security in these regions and their degree of obstacles. The main obstacles across all regions are the total amount of groundwater supply (R5) and the completed investment in wastewater treatment projects (I2). The indicator of total amount of groundwater supply

Table 5. Obstacle degree of subsystems in China's water ecological security from 2012 to 2022

Year	Driving (D)	Pressure (P)	Status (S)	Impact (I)	Response (R)
2012	4.24%	3.11%	26.28%	32.21%	34.16%
2013	4.17%	2.98%	26.92%	28.90%	37.03%
2014	3.97%	2.89%	26.96%	31.08%	35.10%
2015	4.04%	2.97%	25.93%	31.94%	35.13%
2016	4.13%	2.20%	25.62%	33.79%	34.26%
2017	4.21%	2.03%	27.26%	33.75%	32.75%
2018	4.17%	1.98%	26.27%	36.06%	31.53%
2019	4.60%	1.96%	27.55%	34.20%	31.68%
2020	4.96%	4.15%	23.41%	37.89%	29.59%
2021	4.82%	5.20%	22.58%	34.64%	32.75%
2022	4.64%	3.25%	24.08%	38.98%	29.05%

(R5) appears seven times in the table, indicating a widespread problem of groundwater over-exploitation, with groundwater resources under great pressure. This highlights the urgent need to strengthen groundwater management, reasonably control exploitation intensity, and promote water conservation measures. Investment in wastewater treatment projects (I2) also appeared seven times, reflecting the need to further increase investment in water pollution treatment in all regions. Wastewater treatment is a key initiative to safeguard water ecological security, requiring sustained investment to improve pollution control facilities and wastewater treatment capacity.

Table 6. Main obstacles to water ecological security in China's eight economic regions and the degree of obstacles

Region	Sequence 1	Sequence 2	Sequence 3	Sequence 4	Sequence 5
Northeast	I2(10.58%)	S5(8.87%)	R6(8.30%)	S3(7.88%)	I1(6.72%)
Northern coast	S5(10.52%)	S3(9.43%)	R5(9.06%)	I2(8.58%)	R4(7.65%)
Eastern coast	R5(15.14%)	S5(9.87%)	S3(8.87%)	R3(7.37%)	R6(6.99%)
Southern coast	R5(13.75%)	R6(9.13%)	I2(8.43%)	S5(7.40%)	S3(6.72%)
Middle reaches of the Yellow River	S5(9.54%)	I2(8.93%)	S3(8.50%)	R5(7.62%)	I4(7.34%)
Middle reaches of the Yangtze River	R5(13.22%)	I2(9.70%)	R6(9.07%)	S5(7.39%)	S3(6.65%)
Southwest	R5(13.53%)	I2(10.08%)	R6(8.70%)	I4(6.86%)	I1(6.57%)
Northwest	R5(10.24%)	I2(10.12%)	R6(8.82%)	I4(7.68%)	I5(7.02%)

While China's eight major economic regions share commonalities, there are also obvious differences. Water scarcity, over-exploitation of groundwater, inadequate water pollution control, and insufficient water supply capacity are common to all eight regions, but the focus varies. Northeast China, the Northern Coast, and the Middle Reaches of the Yellow River are mainly constrained by the total amount of water resources, while the Eastern Coast, Southern Coast, Southwest Region, and Northwest Region primarily face problems of groundwater over-exploitation and insufficient water-supply capacity. The Middle Reaches of the Yangtze River have obvious shortcomings in groundwater development and water-pollution management. Therefore, each region should consider its water resource endowment and major obstacle factors, taking measures to conserve water, treat pollution, and increase supply according to local conditions to overcome the bottleneck of water ecological security and achieve sustainable water resource use and healthy water environment development.

5 Conclusions

Based on the DPSIR model, a Chinese water ecological security evaluation index system was constructed, and the comprehensive index method and obstacle degree model were used to evaluate China's water ecological security and analyze the obstacles. The study found that: The situation of water ecological security in China is grim, with an overall fluctuating downward trend. Although it has been alleviated at the pressure and impact subsystem levels, the water ecological situation remains worrying due to the combined effects of the driving, state, and response subsystem levels. The water ecological security level of China's eight economic regions varies significantly, showing a stepwise distribution in decreasing order: Eastern Coast > Middle Reaches of the Yangtze River > Northern Coast > Middle Reaches of the Yellow River > Southern Coast > Southwest Region > Northeast Region > Northwest Region. The Eastern Coast performs better than the national average, while the situation in the Southwest, Northeast, and Northwest regions is grim, and the imbalance of development between regions is prominent. The main obstacles to China's water ecological security lie in the impact, state, and response subsystem layers, reflected in the poor state of the water environment, the aggravation of the impact of water problems, and the lack of governance response. The obstacle degree of driving and pressure is relatively low but has rebounded in recent years and should not be ignored. Water scarcity, over-exploitation of groundwater, inadequate water pollution control, and insufficient water supply capacity are prevalent in China's eight economic regions, but the focus varies by region. Water resource endowment and major obstacles should be considered when taking measures to conserve water, treat pollution, and increase supply according to local conditions to break the bottleneck of water ecological security.

In view of the above conclusions, the following three countermeasures are proposed:

Strengthen top-level design and improve the water ecological civilization system. Integrate the concept of water ecological civilization into all aspects of economic and social development, formulate a national master plan for the construction of water ecological civilization, improve the mechanism for dual control of total volume and intensity of water consumption, strictly manage the water resources system, and promote the transformation of water resources allocation from "water supply management" to "water demand management."

Adapt to local conditions and implement regional policies to promote water ecological security. Fully consider the differences in water resources endowment and major obstacles in each region. The Eastern Coast should focus on improving water resource utilization efficiency and strictly controlling groundwater exploitation. The Central and Western regions should vigorously develop water-saving agriculture and industry and accelerate water pollution prevention and control. The Northeast Region should coordinate surface water and groundwater development and strengthen cross-regional water resource allocation.

Strengthen multi-party synergy and build a comprehensive pattern of water ecological civilization construction. Increase investment in water conservancy and upgrade water conservancy infrastructure. Innovate water environment management modes and encourage social capital participation. Strengthen publicity and education on water ecology protection, raise public awareness of water conservation, and form a pattern of water ecology civilization construction built and shared by the whole society.

Acknowledgments. This study receives funding from the National Social Science Fund Project (22BKS143), the Philosophy and Social Sciences Research Project in Jiangsu Province's Universities (2024SJYB1261), and Jiangsu University 'Qinglan Project' funding project (Su Teacher Letter [2024] No. 14).

References

1. Li, Y.Q., Zhang, J., Song, Y.Y.: Comprehensive comparison and assessment of three models evaluating water resource carrying capacity in Beijing, China. Ecol. Indic. **143**, 109305 (2022). https://doi.org/10.1016/j.ecolind.2022.109305
2. Kong, Y., He, W.J., Gao, X., et al.: Dynamic assessment and influencing factors analysis of water environmental carrying capacity in the Yangtze River Economic Belt, China. Ecol. Indic. **142**, 109214 (2022). https://doi.org/10.1016/j.ecolind.2022.109214
3. Silva, R.H., Zwarteveen, M., Stead, D., et al.: Bringing Ecological Urbanism and Urban Political Ecology to transformative visions of water sensitivity in cities. Cities. **145**, 104685 (2024). https://doi.org/10.1016/j.cities.2023.104685
4. Chen, Y., Zhu, M.K., Lu, J.L., et al.: Evaluation of ecological city and analysis of obstacle factors under the background of high-quality development: taking cities in the Yellow River Basin as examples. Ecol. Indic. **118**, 106771 (2020). https://doi.org/10.1016/j.ecolind.2020.106771
5. Ntona, M.M., Busico, G., Mastrocicco, M., et al.: Coupling SWAT and DPSIR models for groundwater management in Mediterranean catchments. J. Environ. Manag. **344**, 118543 (2023). https://doi.org/10.1016/j.jenvman.2023.118543
6. Pang, B., Li, X.X., Fu, Y.: Coupling co-ordination analysis and obstacle factors of water-energy-environment-economy in the Yellow River Basin. J. Clean. Prod. **468**, 143108 (2024). https://doi.org/10.1016/j.jclepro.2024.143108
7. Xie, H.J., Li, K.F., Li, J.Q., et al.: Evaluation of water ecological security in Dianchi Lake basin based on DPSIR model. Environ. Protect. Sci. **47**(6), 94–99 (2021)

Hydrological Change Characteristics of a Lake Connected to Yellow River, Case Study in Lake Ulansuhai, China

Bojun Liu[1,2](✉), Jin Tang[1], Kefei Li[1,2], Yawei Zhao[1,2], and Shuntian Liang[1]

[1] Yellow River Engineering Consulting Co., Ltd., Zhengzhou 450003, China
bojun_1689@126.com

[2] Key Laboratory of Water Management and Water Security for Yellow River Basin, Ministry of Water Resources, Zhengzhou 450003, China

Abstract. Lakes, as an important type of water bodies, play indispensable roles in water regulation and storage, water ecological environmental protection and regional climate regulation. Many countries in the world are still faced with severe challenges of environmental sustainability in lake management, water resources use, and eco-environmental protection. The ecological pattern of lakes in China has also undergone new changes, such as lake water depletion, water ecology deterioration and regional climate revolution, which have brought new issues and challenges to the flood control safety and ecological security of lakes. River-connecting lakes are common all over the world, which are generally connected with big rivers to form water system. The regulation and storage of lake can impose significant effects on the inflow and the outflow between river and lake. The hydrological factors are important indexes for characterizing the change of lake. Analyzing the lake's hydrological variation characteristics can directly reflect the changes of the lake in water level and volume. Accordingly, this study focused on the Lake Ulansuhai in Inner Mongolia, China, and analyzed the variation characteristics of water level and outflow over the past years. The water level of the Lake showed significant increasing trend and may keep increasing in future from 1978 to 2019. The multi-year average annual outflow during the period from 1977 to 2017 was 8333.7 million m^3. On account of artificial regulation, the outflow from the total drainage to the Lake overall increased, without obvious regularity.

Keywords: Environmental sustainability · hydrological factor · change characteristics · water level · outflow · lake Ulansuhai

1 Introduction

Lakes plays an important role in environmental sustainability [1–3]. With the increasing impact of global climate change and human activities, and the acceleration of industrialization, extreme weather events occur frequently. Many countries in the world are still faced with severe challenges of environmental sustainability in lake management, water resources use, and eco-environmental protection. The ecological pattern of lakes

V. Achal et al. (Eds.): EPPC 2024, ESE, pp. 96–101, 2025.
https://doi.org/10.1007/978-3-031-92123-0_9

in China has also undergone new changes, such as lake water depletion, water ecology deterioration and regional climate revolution, which have brought new issues and challenges to the flood control safety and ecological security of lakes.

Climate change and economic and social development will change water circulation process in a lake, which can also lead to a series of problems including the reduction of water volume and the instability of lake area [4]. The hydrological characteristic of lake is an important factor for characterizing lake change. Analyzing the changes of water level and water volume is quite critical for lake management [5, 6]. Especially for the lakes that are connected with big rivers, water from the rivers can enter the lakes and then be discharged into rivers after the regulation and storage by the lakes. The hydrological variation characteristics of river-connected lakes are important basis for integrated lake management. Lake Ulansuhai, located in Inner Mongolia, is a typical inland lake and an important lake for connection in Yellow River basin. Some scholars conducted a great deal of research on the characteristics of climate change, the evolution and evaluation of lake water environment, migration and transformation mechanisms nutrients in freeze-up and non-freeze-up periods, plant community structure and distribution, water balance, and the variation characteristics of lake area [7], but less investigated the change of hydrologic features of Lake Ulansuhai. Accordingly, this study focused on Lake Ulansuhai and investigated the variation characteristics of water level and water discharge with multiple trend analysis methods. The present research results are expected to providing basis for achieving comprehensive management of river-connected lakes.

2 Study Area

Lake Ulansuhai is located in Bayannur City, Inner Mongolia, which is one of the important lakes in the Yellow River basin. The Lake Ulansuhai storage is 3.5×10^8 m^3 and its water mainly comes from Yellow River and the agricultural irrigation backwater of the Hetao irrigation area [8, 9]. The inflows of Lake Ulansuhai are received from the eighth channel, ninth channel, tenth channel, Tongji channel, and lake water flows into the Yellow River by the Wumaoji sluice (seen in Fig. 1).

3 Method and Data

The Mann-Kendall trend analysis test, ordered cluster T-test method and Hurst exponent method are employed to analyze the hydrological change characteristics of Lake Ulansuhai [10–12]. For the Hurst exponent method, if Hurst exponent equals to 0.5, i.e., $H = 0.5$, the past increment of the series at zero moment is irrelevant to the future increment at zero moment; if $H > 0.5$, the past variation trend of the series at zero moment affects the future variation trend in same direction (i.e., positive persistence effect); if $H < 0.5$, the past variation trend of the series at zero moment affects the future variation trend in opposite direction (i.e., negative persistence effect).

The series of lake water level data is from 1978 to 2019, and outflow data is from 1977 to 2017. All data are all collected from the Water Conservancy Development Center of Hetao Irrigation District in Inner Mongolia.

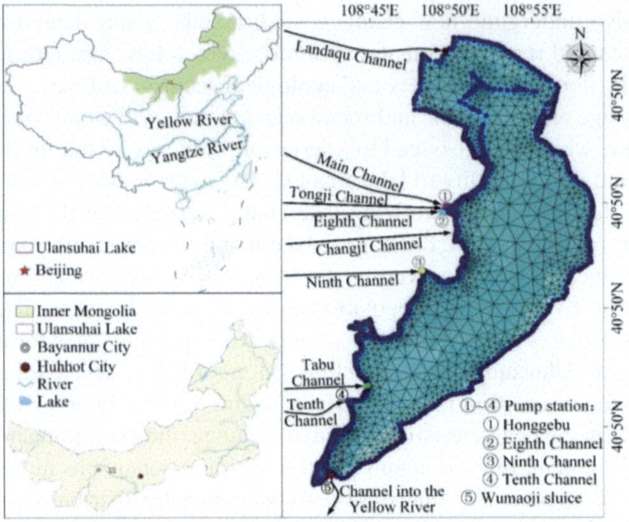

Fig. 1. Lake Ulansuhai location and its inflow and outflow relationship

4 Results and Discussion

4.1 Lake Water Level

Through analysis, the mean annual water level at the Honggebu station (the inlet of Lake Ulansuhai), i.e., the main inlet of the Lake from the total drainage, was 1019.06 m; to be specific, the maximum water level was 1019.47 m, which appeared in 2012 and 2014, respectively, and the minimum water level was 1018.73 m, which appeared in 1981. Over the past 42 years, water levels in 19 years exceeded the mean value, occupying 45.2% of the total statistical years. It suggests that mean annual water level at the inlet from main drainage to the lake did not reach the mean value in more than half of 42 years. It can also be observed from Fig. 2a that the interannual variation of water level at the Honggebu station ranged between 0 and 0.74 m, with uneven and overall increasing trend. According to trend test results, the Z value of water level at the Honggebu station was 5.762, and the trend passed that test at a confidence level of 0.01, i.e., the water level at the station showed significant increase trend. The mutation test results revealed that the mutational point of the water level at the Honggebu station occurred in 1995, 2003 and 2012, and the mutation results all passed the test at a confidence level of 0.01. as shown in Fig. 2a. In 1995, the water level showed the decreasing mutation; to be specific, the water level increased before 1995 and dropped after 1995. 2003 was determined as the increasing mutation point of the water level, suggesting that the water level at the station increased significantly after 2003. 2012 was determined as the decreasing mutation point, i.e., the water level at the station further decreased after 2012. Hurst exponent results demonstrated that the H value of the water level at the Honggebu station was 0.675 (>0.5), suggesting the positive persistence effect of historical water level on the future level. In other words, the water level at the Honggebu station may continue increasing in future.

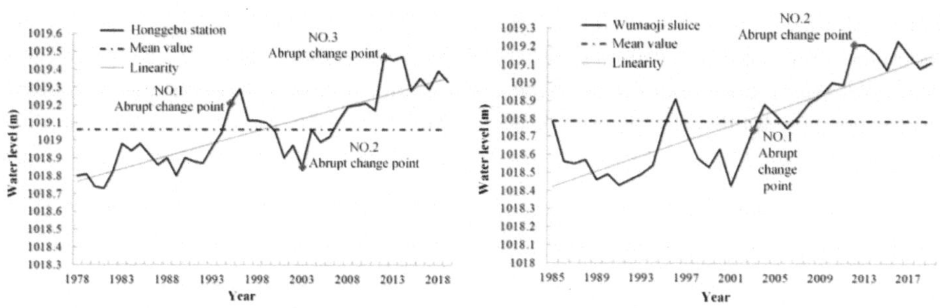

Honggebu station (b) Wumaoji sluice

Fig. 2. Water level changes characteristics.

During the period from 1978 to 2019, the mean annual water level at the Wumaoji gate was 1018.79 m, and the maximum and the minimum were 1019.23 m and 1018.43 m, respectively, which occurred in 1991 and 2001, respectively. Over the past 35 years, the annual water level exceeding the mean value occurred in 15 years, occupying 42.9% of the total statistical years. It suggests that over half of years at the outlet of Lake Ulansuhai did not reach the mean value over the past 45 years. As shown in Fig. 2b, the interannual variation of the water level at the Wumaoji gate ranged from 0 to 0.80 m, with significant change and overall increasing trend. The trend test results revealed that the Z value of the water level at the Wumaoji gate was 5.589, and the trend passed the test at a confidence level of 0.01, suggesting significant rising trend of the water level at the gate. According to the mutation test results, the mutational point of the water level at the gate occurred in 2003 and 2012, and the mutation results passed the test at a confidence level of 0.01, as shown in Fig. 2b. The years with mutation were coincident with the mutation years at the Honggepu station. In 2003, the water level at the Wumaoji gate increased abruptly, suggesting the sudden increase of the water level after 2003. 2012 was determined as the decreasing mutation point of the water level; in other words, the water level at the gate showed decreasing mutation after 2012. The Hurst exponent results also indicated that the H value of the water level at the gate was 0.656 (>0.5), suggesting that the historical water level imposed positive persistence effect on the future water level at the gate. Accordingly, the water level at the gate may keep increasing in the future.

4.2 Lake Outflow

The total drainage is the most important passage into the Yellow River. By analyzing the outflow from the total drainage to the Lake Ulansuhai, the total lake outflow reached up to 8333.7 million m^3, with an average annual outflow of 203.3 million m^3. The maximum and the minimum annual outflows to the Lake Ulansuhai were 660.7 million and 7.6 million m^3, respectively, which occurred in 2016 and 2003. The interannual variation of outflow varied within a range of 0–653 million m^3. Similarly, the outflow from the total drainage to the Lake Ulansuhai showed no irregular variation only increased on the whole. The outflow from the drainage to the lake exceeded 500 million m^3 in 2 years (2016 and 2017), and the outflow exceeded 600 million m^3 only in 2016. Water from the total drainage to the Lake Ulansuhai was used for supplementing the loss of water resources of the Lake. According to the trend test results, the Z value of the total outflow was 0.897, which did not pass the test at any confidence levels. It also suggests that the outflow from the total drainage to the Lake showed no significant increase. The mutation test results revealed that the outflow from the total drainage to the Lake showed mutation in 1983 and 2007. To be specific, the outflow decreased abruptly in 1983 and increased sharply in 2007. This can be regarded as the result of manual storage in accordance with actual water volume.

5 Conclusions

This study focused on the Lake Ulansuhai in inner Mongolia of China, and investigated the variation characteristics of water level and outflow of the Lake. According to the present results, the water level of the Lake Ulansuhai increased significantly during the period from 1978 to 2019, which may keep increasing in future; the average annual outflow in the period from 1977 to 2017 was 8333.7 million m^3 and the outflow from the total drainage to the Lake overall increased but showed no irregular variation rules because of the effects of artificial regulation; the multi-year average annual ecological water supplement during the period from 2008 to 2019 was 229 million m^3, which can play active roles in accelerating water flow and improving water quality of the Lake. In future, it is suggested to analyze the hydrological variation characteristics of the Lake Ulansuhai so as to provide valuable basis for comprehensive management of river-connecting lakes.

Acknowledgments. This work was supported by the National Key R&D Program of China (2023YFC3206703), the Development Funding Project by Water Youth Talent from Beijing Jianghe Water Development Foundation (JHQB202205), and Henan Youth Talent Science and Technology Innovation Program in 2023 (Liu Bojun).

References

1. Karlsson, J., Byström, P., Ask, J., Ask, P., Persson, L., Jansson, M.: Light limitation of nutrient-poor lake ecosystems. Nature. **460**(7254), 506–509 (2009)
2. Vitense, K., Hanson, M.A., Herwig, B.R., Kyle, D., Zimmer, K.D., Fieberg, J.: Predicting total phosphorus levels as indicators for shallow lake management. Ecol. Indic. **96**, 278–287 (2019)
3. Liu, B., Wang, Y., Xia, J., Quan, J., Wang, J.: Optimal water resources operation for Rivers-connected Lake under uncertainty. J. Hydrol. **595**, 125863 (2021)
4. Ayele, H.S., Atlabachew, M.: Review of characterization, factors, impacts, and solutions of Lake eutrophication: lesson for Lake tana, Ethiopia. Environ. Sci. Pollut. R. **28**(12), 14233–14252 (2021)
5. Li, X.Y., Xu, H.Y., Sun, Y.L., Zhang, D.S., Yang, Z.P.: Lake-level change and water balance analysis at Lake Qinghai, West China during recent decades. Water Resour. Manag. **21**(9), 1505–1516 (2007)

6. Chen, Y.D., Zhuge, Y.S., Shi, Y.F., Du, Q., Zhang, X.Y., Nie, R.: Identification of natural eco-hydrological characteristics of lakes in the Qaidam Basin. Water. Resour. Prot. **39**(1), 225–233 (2023)
7. Sun, J.R., Li, X., Wei, J.T.: Study on the area change characteristics of Wuliangsu Sea driven by climate change and human activities in the basin in the past 40 year. Arid. Land. Geography., 1–15 (2024) (in Chinese)
8. Mao, X., Yuan, D., Huang, X., Wei, X., Chen, Q.: Purification-efficient-based hydroecoregion in a lake of irrigation return water. China. Fresen. Environ. Bull. **24**(3b), 1142–1149 (2015)
9. Liu, B., Yang, L., Cui, C., Wan, W., Liang, S.: Is water replenishment an effective way to improve Lake water quality? Case study in Lake Ulansuhai, China. Front. Environ. Sci. **12**, 1392768 (2024)
10. Yue, S., Pilon, P., Cavadias, G.: Power of the mann–kendall and spearman's rho tests for detecting monotonic trends in hydrological series. J. Hydrol. **259**, 254–271 (2002)
11. Ruxton, G.D.: The unequal variance t-test is an underused alternative to student's t-test and the mann–Whitney u test. Behav. Ecol. **17**(4), 688–690 (2010)
12. Chamoli, A., Bansal, A.R., Dimri, V.P.: Wavelet and rescaled range approach for the Hurst coefficient for short and long time series. Comput. Geosci-UK. **33**(1), 83–93 (2007)

Analysis of River Ecological Flow and Safeguard Measures in Ungauged Basins: A Case Study of Jincheng City

Rongkang Yu[1,2], Weiwei Shao[1](✉), Guang Han[1], Zimeng Zhao[3], and Rui Shao[1]

[1] State Key Laboratory of Basin Water Cycle Simulation and Regulation, China Institute of Water Resources and Hydropower Research, Beijing 100038, China
shaoww@iwhr.com

[2] College of Landscape Architecture and Forestry, Qingdao Agricultural University, Qingdao 266109, China

[3] Academy of Eco-civilization Development of Jing-Jin-Ji Megalopolis, Tianjin Normal Universtiy, Tianjin 300387, China

Abstract. In order to solve the problems of lack of runoff data and incomplete data in the process of ecological flow research in some areas, this study took Jincheng City as a typical study area, used the method of global natural runoff simulation data, and used the vector river network confluence model to simulate the runoff data. On the basis of this data, the Tennant method in the hydrology method was used to calculate the ecological water demand of the Jincheng River, and the design guarantee rate of the ecological flow (water volume) target was scientifically and reasonably determined according to the water resources conditions, engineering regulation and control ability, and the importance of ecological protection in the basin, and the analysis of the ecological flow guarantee rate and satisfaction of the river was made. The future implementation direction of the optimal allocation of water resources in Jincheng City was determined, and new research methods and ideas were provided for the study of ecological flow in other ungauged basins.

Keywords: Ungauged Basins · Ecological Flow · RAPID Model · Tennant Method

1 Introduction

The study of river ecological flow is the basic work for the optimal allocation and rational development of water resources, and the improvement of river water resource utilization rate and sustainable development of water resources [1]. It is particularly important to scientifically establish the demand for river ecological flow for the river supervision and guarantee system, and can also lay a good foundation for high-level water system ecological goals such as water environment and water ecology [2, 3]. Therefore, it is very important to study the ecological flow of rivers. At present, there are four main methods to solve river ecological flow, namely hydrology, hydraulics, habitat simulation and holistic [4]. Due to its convenience and low cost, hydrology is the most common

V. Achal et al. (Eds.): EPPC 2024, ESE, pp. 102–107, 2025.
https://doi.org/10.1007/978-3-031-92123-0_10

calculation method in ecological flow. The Qp method and the Tennant [5] method, which are mainly involved in this paper, are common calculation methods in hydrology [4]. However, due to the difficulty of calculating and analyzing river ecological flow in ungauged basins, it is an effective way to solve this problem by verifying its rationality through global runoff simulation data. In this study, this method was used to effectively solve the problem of lack of runoff data in ungauged basins. Taking Jincheng City as an example, the method has been effectively verified, and after obtaining the data, the Tennant method was used to calculate the ecological flow guarantee of Jincheng River, and the ecological water demand of 67 rivers in Jincheng City was calculated and analyzed.

2 Study Area

Jincheng City is located in the southeast of Shanxi [6], and the whole territory is in the Jincheng Basin, that is, the basin of the middle and lower reaches of the Dan River and the Qin River. The plains, hills and mountains of Jincheng City accounted for 12.9% (1221.6 km^2), 28.5% (2704 km^2) and 58.6% (5564.4 km^2) of the total area, respectively. The average altitude of the whole territory is 600–700 m, the highest point is Shunwangping (2322 m above sea level) in Zhongtiao Mountain, and the lowest is in the valley of the lower reaches of the Dan River and Qin River (nearly 300 m above sea level). The Jincheng River belongs to the two major basins of the Yellow River and the Haihe River, According to the statistics of Jincheng Meteorological Station, the average rainfall for many years is 573.6 mm, the maximum rainfall is 1010.4 mm (1956), and the minimum rainfall is 265.7 mm (1997).

3 Data and Methodology

With the goal of meeting the basic needs of river ecology and maintaining the functions of river and lake ecosystems, the ecological flow guarantee scheme of different types of rivers is proposed, the ecological flow guarantee objectives of 20 rivers in the city are determined, the river ecological flow, safeguard measures and responsibility assessment mechanism are determined, and the ecological flow guarantee plan of key rivers is formulated. According to the analysis and comparison of several methods [7], suitable calculation methods for ecological flow were selected.

3.1 Methods for Calculating Ecological Flow in Basins with Data

Based on an adequate long series of runoff data, methods such as the Tennant method, the Qp method, and the average flow of the driest month in the past decade can be chosen for calculating ecological flow. These methods are applicable to different scenarios and have their own merits and demerits.

The calculation using the Tennant method is more in line with the annual runoff variation pattern of northern rivers, where the runoff is mainly concentrated in the flood season and the proportion of runoff in the non-flood season is relatively small. Moreover, different classifications of river ecological flow values are delineated based on different regions, types, and degrees of development and utilization of rivers. This method is simple and convenient. After addressing the issue of lacking data, the Tennant method was employed for calculation in this study.

3.2 Calculation of Ecological Flow in Ungauged Basins

VIC (Variable Infiltration Capacity) [8] is a large-scale land surface hydrological model developed jointly by the University of Washington, the University of California, Berkeley, and Princeton University. It is a grid-based distributed hydrological model that is spatially distributed, which makes it easy to embed with climate models and water resources models to evaluate the impact of climate change on water resources. RAPID is a numerical model for simulating the water flow propagation in tens of thousands to hundreds of thousands of rivers forming a river network. Its main applications include river-aquifer interaction studies, large-scale high-resolution flow modeling, and other aspects.

3.3 Model Accuracy Verification

By comparing the actual measurement results of some stations with the VIC model, the credibility of the model is reflected. According to the runoff observation data of Runcheng station and Zhangfeng Station, the natural runoff simulation data product of Jincheng City is verified, and the results show that the simulation accuracy of the natural runoff simulation data product of Jincheng City is high, and the correlation coefficients with the monthly value of runoff observation data of Runcheng Station (see Fig. 1) and Zhangfeng Station (see Fig. 2) are 0.63 (p-value > 0.01) and 0.86 (p-value > 0.01, respectively), which can reasonably quantify the runoff, and further calculate the ecological flow of the undocumented rivers in Jincheng City based on this data.

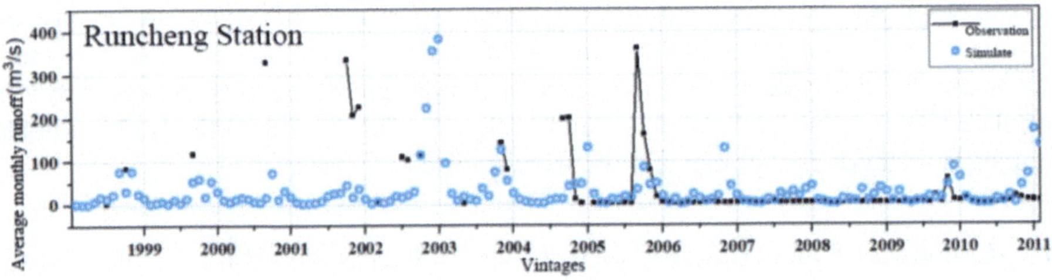

Fig. 1. Comparison chart of the monthly value of Jincheng natural runoff simulation data product and Runcheng station runoff

The monitoring results of 14 rivers in the dry season of Jincheng City were selected (see Fig. 3). The results were used to verify the accuracy of the runoff simulations from late April to early May and mid-to-late November 2013. By comparing with the natural runoff simulation data products of Jincheng City, the results show that the correlation

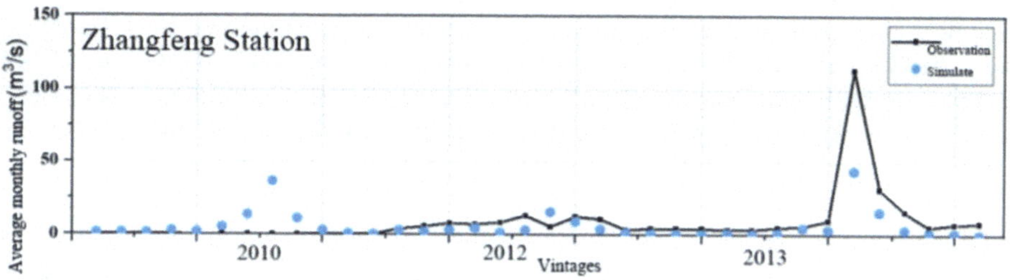

Fig. 2. Comparison chart of the monthly value of Jincheng natural runoff simulation data product and Zhangfeng station runoff

coefficient between the observed runoff and the simulated runoff of 14 rivers is 0.90, and the accuracy of simulated runoff is high, which can be used for the calculation of ecological runoff in ungauged basins.

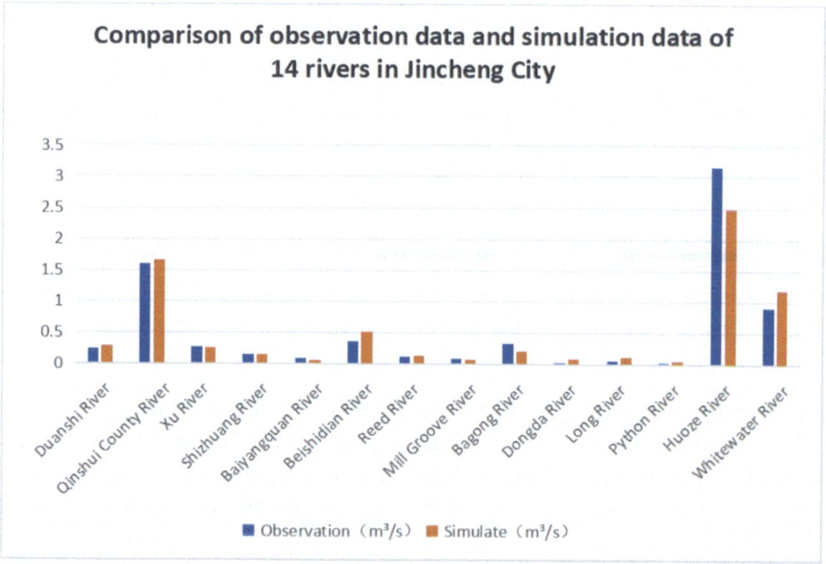

Fig. 3. Comparison of observation data and simulation data of 14 rivers in Jincheng City

4 Discussions and Conclusions

Based on the above method, the ecological flow of the river was calculated according to the collected measured data of the station and the simulation data product of natural runoff in Jincheng City. According to the analysis of the ecological flow calculation results, 33 of the 67 rivers in Jincheng City have an annual ecological water demand of less than 50,000 m^3. There are 9 rivers with an annual ecological water demand of

5–10 m^3; There is 1 river with an annual ecological water demand of 10–150,000 m^3; There are 2 rivers with an annual ecological water demand of 15–200,000 m^3; There are 15 rivers with an annual ecological water demand of 20–1 million m^3; There are 7 rivers with an annual ecological water demand of more than 1 million m^3 (see Fig. 4).

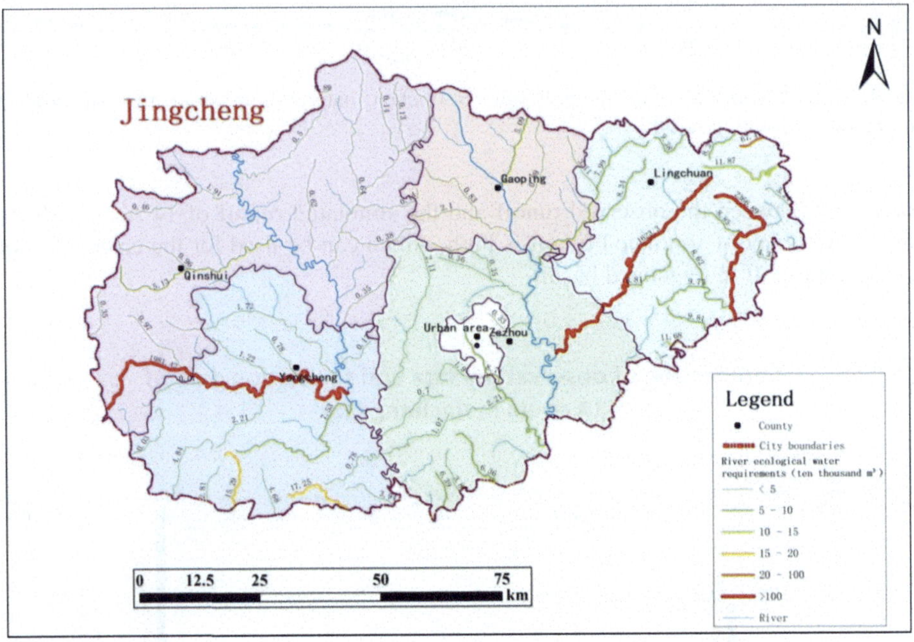

Fig. 4. Calculation results of river ecological flow in Jincheng City

In this study, the simulation data were simulated using natural runoff simulation data in the ungauged basins, and the simulation data were obtained based on the VIC hydrological process model and the RAPID vector river network confluence model, and then the runoff simulation data in the Runcheng station and Zhangfeng station in the typical study area were verified and analyzed through the runoff observation data of Runcheng Station and Zhangfeng Station in Jincheng City, and it was believed that the data obtained from the simulation data of the natural runoff simulation data product in Jincheng City had high accuracy. This method can be used in areas where runoff data are lacking. After solving the problem of lack of data, the ecological flow calculation of Jincheng rivers was completed through the Tennant method in hydraulics, and relevant guarantee plans were formulated for 20 rivers in the ecological flow guarantee list, and the ecological flow compliance status of Jincheng rivers was further guaranteed and implemented on the basis of data calculation, and the ecological flow satisfaction analysis of 67 rivers in Jincheng City was completed, and the overall satisfaction was high, which improved the optimal allocation of water resources and the ecological water demand efficiency of Jincheng City.

Acknowledgments. This research was partially supported by the Natural Science Foundation of Beijing Municipality (grant no. 8222036), the National Natural Science Foundation of China (grant no. 51979285), the Research Foundation of the State Key Laboratory of Simulation and Regulation of Water Cycle in River Basin of China Institute of Water Resources and Hydropower Research (grant no. SKL2022TS11), and the Major Science and Technology Project of the Ministry of Water Resources (grant no. SKS-2022126, WR110145B0042023).

References

1. Wu, M.H., Liu, Z.C., Li, H., Wu, H.: Urban River ecological restoration from the landscape hydrology perspective: a case study of the Los Angeles River revitalization. Landsc. Archit. **27**(08), 35–41 (2020)
2. Wang, X.F., Miao, L., Zhai, C.W.: Research on water resources dispatch based on the guarantee of ecological flow in Wuding River basin. Shaanxi Water Resour. **2024**(04), 32–35 (2024)
3. Dong, Z.R., Zang, J., Zhao, J.Y.: Scientific connotation of ecological flow. China Water Resour. **2020**(15), 15–19 (2020)
4. Zhu, C.H., Yu, H.T., Song, B.: Calculation and analysis of the ecological flow of Wanmipo hydropower station. Water Res. Develop. Manage. **10**(03), 11–15+31 (2024)
5. Tennant, D.L.: Instream flow regimens for fish, wildlife, recreation and related environmental resources. Fisheries. **1**(4), 6–10 (1976)
6. Shu, X.Y., Xu, Z.X., Ye, C.L., et al.: Analysis of flood driving factors in different regions of Jincheng City. J. Beijing Norm. Univ. (Nat. Sci.) **60**(02), 203–217 (2024)
7. Lin, J.Q., Li, Y., et al.: Recent progress in ecological operation and adaptive management for stimulating fish natural spawning. Shuili Xuebao **53**(04), 483–495 (2022)
8. Liang, X., Xie, Z., Huang, M.: A new parameterization for surface and groundwater interactions and its impact on water budgets with the variable infiltration capacity (VIC) land surface model. J. Geophys. Res. Atmos. **108**(D16) (2003)

Windward Self-organizing Ecological Restoration Technology for Bank-Collapsed Lakeside Zone

Qun Liu[1(✉)] and Xiqiang Zheng[2]

[1] Anhui Sanlian University, Hefei 230601, China
liuqunxq@163.com

[2] Anhui Institute of Environmental Sciences, Key Laboratory of Wastewater Treatment Technology in Anhui Province, Hefei 230601, China

Abstract. The ecological restoration of lakeside zones is not only transplanting aquatic plants but also requires the provision of external physical conditions suitable for their survival, such as matrix and hydrology. As a carrier of aquatic vegetation, the matrix must possess appropriate flexibility and moderate stiffness. Based on a comprehensive survey of the bank-collapsed lakeside zone of Chaohu Lake, this study systematically analyzed the causes of ecological degradation in the bank-collapsed lakeside zone and developed an integrated matrix–hydrology–biology restoration technology. This helped to solve the problems of matrix loss and hydraulic cutting on aquatic plants during the recovery growth period, creating an appropriate environment for aquatic plants. The demonstration project of the Wannianbu lakeside zone on the northwest coast of Chaohu Lake achieved a sound governance effect, offering an eco-friendly technology that can be implemented for the ecological restoration of the bank-collapsed lakeside zone of Chaohu Lake.

Keywords: Lakeside zone · Bank collapse · Ecological restoration · Physical matrix · Chaohu Lake

1 Introduction

Chaohu Lake is one of the five major freshwater lakes in China was renowned for its rich biodiversity, multi-level hierarchy, and favorable ecological environment. However, with the economic development and the intensification of unreasonable human development activities, more and more lakeside zones are being seriously degraded under human interference. Lakeside zones, also known as aquatic-terrestrial ecotones, are important impoundment and accommodation sites for surface watersources entering the lake and special ecosystems with rich biodiversity. They also serve as essential gene pools and seed banks for the lake. The spatial scope of a lakeside zone mainly depends on the spatial structure of the lakeside dry-wet cycles caused by water level fluctuations and the spatial environmental characteristics of adjacent aquatic and terrestrial systems,Studies from University Pablo de Olavide Add New Findings in the Area of Nature Ecology and Evolution (Multiple elements of soil biodiversity drive ecosystem

V. Achal et al. (Eds.): EPPC 2024, ESE, pp. 108–117, 2025.
https://doi.org/10.1007/978-3-031-92123-0_11

functions across biomes) [1–4]. Lakeside zones play important roles in water conservation, flood storage & drought prevention, siltation promotion & land reclamation, water purification, and maintenance of biodiversity and ecological balance,such as the role of macrophyte-associated microbiomes in lacustrine wetlands [5–7].

In recent years, with the deepening of research on lake governance, technologies involved in ecological restoration and reconstruction of lakeside zones have gradually become a focus of environmental ecology [8]. The research conducted by Canada on its Great Lakes Region has covered the demarcation of geographical scopes, the development and benefits of restoration measures, and so forth [9]. The ecological restoration plan implemented by the US around Lake Washington has effectively controlled its water eutrophication [10]. Some domestic scientific researchers have also investigated the ecological restoration and reconstruction of lakes with severe water eutrophication and made certain achievements in theoretical research and engineering practice, with the gradient change in water environmental factors, the spatial structure of the lakeside zone exhibits distinct circular characteristics in its transverse, longitudinal, and axial distributions. The transverse structure comprises the landward radiation zone, the water level fluctuation zone, and the aquatic radiation zone. The longitudinal structure consists of atmospheric, aquatic, and sedimentary components. The axial structure represents the landscape changes and hierarchical organization of the lakeside zone across different shoreline segments, encapsulating the overall composition and typological integration of the entire lakeside zone [3, 11].

Focusing on the northwest coast of Chaohu Lake (where Lakeside New District of Hefei is located), this study investigated ecological restoration technology and engineering demonstration for the bank-collapsed lakeside zone of Chaohu Lake, with a view to providing technical support for the bank-collapsed governance of Chaohu Lake.

2 Materials and Methods

2.1 Comprehensive Survey on Bank-Collapsed Lakeside Zone

2.1.1 Matrix

The texture of the lakeside zone on the northwest coast of Chaohu Lake is dominated by expansive soil, with a thickness of 10–40 m, a free expansion rate of 50–60%, an expansion force of 86.2 kPa, a linear shrinkage of 1.25%, volumetric shrinkage of 9.5%, a shrinkage coefficient of 0.28, a shrinkage limit of 12.0%, and a total expansion-shrinkage ratio of 2.1%. The soil layer is highly expansive after wetting and shrinkable after drying [12]. It can be seen that the matrix distributed on the northwest coast of Chaohu Lake has excess flexibility but insufficient stiffness. That is, it is suitable for the growth of aquatic plants but vulnerable to invasion by lake waves, resulting in an unstable basement and making aquatic plants susceptible to lodging and death. In this sense, the key to matrix restoration is adding a stiff skeleton to the flexible matrix and maintaining basement stability.

2.1.2 Hydrology

The perennial prevailing wind direction in the Chaohu Lake basin is the southeasterly wind, with a multi-year average wind speed of 3.2 m/s and a maximum lake breeze speed of 18 m/s. The lakeside zone on the northwest coast of Chaohu Lake is located on the windward side of the perennial prevailing wind direction. In the early growth stages (March to April) of transplanted aquatic plants, the winds were harsh, and the waves roughed on the lake surface. The calculated wave run-up and altitude of wind banked-up added up to 2.5–3 m, making it almost impossible for aquatic plants to survive hydraulic cutting. Therefore, the key to hydrological restoration lies in creating hydraulic buffering conditions, ensuring the survival of transplanted aquatic plants in their early growth stages, and constructing a natural barrier to hydraulic erosion.

2.1.3 Biology

Reeds are an in situ dominant emergent plant species in the Chaohu Lake basin. Currently, reeds are thriving in the polder areas along the lake. Even after the degradation of the bank-collapsed lakeside zone, the remaining scattered reeds were dominant. The key to biological restoration lies in formulating a transplanting plan based on the growth period, plant behavior, ecological habits, and other traits of reeds.

2.2 Causes of Ecological Degradation

The ecological degradation of the northwest coast of Chaohu Lake began with the construction of embankments in the 1970s and 1980s. Before that, the shoreline was well covered by vegetation, and the depth of emergent vegetation (dominated by reeds) in the lakeside zone reached 50–200 m. During the construction, stretches of reeds were manually pulled out, resulting in the gradual loss of effective vegetation protection over the expansive soil shoreline, the intensification of hydraulic erosion, and the aggravation of shoreline collapse. Shoreline collapse, in turn, caused the matrix of aquatic plants (such as reeds) to be unstable. Hydraulic erosion also intensified the hydraulic cutting of aquatic vegetation, posing greater difficulties to the recovery growth of reeds and other aquatic plants. At this point, the ecosystem of the northwest coast of Chaohu Lake entered a vicious cycle of degradation.

The ecological restoration of the bank-collapsed lakeside zone on the northwest coast of Chaohu Lake requires a comprehensive addressing of three interrelated, interdependent problems: the disappearance of aquatic vegetation, the instability of the lakeshore matrix, and the intensification of hydraulic erosion. It is difficult for emergent plants planted without supporting measures to survive here. The restoration of aquatic vegetation communities can be effective only through adequate lakeshore geological restoration and lakeside hydrological restoration.

3 Results and Discussion

3.1 Ecological Restoration Technologies

Depending on the gradient of the bank collapse line, necessary slope cutting is provided for the landward lakeside zone with high ridges or steep slopes above the highest water level, and trees, shrubs, and grasses are planted together to strengthen slope protection by vegetation. Emergent plants are transplanted to the riparian and waterward lakeside zones below the highest water level to restore the revetment system. The specific plan is as follows:

(1) During the winter's low-water level period of Chaohu Lake, a dredger is used to conduct excavation operations perpendicular to the shoreline. The excavated lakebed soil is piled up on both sides of the dredger, forming multiple earth dams above the water surface perpendicular to the shoreline and parallel to each other (Fig. 1). The striped earth dams, perpendicular to the shoreline, can weaken wind wave erosion to the largest extent, creating a recovery growth area for emergent vegetation. In the recovery growth area of emergent vegetation, emergent vegetation is recovered by transplanting rhizomes. Wave-dissipating earth dams protect the nearshore from wind wave scouring in the lake area. In addition to maintaining the stability of the shoreline matrix, they also prevent emergent plants from being cut by wind waves during the spring germination period. In addition, wave-dissipating earth dams also play a role in water conservation, providing an appropriate aquatic environment for germinating and growing emergent plants.
(2) The strong rhizomes of emergent plants that are beginning to germinate in the polder areas along Chaohu Lake are selected around the spring and transplanted to the recovery growth area.
(3) Transplanted emergent plants are then allowed to germinate further and grow. At the same time, after the wind wave erosion of two wave-dissipating earth dams, the soil gradually flows back into the channel left by dredger excavation, causing the heights of earth dams to gradually decline. With the increase in spring rainfall, the outer lake's water level gradually rises
(4) After entering the summer, the stems, leaves, and roots of emergent plants have initially grown up, and the ability of the plants to withstand wind waves has gradually improved. When the water level of the outer lake exceeds the heights of wave-dissipating earth dams, the recovery growth area joins with the outer lake area, and the rhizomes of emergent plants grow to be even thicker and stronger under wind wave erosion.
(5) In the autumn, the water level of the outer lake gradually drops, and wave-dissipating earth dams have basically disappeared under wind wave erosion. The area around the water level has been covered by emergent plants, forming effective vegetation protection.
(6) In subsequent years, the recovered emergent plant communities continue towards waterward and landward directions, giving rise to a shoreline vegetation revetment system with wider coverage, tighter root entanglement, and stronger anti-interference ability.

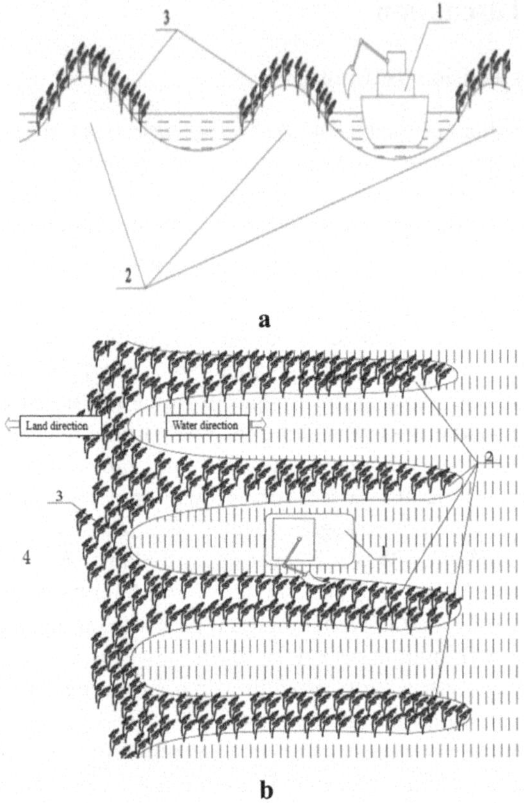

Fig. 1. Method 2 for recovery of emergent vegetation in the bank-collapsed lakeside zone of Chaohu Lake (**a**: section view; **b**: top view; 1—dredger, 2—wave-dissipating earth dams, 3—recovery growth area of emergent vegetation, 4—shoreline)

3.2 Restoration Conditions and Strategies

3.2.1 Matrix Restoration

Matrix is a carrier of ecosystem development and existence. The first step in the matrix restoration of lakeshore wetlands is to balance the sediments provided by rivers or surrounding lakeshore and the eroded sediments. That is, the sediments along the lakeshore are constantly changing, while the landscape of the entire lakeshore zone remains stable. The bank-collapsed soil layer of Chaohu Lake is highly expansive after wetting and shrinkable after drying. After embankment construction, water level fluctuations are directly related to local rainfall and evaporation. The unevenness of rainfall and the difference in evaporation cause the water level of Chaohu Lake to vary by about 3–4 m between rainy and dry seasons. Under different water level fluctuations and infiltration erosion, bank collapse frequently occurs, especially during rainy seasons.

An analysis of the matrix of the lakeside zone reveals that it has excess flexibility but insufficient stiffness. Clay is suitable for the growth of aquatic plants but is vulnerable to invasion by lake waves, resulting in unstable basements and making aquatic plants susceptible to lodging and death after recovery. In this sense, the key to matrix restoration is maintaining basement stability.

3.2.2 Hydrological Restoration

Wind-driven currents constitute a hydrological characteristic of Chaohu Lake, and the impacts of wind force and direction on lake waves are undeniable. Due to the artificial removal of lakeside vegetation during embankment construction against flood, there is no hydraulic buffering in the lakeside zone. Lake waves directly beat against the lakeshore, causing frequent bank collapse. Aquatic vegetation provides a natural lakeside wave-dissipating zone but requires effective hydraulic buffering during its growth period. During the growth period of aquatic vegetation (from March to July), the winds are harsh, and the waves rough on the lake's surface. The calculated wave run-up and altitude of wind banked-up add up to 2.5–3 m, posing great difficulties in recovering aquatic vegetation under hydraulic cutting.

Currently, extensive research has been conducted to explore how to dissipate waves. During embankment construction against flood, aquatic plants (especially emergent plants) have been removed on a large scale, causing the natural wave-dissipating effect to disappear. As a result, the originally quiet lakeside zone is now exposed to harsh winds and rough waves, making it almost impossible for transplanted aquatic plants to survive. Therefore, the key to hydrological restoration lies in creating hydraulic buffering conditions, ensuring the recovery of aquatic vegetation during the growth period, and constructing a natural barrier to hydraulic erosion.

3.2.3 Biological Restoration

If the above two restoration processes are brought into full play, the ecosystem will naturally be recovered step by step. However, if the organisms to be restored no longer exist (i.e., extinct locally) or are greatly impacted by alien species, artificial intervention measures must be taken. In this regard, the operational biological restoration measures include selecting species and communities and adopting vigilance measures against alien species. Reeds and cattails are dominant emergent plant species in the Chaohu Lake basin. Notably, stretches of reeds were manually pulled out during the embankment construction against flood years ago, but the remaining scattered reeds can still show their dominance. The comprehensive ecological benefits of recovering lakeside reeds are beyond doubt.

3.3 Ecological Restoration Strategies

The ecological restoration of lakeside zones is not simply a matter of transplanting aquatic plants but also requires the provision of external physical conditions suitable for their survival, such as matrix and hydrology, and the effects of spatially heterogeneous lakeside development on nearshore biotic communities in eutrophic lake [13–15].

As a carrier of aquatic vegetation, the matrix must possess appropriate flexibility and moderate stiffness. Besides being suitable for the growth of aquatic plants, it must also be able to withstand wave erosion and protect vegetation from lodging. Moderate hydrology is undoubtedly a critical condition for the survival of aquatic plants. High flow velocities and high wind waves scour not only the matrix but also subject aquatic plants in their early growth stages to fatal hydraulic cutting. The transplanting of aquatic plants requires a reasonable selective breeding of predominantly local species and a temporal matching with their growth period. The researchers investigated the various stages and management approaches associated with lakeside ecosystem restoration [8, 16]. In fact, the ecological restoration of lakeside zones is a systematic project that requires optimization and reorganization from three aspects: matrix, hydrology, and biology. Based on the above analysis, the ecological restoration of the bank-collapsed lakeside zone on the northwest coast of Chaohu Lake obeys the following strategies:

3.3.1 Banded Restoration

In this respect, efforts are made to increase the depth of the lakeside zone as much as possible, restore the topographic structure of complete succession series (such as the waterward lakeside zone, the riparian zone, and the landward lakeside zone), create gentle-slope entry conditions conducive to vegetation growth, and construct aquatic-terrestrial alternate wetlands with varying depths of 50–100 m.

3.3.2 Wave Dissipation First, Restoration Afterward

Wave dissipation is the primary task in implementing the ecological restoration of the lakeside zone on the northwest coast of Chaohu Lake. Effective wave dissipation can alleviate the hydraulic cutting of aquatic plants in early growth stages while reducing the occurrence of bank collapse, providing a stable matrix for aquatic plants.

3.3.3 Eco-Friendly Project Implementation

The bank collapse line of Chaohu Lake stretches for 88 km, and its large-scale ecological restoration requires the development of eco-friendly technologies that can be implemented in engineering practice. Attention should also be paid to the eco-friendliness of the project implementation process, including compressing the working face, reducing the work amount, and avoiding the squeezing of storage capacity. The key is to provide a turning point for the reverse recovery of degraded ecosystems, restoring their original ecological state through natural forces and minimizing human traces.

3.4 Engineering Case

In 2021, a demonstration project for the ecological restoration of bank-collapsed lakeside zones was constructed in the Wannianbu lakeside zone on the northwest coast of Chaohu Lake. The project produced desirable results. The project was located in the 1 km lakeside zone on the windward side of Chaohu Lake, where ecological degradation was severe and aquatic vegetation had completely disappeared (Fig. 2). During the dry season, a dredger was used to conduct excavation operations perpendicular to the shoreline. From March to April, the rhizomes of local reeds were collected from nearby ponds. Aboveground rhizomes were kept at 20–30 cm, and underground rhizomes remained intact and began to germinate. The collected rhizomes were transplanted to the recovery growth area of emergent vegetation by the density of 5–7 plants/m^2.

Fig. 2. Original appearance of the bank-collapsed lakeside zone

By early April, the germination rate of reeds exceeded 80%, and new ones replaced the non-germinated reeds. By late June, the heights of reed plants reached 0.8–1.2 m, and local gaps began to appear in wave-dissipating earth dams under wind wave scouring. By early August, reed plants were as high as 1.6 m, and wave-dissipating earth dams had disappeared under the combined action of wind wave erosion and the rising water level of the outer lake. The recovery growth area of emergent vegetation joined together with the outer lake area (Fig. 3).

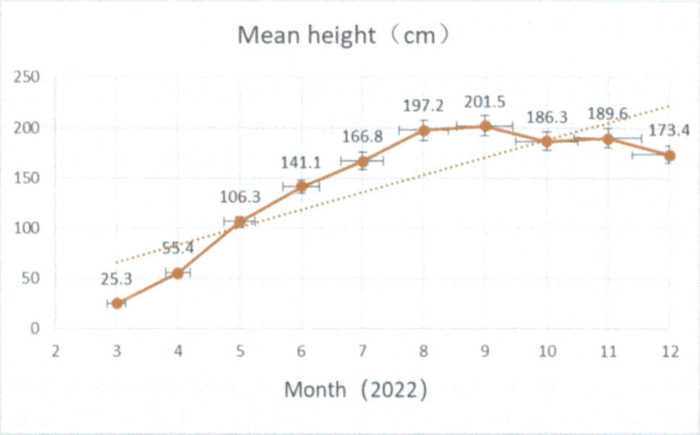

Fig. 3. Trend chart of plant growth changes

By late September, reed plants reached the heights of 1.5–2.0 m, and there was no plant lodging or death. By late November, the flowering rate of reeds exceeded 50%. At this point, vegetation had effectively protected the shoreline, and there was no shoreline collapse or retreat (Fig. 4).

Fig. 4. Demonstration project results: Reed after initial grow-up

4 Conclusions

The windward self-organizing integrated restoration technology is based on constructing wave-dissipating earth dams. It not only solves the problems of matrix loss and hydraulic cutting but also creates an appropriate aquatic environment for the recovery of aquatic plants. The soil used for constructing wave-dissipating earth dams was taken from and used for the lake without squeezing storage capacity. In addition, using a dredger for excavation operations solved the problem of stranding large machinery in nearshore construction. The governance effect of the Wannianbu lakeside zone on the northwest coast of Chaohu Lake demonstrates that the windward self-organizing integrated restoration technology is an eco-friendly bank collapse governance technology that can be implemented in engineering practice, offering a beneficial technical reserve for the bank collapse governance of Chaohu Lake.

Acknowledgments. This study was supported by the Anhui Province Outstanding young talent projects "Reform and Innovation of Nursing Teaching Model from the Perspective of New Medical Sciences" (Project Number: gxyq2022137), Anhui Province Quality Engineering Project Blended online and offline courses "Biochemistry" (Project Number: 2023xsxx364).

References

1. Delgado-Baquerizo, M., Reich, P.B., Trivedi, C., et al.: Multiple elements of soil biodiversity drive ecosystem functions across biomes. Nat. Ecol. Evol. **4**, 210–220 (2020)
2. Haskell, D.E., Webster, C.R., Bales, A.L., Meyer, M.W., Flaspholer, D.J.: Assessment of wildlife habitat attributes at restoration projects on Northern Wisconsin Lakeshores. N. Nat. **24**, 391–412 (2017)
3. Ye, C., Li, C., Deng, T.: Structures and ecological functions of lake littoral zones. Res. Environ. Sci. **28**, 171–181 (2015)
4. Huang, G.H., Xi, J.: Barriers to sustainable water-quality management. J. Environ. Manage. **61**, 1–23 (2001)
5. de Paula, C.C., Bárta, J., Rejmánková, E., Komárková, J., Lin, Q., Kohoutová, I., Sirová, D.: The role of macrophyte-associated microbiomes in lacustrine wetlands: an example of the littoral zone of lake Atitlan Guatemala. Hydrobiologia. **851**, 1637–1655 (2022)

6. Sandau, N., Naisbit, R.E., Fabian, Y., Bruggisser, O., Kehrli, P., Aebi, A., Rohr, R., Bersier, L.: Understanding negative biodiversity-ecosystem functioning relationship in semi-natural wildflower strips. Oecologia. **189**, 185–197 (2019)
7. Ali, A., Lin, S.L., He, J.K., Kong, F.M., Yu, J.H., Jiang, H.S.: Climate and soils determine aboveground biomass indirectly via species diversity and stand structural complexity in tropical forests. For. Ecol. Manage. **432**, 823–831 (2019)
8. Chai, P., Dai, Y., Liang, W., Cheng, S., Wu, Z.: Advances of ecological restoration of lakeside zone. Eng. Sci. **12**, 32–35 (2010)
9. Hartig, J.H., Thomas, R.L.: Development of plans to restore degraded areas in the Great Lakes. Environ. Manage. **12**, 327–347 (1988)
10. Lane, R., Day, J., Kemp, G.: The 1994 experimental opening of the Bonnet Carre spillway to divert Mississppi River water into Lake Pontchartrain, Louisana. Ecol. Eng. **17**, 411–422 (2001)
11. Yang, H., Zhu, S., Shen, Z., Wang, W.: Theories and technologies on restoration and reconstruction for ecological zone of lakeside. J. Agro-Environ. Sci. **25**, 819–824 (2006)
12. Sun, F.: The features of bank collapse in the Chaohu Lake. J. Geol. Haz. Environ. Preserv. **11**, 287–291 (2000)
13. Meyer, M.F., Ted, O., Woo Kara, H., Kirill, S., Galloway Aaron, W.E., Schram Julie, B., Hampton Stephanie, E.: Effects of spatially heterogeneous lakeside development on nearshore biotic communities in a large, deep, oligotrophic lake. Limnol. Oceanogr. **67**, 2649–2664 (2022)
14. Nakamura, K., Tockner, K.: River and wetland restoration in Japan. In: Zagreb: Proceedings of the 3rd European Conference on River Restoration, pp. 211–220 (2004)
15. Nishihiro, J., Miywaki, S., Fujiwara, N., Washitani, I.: Regeneration failure of lakeshore plants under an artificially altered water regime. Ecol. Res. **19**, 613–623 (2004)
16. Lu, H., Zeng, G., Jin, X., Jiao, S.: Theories, technologies and applications of ecological restoration and reconstruction on aquatic-terrestrial ecotone. Urban Environ Urban Ecol. **16**, 91–93 (2003)

Study on the Ecological Compensation Mechanism for Watersheds with Water Diversion Projects from the Perspective of Game Theory

Guang Han, Weiwei Shao(✉), Jiaqi Li, and Rongkang Yu

State Key Laboratory of Simulation and Regulation of Water Cycle in River Basin, China Institute of Water Resources and Hydropower Research, Beijing 100038, China
shaoww@iwhr.com

Abstract. Due to geographical disparities in economic development, population distribution, and water resources, many watersheds have implemented water diversion projects. These projects have increased the number of stakeholders in watershed ecological compensation, making game theory characteristics more pronounced and complex. However, analysis often overlooks the natural upstream-downstream relationships of watersheds and the human-induced water diversion projects affecting water intake and receiving. Therefore, this study, from a theoretical perspective, focuses on ecological compensation in watersheds with water diversion projects. It develops an evolutionary game model involving four key stakeholders: the central government, upstream water-taking area governments, downstream governments, and water-receiving area governments. The study also explores potential equilibrium points in this model.

Keywords: Watershed ecological compensation · Water diversion project · Evolutionary game theory

1 Introduction

Watershed ecological compensation is a derivative of ecological compensation, encouraging stakeholders to collectively achieve environmental protection through voluntary transactions between ecosystem service providers and consumers, as well as government transfer payments. This approach is an effective means to address water pollution and has been implemented in many countries worldwide, with approximately 387 watershed service payment programs globally. Typical countries include Ecuador, Ethiopia, Mexico, Costa Rica, and Brazil [1–6]. In China, research on watershed ecological compensation began in the early twenty-first century, focusing primarily on concepts of ecological compensation, stakeholder identification, compensation standard calculation, and compensation methods. Since the twenty-first century, China has implemented watershed ecological compensation in regions like the Xin'an, Jiuzhou, and Tingjiang River Basins, as well as the Luan River Diversion Project, effectively mitigating cross-basin water pollution.

V. Achal et al. (Eds.): EPPC 2024, ESE, pp. 118–127, 2025.
https://doi.org/10.1007/978-3-031-92123-0_12

However, watershed ecological compensation differs from other types of ecological compensation. In recent years, numerous water diversion projects have altered natural river systems, complicating stakeholder relationships beyond traditional upstream-downstream or source-receiving area interactions. Firstly, upstream and downstream regions share water resource rights. However, to ensure high-quality water for middle and downstream areas, upstream regions must prioritize environmental protection during economic development, restricting polluting industries and incurring opportunity costs. Downstream areas should compensate for these sacrifices. Water diversion projects, addressing water scarcity and declining groundwater levels in receiving areas, impose higher quality and quantity demands on upstream sources. Unlike natural downstream areas, water diversion receiving areas lack inherent rights to river resources, necessitating compensation for a stable water supply. The introduction of water diversion projects also reduces and destabilizes downstream flow, shrinking the water environment's capacity and heightening risks of water degradation and scarcity. Thus, receiving areas should compensate downstream areas to maintain sustainable water intake and mitigate potential conflicts.

In summary, water diversion projects have created distinct and complex game dynamics in watershed ecological compensation. While much research has examined game dynamics within watershed ecological compensation, most studies focus on either upstream-downstream or source-receiving interactions, without jointly considering natural and diversion-induced relationships [7, 8]. Accordingly, as shown in Fig. 1, this study focuses on the central government, upstream water-taking area governments, downstream governments, and water-receiving area governments, and develops a theoretical model of evolutionary game dynamics among these four stakeholders. Figure 1 defines the upstream water-taking area as the region above the diversion project source, encompassing both the river's upstream area and the water source for the diversion. The downstream area is defined as the river section below the source, while the water-receiving area refers to the region supplied by the diversion project.

2 Evolutionary Game Modeling

Unlike traditional static games, evolutionary game theory assumes that players are boundedly rational and do not make decisions simultaneously. The game process is dynamic, which better reflects real-world scenarios. Evolutionary game theory consists of three main components: the payoff matrix, the replicator dynamics equation, and the evolutionarily stable strategy.

2.1 Initial Condition Assumptions

1. Assume that all subjects are finitely rational and all have two strategy choices.
2. In the evolutionary game model, the central government aims to promote the establishment of ecological compensation to maximize environmental and social benefits, and the governments of the upstream water-taking area, the downstream and the water-receiving area all aim to maximize their own economic interests.

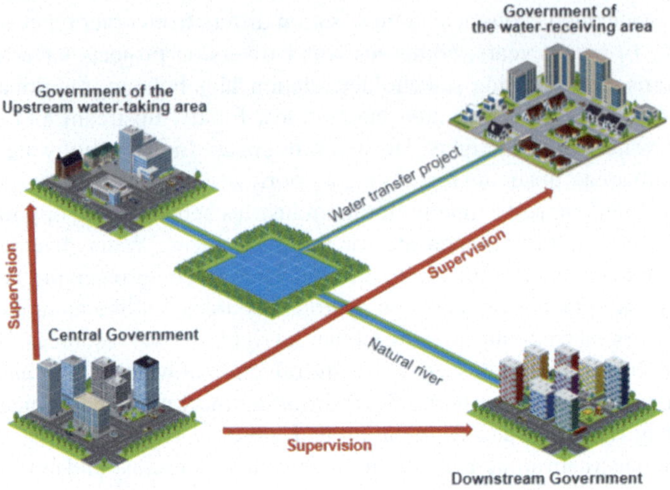

Fig. 1. Game relations between the subjects

3. The probability of strict supervision by the central government is x and the probability of lax supervision is 1-x. The probability that the upstream water-taking area protects the water environment is y and the probability that it does not is 1-y. The probability that the downstream pays the compensation fee is z and the probability that it does not is 1-z. The probability that the receiving area pays the compensation fee is w and the probability that it does not is 1-w.
4. The central government promotes the establishment of an ecological compensation system through the use of penalties and incentives from the perspective of cooperation and fairness. When the principle of equity is violated by the upstream water-taking area, downstream, or water-receiving area, such as when the downstream or water-receiving area grants compensation and the upstream water-taking area does not protect the water environment, or when the upstream water-taking area protects the water environment and the downstream or water-receiving area does not compensate for it, the central government will penalize the beneficiaries, and at the same time, reward those whose interests have been compromised.
5. This paper assumes that watershed ecological compensation is based on the establishment of a horizontal ecological compensation mechanism, so when none of the parties have the will to establish an ecological compensation mechanism, the central government will not take punitive measures. In addition, this paper hypothesizes that the central government's incentives need to be based on the premise that it is able to grasp the actual work of each party and the social and environmental benefits. When the central government chooses loose regulation, it is difficult to grasp the actual work of the parties and judge whether the social and environmental benefits are generated, so this paper assumes that the central government only adopts incentives in strict regulation [9].

2.2 Game Strategies and Parameter Setting

When the central government adopts a strict supervision strategy, its total environmental and social benefits is A_1, but it incurs supervision costs B. When the central government adopts a lenient supervision strategy, its total environmental and social benefits is A_2. The compensations provided to the upstream water-taking area, downstream, and receiving areas are C_1, D_1, and E_1, respectively, while the penalties are C_2, D_2, and E_2. The total payoff for the downstream area is H_1 when it pays compensation and H_2 when it does not. The compensation paid by the downstream area to the upstream water-taking area is I, and the cost of water quality treatment required by the downstream area is J. The total payoff for the upstream water-taking area when protecting the water environment is F_1 and it sacrifices opportunity costs G_1 for development. The total payoff when the upstream water-taking area does not protect the environment is F_2 and water quality treatment costs incurred from environmental degradation. The total payoff for the downstream area when paying compensation is F_2, costs of water quality treatment due to environmental degradation G_2 need to be paid. When the receiving area pays compensation, its total payoff is K_1, with a compensation payment of L_1 to the upstream water-taking area and L_2 to the downstream area. If the receiving area does not pay compensation, its total payoff is K_2. The cost incurred by the receiving area for water quality treatment is M. Specific parameter settings and the game relationships among the stakeholders are shown in Fig. 2.

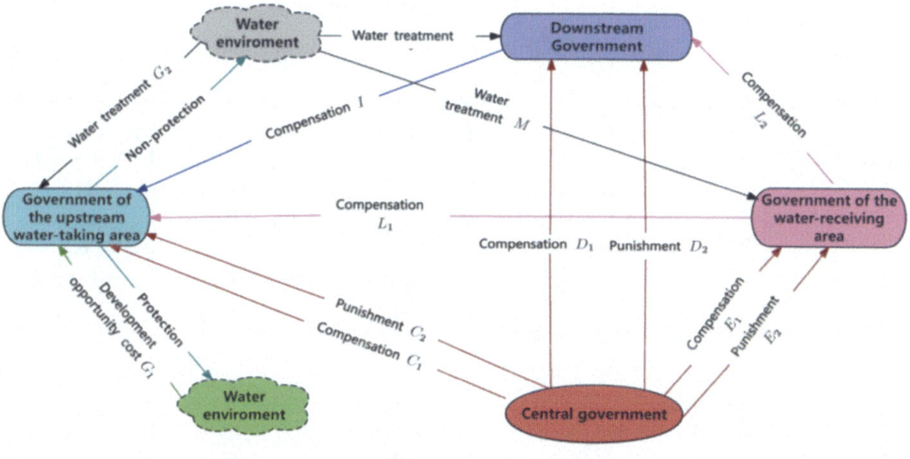

Fig. 2. Game relations and parameter setting

2.3 Payoff Matrix

Based on the assumptions and parameter settings above, the payoff matrix for each participant in the game model is shown in Table 1.

Table 1. Payoff matrix for each participant

		Central Government			
		Strict supervision x		Lenient supervision 1-x	
		Upstream water-taking area		Upstream water-taking area	
		Protection y	Non-protection 1-y	Protection y	Non-protection 1-y
(Downstream) Compensate z	(Receiving area) Compensate w	$a_{11} = A_1 - B$ $b_{11} = F_1 + C_1 + J + K_1 - E_1$ $c_{11} = H_1 - I + L_2$ $d_{11} = K_1 - L_1 - L_2$	$a_{12} = A_1 + C_2 - D_1 - E_1$ $b_{12} = F_2 + I + L_1 - C_2 - G_2$ $c_{12} = H_1 + D_1 + L_2 - J$ $d_{12} = K_1 + E_1 - L_1 - L_2 - M$	$a_{13} = A_2$ $b_{13} = F_1 + I + L_1 - G_1$ $c_{13} = H_1 - I + L_2$ $d_{13} = K_1 - L_1 - L_2$	$a_{14} = A_2 + C_2$ $b_{14} = F_2 + I + L_1 - C_2 - G_2$ $c_{14} = H_1 - I + L_2 - J$ $d_{14} = K_1 - L_1 - L_2 - M$
	(Receiving area) Not compensate $1 - w$	$a_{21} = A_1 + E_2 - C_1 - B$ $b_{21} = F_1 + C_1 + I - G_1$ $c_{21} = H_1 - I$ $d_{21} = K_2 - E_2$	$a_{22} = A_1 + C_2 - D_1 - B$ $b_{22} = F_2 + I - C_2 - G_2$ $c_{22} = H_1 + D_1 - I - J$ $d_{22} = K_2 - M$	$a_{23} = A_2 + E_2$ $b_{23} = F_1 + I - G_1$ $c_{23} = H_1 - I$ $d_{23} = K_2 - E_2$	$a_{24} = A_2 + C_2$ $b_{24} = F_2 + I - C_2 - G_2$ $c_{24} = H_1 - I - J$ $d_{24} = K_2 - M$
(Downstream) Not compensate $1 - z$	(Receiving area) Compensate w	$a_{31} = A_1 + D_2 - C_1 - B$ $b_{31} = F_1 + C_1 + L_1 - G_1$ $c_{31} = H_2 + L_2 - D_2$ $d_{31} = K_1 - L_1 - L_2$	$a_{32} = A_1 + C_2 - E_1 - B$ $b_{32} = F_2 + L_1 - C_2 - G_2$ $c_{32} = H_2 + L_2 - J$ $d_{32} = K_1 + E_1 - L_1 - L_2 - M$	$a_{33} = A_2 + D_2$ $b_{33} = F_1 + L_1 - G_1$ $c_{33} = H_2 + L_2 - D_2$ $d_{33} = K_1 - L_1 - L_2$	$a_{34} = A_2 + C_2$ $b_{34} = F_2 + L_1 - C_2 - G_2$ $c_{34} = H_2 + L_2 - J$ $d_{34} = K_1 - L_1 - L_2 - M$
	(Receiving area) Not compensate $1 - w$	$a_{41} = A_1 + D_2 + E_2 - C_1 - B$ $b_{41} = F_1 + C_1 - G_1$ $c_{41} = H_2 - D_2$ $d_{41} = K_2 - E_2$	$a_{42} = A_1 - B$ $b_{42} = F_2 - G_2$ $c_{42} = H_2 - J$ $d_{42} = K_2 - M$	$a_{43} = A_2 + D_2 + E_2$ $b_{43} = F_1 - G_1$ $c_{43} = H_2 - D_2$ $d_{43} = K_2 - E_2$	$a_{44} = A_2$ $b_{44} = F_2 - G_2$ $c_{44} = H_2 - J$ $d_{44} = K_2 - M$

2.4 Replicator Dynamics Equation

(1) Central Government

Assuming that the expected payoff for the central government when adopting a strict supervision strategy is O_1 and the expected payoff when adopting a lenient supervision strategy is O_2, the replication dynamic equation for the central government F(x) is as follows:

$$F(x) = x(1-x)(O_1 - O_2)$$

$$\begin{aligned}
&= x(1-x)[yzw(a_11 - a_13) + yz(1-w)(a_21 - a_23) + yw(1-z)(a_31 - a_33) \\
&+ y(1-z)(1-w)(a_41 - a_43) + wz(1-y)(a_12 - a_14) \\
&+ z(1-w)(1-y)(a_22 - a_24) + w(1-z)(1-y)(a_32 - a_34) \\
&+ (1-y)(1-z)(1-w)(a_42 - a_44)]
\end{aligned}$$

$$\begin{aligned}
&= x(1-x)[yzw(A_1 - A_2 - B) + yz(1-w)(A_1 - A_2 - C_1 - B) \\
&+ yw(1-z)(A_1 - A_2 - C_1 - B) + y(1-z)(1-w)(A_1 - A_2 - C_1 - B) \\
&+ wz(1-y)(A_1 - A_2 - D_1 - E_1) + z(1-w)(1-y)(A_1 - A_2 - D_1 - B) \\
&+ w(1-z)(1-y)(A_1 - A_2 - E_1 - B) + (1-y)(1-z)(1-w)(A_1 - A_2 - B)]
\end{aligned}$$

(2) Upstream Water-taking Area

Assuming the expected payoff for the upstream water-taking area government when protecting the water environment is P_1 and when not protecting is P_2, the replication dynamic equation F(y) for the (2) Upstream Water-taking Area government is expressed as follows:

$$F(y) = y(1-y)(p_1 - p_2)$$

$$\begin{aligned}
&= y(1-y)[xzw(b_{11} - b_{12}) + xz(1-w)(b_{21} - b_{22}) + xw(1-z)(b_{31} - b_{32}) \\
&+ x(1-z)(1-w)(b_{41} - b_{42}) + wz(1-x)(b_{13} - b_{14}) + z(1-w)(1-x)(b_{23} - b_{24}) \\
&+ w(1-z)(1-x)(b_{33} - b_{34}) + (1-x)(1-z)(1-w)(b_{43} - b_{44})]
\end{aligned}$$

$$\begin{aligned}
&= y(1-y)[xzw(F_1 + C_1 - G_1 - F_2 + C_2 + G_2) + xz(1-w)(F_1 + C_1 - G_1 - F_2 + C_2 + G_2) \\
&+ xw(1-z)(F_1 + C_1 - G_1 - F_2 + C_2 + G_2) + x(1-z)(1-w)(F_1 + C_1 - G_1 - F_2 + G_2) \\
&+ zw(1-x)(F_1 + C_2 - G_1 - F_2 + G_2) + z(1-x)(1-w)(F_1 + C_2 - G_1 - F_2 + G_2) \\
&+ w(1-x)(1-z)(F_1 + C_2 - G_1 - F_2 + G_2) + (1-x)(1-z)(1-w)(F_1 - G_1 - F_2 + G_2)]
\end{aligned}$$

(3) Downstream Area

Assuming the expected payoff for downstream area paying compensation is$Q_1$and not paying isQ_2, the replication dynamic equation for the downstream area is:

$$F(z) = z(1 - z)(Q_1 - Q_2)$$

$$\begin{aligned}
&= z(1 - z)[xyw(c_{11} - c_{31}) + xw(1 - y)(c_{12} - c_{32}) + yw(1 - x)(c_{13} - c_{33}) \\
&+ w(1 - x)(1 - y)(c_{14} - c_{34}) + xy(1 - w)(c_{21} - c_{41}) + x(1 - y)(1 - w)(c_{22} - c_{42}) \\
&+ y(1 - x)(1 - w)(c_{23} - c_{43}) + (1 - x)(1 - y)(1 - w)(c_{24} - c_{44})]
\end{aligned}$$

$$\begin{aligned}
&= z(1 - z)xyw(H_1 - I - H_2 + D_2) + xw(1 - y)(H_1 + D_1 - H_2) \\
&+ yw(1 - x)(H_1 - I - H_2 + D_2) + w(1 - x)(1 - y)(H_1 - I - H_2) \\
&+ xy(1 - w)(H_1 - I - H_2 + D_2) + x(1 - y)(1 - w)(H_1 + D_1 - I - H_2) \\
&+ y(1 - x)(1 - w)(H_1 - I - H_2 + D_2) + (1 - x)(1 - y)(1 - w)(H_1 - I - H_2)
\end{aligned}$$

(4) Receiving Area

Assuming that the expected payoff for water-receiving area choosing to pay compensation is R_1 and the expected payoff for choosing not to pay compensation is R_2, the replication dynamic equation for the receiving area government can be expressed as:

$$F(w) = w(1 - w)(R_1 - R_2)$$

$$\begin{aligned}
&= w(1 - w)[xyz(d_{11} - d_{21}) + xz(1 - y)(d_{12} - d_{22}) + yz(1 - x)(d_{13} - d_{23}) \\
&+ z(1 - x)(1 - y)(d_{14} - d_{24}) + xy(1 - z)(d_{31} - d_{41}) + x(1 - y)(1 - z)(d_{32} - d_{42}) \\
&+ y(1 - x)(1 - z)(d_{33} - d_{43}) + (1 - x)(1 - y)(1 - z)(d_{34} - d_{44})]
\end{aligned}$$

$$\begin{aligned}
&= w(1 - w)xyz(K_1 - l_1 - l_2 - K_2 + E_2) + xz(1 - y)(K_1 + E_1 - L_1 - L_2 - K_2) \\
&+ yz(1 - x)(K_1 - L_1 - L_2 - K_2 + E_2) + z(1 - x)(1 - y)(K_1 - L_1 - L_2 - K_2) \\
&+ xy(1 - z)(K_1 - L_1 - L_2 - K_2 + E_2) + x(1 - y)(1 - z)(K_1 + E_1 - L_1 - L_2 - K_2) \\
&+ y(1 - x)(1 - z)(K_1 - L_1 - L_2 - K_2 + E_2) + (1 - x)(1 - y)(1 - z)(K_1 - L_1 - L_2 - K_2)
\end{aligned}$$

2.5 Evolutionary Stabilization Strategies

Based on the replication dynamic equations for the four subjects, the Jacobian matrix can be formulated as follows

$$J = \begin{bmatrix}
\frac{\partial F(x)}{\partial x} & \frac{\partial F(x)}{\partial y} & \frac{\partial F(x)}{\partial z} & \frac{\partial F(x)}{\partial w} \\
\frac{\partial F(y)}{\partial x} & \frac{\partial F(y)}{\partial y} & \frac{\partial F(y)}{\partial z} & \frac{\partial F(y)}{\partial w} \\
\frac{\partial F(z)}{\partial x} & \frac{\partial F(z)}{\partial y} & \frac{\partial F(z)}{\partial z} & \frac{\partial F(z)}{\partial w} \\
\frac{\partial F(w)}{\partial x} & \frac{\partial F(w)}{\partial y} & \frac{\partial F(w)}{\partial z} & \frac{\partial F(w)}{\partial w}
\end{bmatrix}$$

The stability of the evolutionary game model can be determined using the Lyapunov method. For a multi-agent evolutionary game model, the equilibrium points are pure strategy equilibrium points, and the condition for them to be stable equilibrium points is that all eigenvalues of the Jacobian matrix of the game model are negative [10]. Since each of the four entities in the model has 2 strategies, the model has a total of 16 pure strategy equilibrium points. By substituting all pure strategy equilibrium points of the model into the Jacobian matrix based on the replication dynamic equations of each entity, and using the method for determining stable equilibrium points, the possible stable equilibrium points of the model are obtained as shown in Table 2.

Table 2. Possible stable equilibrium points

Equilibrium points	Eigenvalues	Positive or negative	Stability
$\theta_1(0,0,0,0)$	$A_1 - A_2 - B, F_1 - G_1 - F_2 + G_2, G_1 - I - H_2, K_1 - L_1 - L_2 - K_2$	U − − −	Conditional ESS
$\theta_2(0,1,1,0)$	$A_1 - A_2 - C_1 - B, G_1 + F_2 - F_1 - G_2 - C_2, H_2 + I - H_1 - D_2, K_1 + E_2 - L_1 - L_2 - K_2$	UUUU	Conditional ESS
$\theta_3(0,1,0,1)$	$A_1 - A_2 - C_1 - B, G_1 + F_2 - F_1 - G_2 - C_2, H_1 + D_2 - I - H_2,\ K_2 + L_1 + L_2 - K_1 - E_2$	UUUU	Conditional ESS
$\theta_4(0,1,1,1)$	$A_1 - A_2 - B, G_1 + F_2 - F_1 - C_2 - G_2, G_2 + I - H_1 - D_2, K_2 + L_1 + L_2 - K_1 - E_2$	UUUU	Conditional ESS
$\theta_5(1,1,1,1)$	$B - A_1 + A_2, G_1 + F_2 - F_1 - C_2 - C_1 - G_2, H_2 + I - H_1 - D_2, K_2 + L_1 + L_2 - K_1 - E_2$	U − − −	Conditional ESS

Annotation: U-Uncertainty of Positive or negative.

3 Results

Based on the model results, the evolutionary game model involving the four subjects in the watershed, under the given assumptions, has 5 possible stable equilibrium points. The first possible equilibrium point is $\theta_1(0,0,0,0)$, which is a stable equilibrium point if the eigenvalue condition is $A_1 - A_2 - B<0$. In this scenario, none of the upstream water-taking areas, downstream, or water-receiving regions have the initiative to establish an ecological compensation mechanism. However, the central government benefits more from a lenient supervision strategy, leading all entities to stabilize in a state where ecological compensation is not established. Conversely, if this eigenvalue condition is not met, the central government will choose stricter supervision for higher returns, forcing the other entities to be unable to stabilize in the strategy combination of no protection and no compensation.

For the second possible stable equilibrium point $\theta_2(0,1,1,0)$, the condition for stability is that all four eigenvalues are negative. In this scenario, the central government's lenient supervision strategy yields higher returns. The compensation amount from the downstream can offset the losses incurred by the upstream water-taking areas due to free-riding by the water-receiving areas. Downstream entities choose to pay compensation to avoid more severe penalties, while the water-receiving areas opt for free-riding, which offers greater benefits despite the risk of being penalized. The stability condition for this equilibrium point requires all four eigenvalues to be negative; any single eigenvalue not meeting this requirement will not satisfy the stability condition.

For the third equilibrium point $\theta_3(0,1,0,1)$, similar to the second, the central government chooses lenient supervision when the eigenvalue condition is met. Downstream entities adopt a riskier free-riding strategy, while upstream water-taking areas and receiving areas establish a horizontal ecological compensation mechanism. All stakeholders eventually achieve a stable equilibrium.

For the fourth equilibrium point $\theta_4(0,1,1,1)$, there is a consensus among the upstream water-taking areas, downstream, and water-receiving areas regarding the establishment of a horizontal ecological compensation mechanism, aligning with their consistent assessment of benefits. Given the strong protection and compensation awareness of the other entities, strict supervision by the central government does not increase the overall effective system's returns but instead incurs high regulatory costs. Consequently, the central government opts for lenient supervision.

For the fifth equilibrium point $\theta_5(1,1,1,1)$, the upstream water-taking areas, downstream, and water-receiving areas all exhibit strong compensation and protection awareness, but under the condition of strict supervision by the central government. If the central government adopts lenient supervision, the horizontal ecological compensation mechanism will not be sustainable. Therefore, under the long-term strict supervision of the central government, the other entities will establish a sustainable ecological compensation mechanism.

4 Conclusion

Taking the basin containing the water transfer project as the research object, this study clarifies the cooperative relationship among the subjects and establishes the theoretical game model of the four parties. The results of the model show that there are five possible stable equilibrium points, but different parameter conditions need to be satisfied. In the future research, the authors consider further analyzing the influence of different parameter changes on the establishment of ecological compensation mechanism of water transfer project containing, and then provide reference for optimizing the ecological compensation mechanism of the basin.

Acknowledgements. This research was funded by the Chinese National Natural Science Foundation (grants no. U2240223, 51,979,285), the Beijing Municipal Natural Science Foundation (grant no. 8222036), the Research Fund of the State Key Laboratory of Simulation and Regulation of Water Cycle in River Basin of China Institute of Water Resources and Hydropower Research (grant no. SKL2022TS11), and the Major Science and Technology Programs of the Ministry of Water Resources (grants no. SKS-2022126, WR110145B0042023).

References

1. Salzman, J., Bennett, G., Carroll, N., Goldstein, A., Jenkins, M.: The global status and trends of payments for ecosystem services. Nat. Sustain. **1**, 136–144 (2018)
2. Cordero Camacho, D.: Payment schemes for environmental services in watersheds in Ecuador. For. Syst. **17**, 54–66 (2008)
3. Feyera, S., Abi, M., Tolossa, D., Abate, E., Terefe, G.: Sustainable utilization and conservation through compensation for ecosystem services: experiences from the Koga watershed in Northern Ethiopia. Land. **13**(8), 1287 (2024)
4. Jones, K.W., et al.: The emergence and persistence of payments for watershed services programs in Mexico. Water Resour. Econ. **42**, 100217 (2023)
5. Ricardo, R.: Water and payment of environmental services in Costa Rica. In: Conference: 21st Century Watershed Technology: Improving Water Quality and Environment, (2010)
6. Young, C., Bakker, L.: Payments for ecosystem services from watershed protection: a methodological assessment of the oasis project in Brazil. Natureza Conservacao. **12**, 71–78 (2014)
7. Ding, J., Chen, L., Deng, M., Chen, J.: A differential game for basin ecological compensation mechanism based on cross-regional government-enterprise cooperation. J. Clean. Prod. **362**, 132335 (2022)
8. Zhu, K., Zhang, Y., Wang, M., Liu, H.: The ecological compensation mechanism in a cross-regional water diversion project using evolutionary game theory: the case of the Hanjiang River Basin, China. Water. **14**(7), 1151 (2022)
9. Gao, X., Song, R., et al.: Ecological compensation of major water transfer project based on stochastic evolutionary game: a case study of south-to-north water transfer project [J] (in Chinese). Front. Sci. Technol. Eng. Manage. **41**(01), 26–34 (2022)
10. Franco, B., Giulia, G.: Polyhedral Lyapunov functions structurally ensure global asymptotic stability of dynamical networks iff the Jacobian is non-singular. Autom. **86**, 183–191 (2017)

The Trend and the Influencing Factors of Pan Evaporation in China for Long Time Series

Jiaqi Li[1], Bing Gao[2,3], Weiwei Shao[1](✉), Guang Han[1], and Rongkang Yu[1]

[1] State Key Laboratory of Simulation and Regulation of Water Cycle in River Basin, China Institute of Water Resources and Hydropower Research, Beijing 100038, China
shaoww@iwhr.com

[2] Key Laboratory of Groundwater Conservation of MWR, China University of Geosciences, Beijing 100083, China

[3] China School of Water Resources and Environment, China University of Geosciences, Beijing 100083, China

Abstract. Pan evaporation (E_{pan}) is a critical indicator of atmospheric evaporative demand and a key input for hydrological models. This study explores the changes of E_{pan} and its dominant factors across China from 1963 to 2017 using Mann-Kendall analysis. The results show that the national E_{pan} exhibited a decline followed by an increase during the study period, with the majority of trend reversals occurring around 1996. The decline in net radiation and wind speed may have contributed to the decrease in E_{pan}, while increases in air temperature and vapor pressure deficit were associated with the rise in E_{pan}. VPD is the primary factor influencing E_{pan} across the meteorological stations in China, with wind speed playing a significant role before 1996. Air temperature and net radiation had a relatively minor contribution to the changes in E_{pan}.

Keywords: Pan evaporation · Vapor pressure deficit · Climate change

1 Introduction

E_{pan} is frequently used to estimate open water evaporation or potential evapotranspiration [1], serving as a key indicator of atmospheric evaporative demand and a foundational data for assessing drought conditions [2].

E_{pan} is highly sensitive to climate variables [3]. Previous studies showing that factors such as wind speed, relative humidity, air temperature, vapor pressure deficit (VPD), and solar radiation can influence E_{pan}. The effects of these factors on E_{pan} are not uniform [4]. For instance, an increase in wind speed or air temperature typically leads to an increase in E_{pan}, while a rise in relative humidity generally results in a decrease [5]. Therefore, analysing the change of E_{pan} and understand the influence of meteorological factors on E_{pan} [6] is a hot research area in the field of hydro-climatology [7].

While most studies on E_{pan} trends and influencing factors have focused on regional scales and short time series, this study aims to fill this gap by analysing the E_{pan} trends and dominant factors (net radiation, wind speed, air temperature and VPD) across China from 1963 to 2017, providing a theoretical basis for understanding the response of the hydrological cycle to climate change.

V. Achal et al. (Eds.): EPPC 2024, ESE, pp. 128–137, 2025.
https://doi.org/10.1007/978-3-031-92123-0_13

2 Data and Methods

2.1 Data

The E_{pan} data used in this study was a dataset which reconstructed based on an improved PenPan model [8] proposed by Yang et al. [9], which corrects the Class A E_{pan} data [10]. The dataset demonstrates good simulation performance within the China.

Meteorological data were obtained from the China Surface Climate Daily Dataset (V3.0), available for download at http://data.cma.cn, which includes daily records of sunshine hours, atmospheric pressure, air temperature, relative humidity, and wind speed.

2.2 Methods

2.2.1 Mann-Kendall Analysis

Mann Kendall analysis is a non-parametric test method, which is widely used in the analysis of hydrological and meteorological elements, and is less disturbed by outliers and missing values. By Mann Kendall analysis, the change of the E_{pan} trend can be detected.

$$\mathrm{s_k} = \sum\nolimits_{\mathrm{i}=1}^{\mathrm{k}} \mathrm{r_i} \left(\mathrm{x_i} > \mathrm{x_j}, \mathrm{r_i} = 1; \mathrm{x_i} < \mathrm{x_j}, \mathrm{r_i} = 0 \right) \tag{1}$$

$$\mathrm{UF} = \frac{\mathrm{s_k} - \mathrm{E(s_k)}}{\sqrt{\mathrm{Var(s_k)}}} \ (\mathrm{k} = 2, 3, 4, \cdots, \mathrm{n}) \tag{2}$$

$$\mathrm{UB_i} = -\mathrm{UF_i}(\mathrm{i} = \mathrm{n}, \mathrm{n} - 1, \cdots, 1) \tag{3}$$

For a time series X (x_1, x_2, ..., x_n), the situation of UF > 0 means he trend of E_{pan} is increased, otherwise, it means the decreased trend. If the line of UF intersects the line of UB, it means that the trend is abrupt. In this paper, we set the significance level as 0.05, that is, if UF is within the range of [−1.96, 1.96], it can be determined that the change trend and mutation are significant.

2.2.2 Partial Differential Derivation Attribution Method

In order to reveal the influence of meteorological factors such as wind speed, net radiation, air temperature and VPD on E_{pan} variation, partial differential derivation attribution method was used to further analyse [11]. Based on the PenPan model, this method quantifies the contribution of different meteorological factors to E_{pan} changes through partial differential derivation.

$$\frac{dE_{pan}}{dt} \approx R_n{}^* \frac{dR_n}{dt} + T^* \frac{dT}{dt} + U^* \frac{dU}{dt} + D^* \frac{dD}{dt} \tag{4}$$

$$R_n{}^* = \left. \frac{\partial Epan}{\partial R_n} \right|_{R_n = \overline{R_n}} \tag{5}$$

$$T^* = \left. \frac{\partial Epan}{\partial T} \right|_{T = \overline{T}} \tag{6}$$

$$U^* = \left. \frac{\partial Epan}{\partial U} \right|_{U=\overline{U}} \tag{7}$$

$$D^* = \left. \frac{\partial Epan}{\partial D} \right|_{D=\overline{D}} \tag{8}$$

Where, $R_n{}^*\frac{dR_n}{dt}$, $T^*\frac{dT}{dt}$, $U^*\frac{dU}{dt}$, and $D^*\frac{dD}{dt}$ represents the contribution of net radiation, air temperature, 2 m wind speed and VPD to E_{pan}, respectively. The meteorological factor with the largest absolute contribution has the greatest impact on E_{pan} and is regarded as the leading factor causing E_{pan}. $\overline{R_n}$, $\overline{T}$, $\overline{U}$, $\overline{D}$ represents the annual net radiation, annual air temperature, annual windspeed, and annual VPD.

2.2.3 The Division of Region

To better understand the spatial distribution of E_{pan} changes and their influencing factors, China was divided into Northern region and Southern region, as shown in Fig. 1.

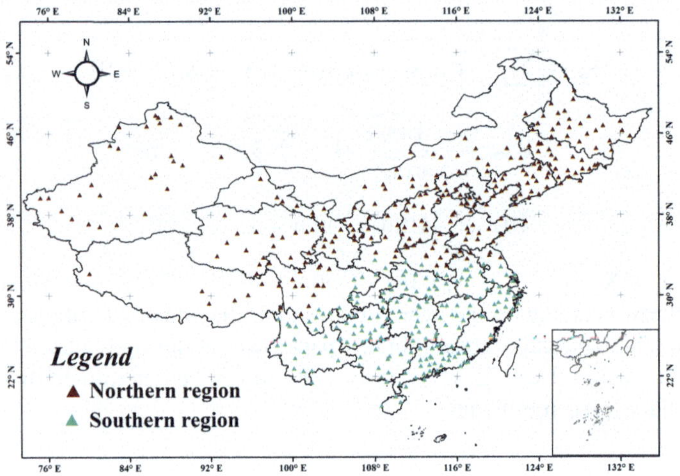

Fig. 1. The division of Northern region and Southern region

3 Results

3.1 The Trend of E_{pan}

Mann-Kendall trend analysis was conducted for the E_{pan} data across meteorological stations in China. The analysis revealed that, from 1963 to 2017, most stations exhibited a declining trend in E_{pan} until 1996 (Fig. 2a), followed by an increase (Fig. 2b). Moreover, 35.9% of stations experienced a trend reversal in 2002. For other stations, the trend turned around 1999 Regionally, stations with trend reversals in 1999 were mainly located in the Northern region and Guangxi Province, while those in 2002 were more common in inland southern areas. For the period from 1963 to 1996, the average E_{pan} trend for all 462 stations was −3.61 mm per decade. As shown in Fig. 2b, after 1996, most stations showed an increasing trend in E_{pan}, with only 138 stations experiencing a decline, predominantly located in the northeast and southern coastal regions.

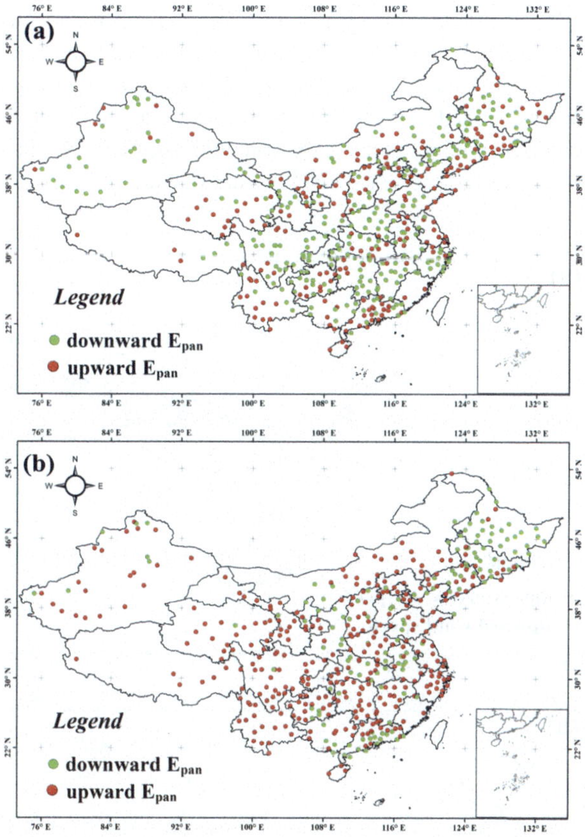

Fig. 2. The trend of E_{pan} in 1963–1996 and1997–2017 (a) represents 1963–1996 and (b) represents 1997–2017.

By analysing the average E_{pan} change trend of the two regions, we can see that the change trend of E_{pan} in the Southern region is greater than that in the Northern region, whether it is before 1996 (−0.01 mm per decade in the Southern region and − 0.009 mm per decade in the Northern region) or after the trend turns from downward to upward (0.133 mm per decade in the Southern region and 0.111 mm per decade in the Northern region).

3.2 Trend of Meteorological Factors

In order to better understand the influence of different meteorological factors on E_{pan}, we also analysed the trend of each meteorological factor in different periods.

Before 1996, wind speed showed a decreased trend at a rate of -0.1 m s^{-1} per decade, net radiation also showed a downward trend, while air temperature and VPD showed an upward trend.

From 1996 to 2017, the decreasing trend of wind speed is more obvious (Fig. 3), which is -0.2 m s^{-1} per decade. The upward trend of VPD also changed, from 0.003 Kpa per decade to 0.022 Kpa per decade (Fig. 4). The warming effect continued (Fig. 5), and the trend from 1997 to 2017 (0.18 °C per decade) was similar to that before 1996 (0.17 °C per decade). The decline in net radiation continued (Fig. 6), but at a slower rate, from -4.29 MJ m^2 per decade to -1.87 MJ m^2 per decade.

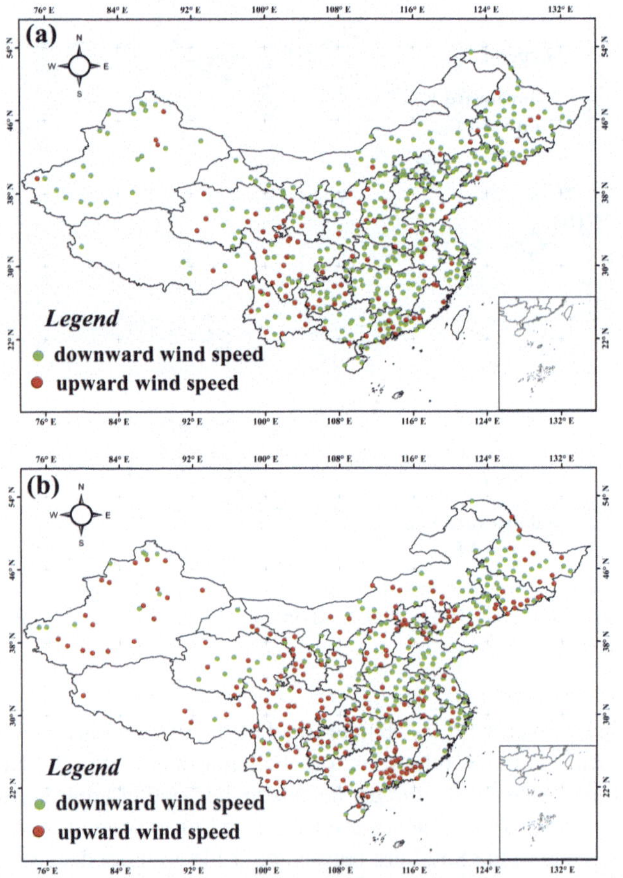

Fig. 3. The trend of wind speed in 1963–1996 and 1997–2017 (**a**) represents 1963–1996 and (**b**) represents 1997–2017.

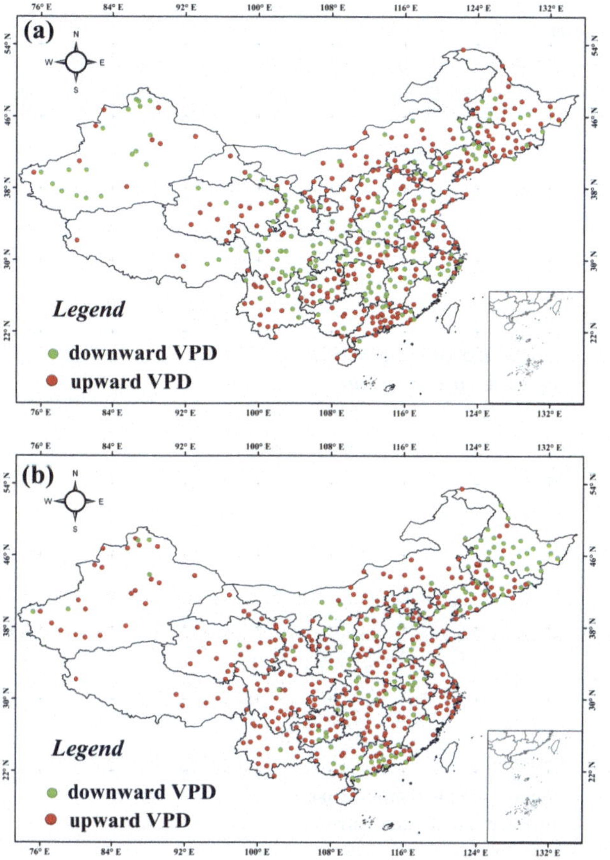

Fig. 4. The trend of VPD in 1963–1996 and1997–2017 (**a**) represents 1963–1996 and (**b**) represents 1997–2017.

According to the result of the spatial distribution, in the Northern region, the wind speed showed a decreasing trend both before and after 1996 (Fig. 3), but the decreasing trend before 1996 was greater. The downward trend of wind speed in the Southern region turned into an upward trend after 1996 (Fig. 3b). In the whole country, there is a downward trend throughout the period, which indicates that even though the wind speed in the Southern region has weakened to a certain extent, the downward trend in the Northern region after 1996 is more obvious, so the whole country still shows a decline in wind speed.

For VPD, both the Southern region and Northern region experienced an increase in both periods, and the increase was greater in the second period. Especially in the Southern region, the trend of VPD is greatly increased (Fig. 4b). The air temperature also rose in both periods and in both regions, but in the Southern region after 1996 (Fig. 5b), the warming speed was faster than that before 1996, while in the Northern region, it was the opposite. In contrast to VPD, net radiation decreased in both Southern region and Northern region (Fig. 6), and the rate of decline in the second period was less than that in the first period.

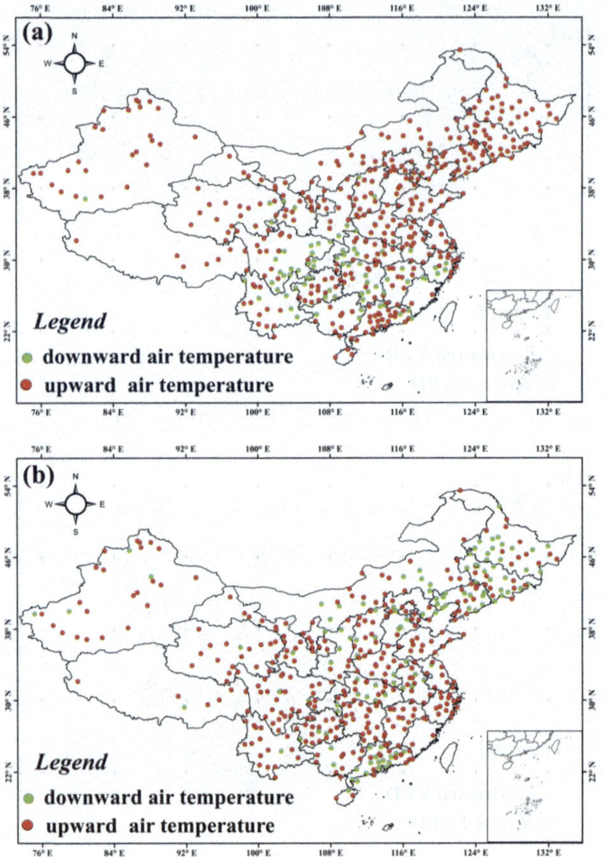

Fig. 5. The trend of air temperature in 1963–1996 and1997–2017 (**a**) represents 1963–1996 and (**b**) represents 1997–2017.

3.3 Attribution Analysis of E_{pan} Changes

As shown in Fig. 7a, using the partial derivative attribution method, we found that, from 1963 to 1996, the dominant factor influencing E_{pan} changes was VPD, with 334 stations showing its influence. Wind speed was the second most significant factor, affecting 74 stations. As shown in Fig. 7b, after 1996, the number of stations where VPD was the dominant factor increased to 400, while wind speed remained the second most influential factor.

Regional analysis revealed that the decrease in wind speed and net radiation led to a decrease in E_{pan}, while increases in VPD and air temperature resulted in an increase in E_{pan}. Although VPD dominated E_{pan} changes in most stations, in the Northern region before 1996, the decline in wind speed played a more significant role. In the coastal area of Southern regions, the decline in E_{pan} was associated with decreases in VPD. Additionally, net radiation had a minimal effect on E_{pan} changes.

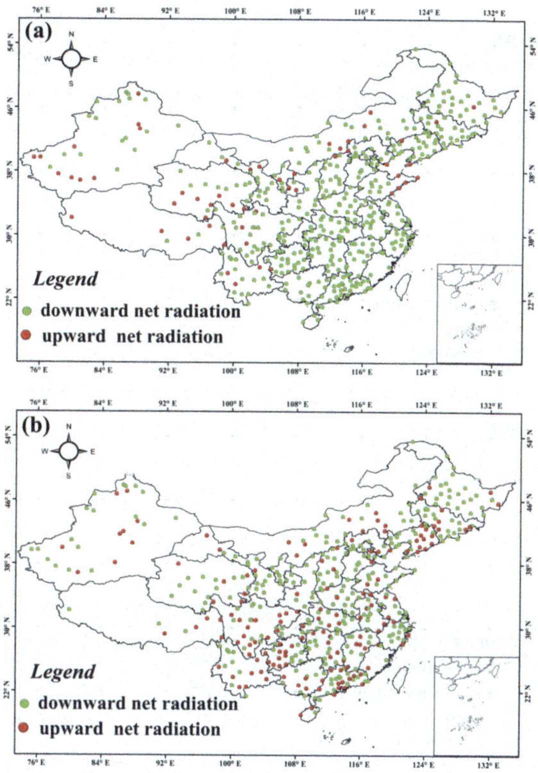

Fig. 6. The trend of net radiation in 1963–1996 and1997–2017 (**a**) represents 1963–1996 and (**b**) represents 1997–2017.

4 Discussion

4.1 The Trend of E_{pan} and the Influence of Meteorological Factors

The results of this study show that, from 1963 to 2017, the annual average E_{pan} at meteorological stations in China decreased and then increased, with a significant reversal around 1996. Both the Northern and Southern region followed this trend, with the Southern region exhibiting more substantial changes than the Northern region.

By analysing the contributions of VPD, wind speed, air temperature, and net radiation to E_{pan} using the PenPan model, we determined the dominant factors influencing E_{pan} changes across different regions and time periods. E_{pan} changes were found to be the result of multiple meteorological factors acting together. In the pre-1996 period, wind speed dominated the declining E_{pan} trend, but after 1996, the increase in VPD counteracted this effect, leading to an increase in E_{pan}. In the second period, the increase in VPD and air temperature was the primary driver of the upward trend in E_{pan}, especially VPD. It was found that from 1963 to 1996, while E_{pan} showed a decreasing trend, the dominant factor was VPD, which was also increasing during that period. However, the change of E_{pan} was the result of a combination of meteorological factors, and the decrease of net radiation and wind speed during this period inhibited the rise of E_{pan}, so E_{pan} showed a downward trend.

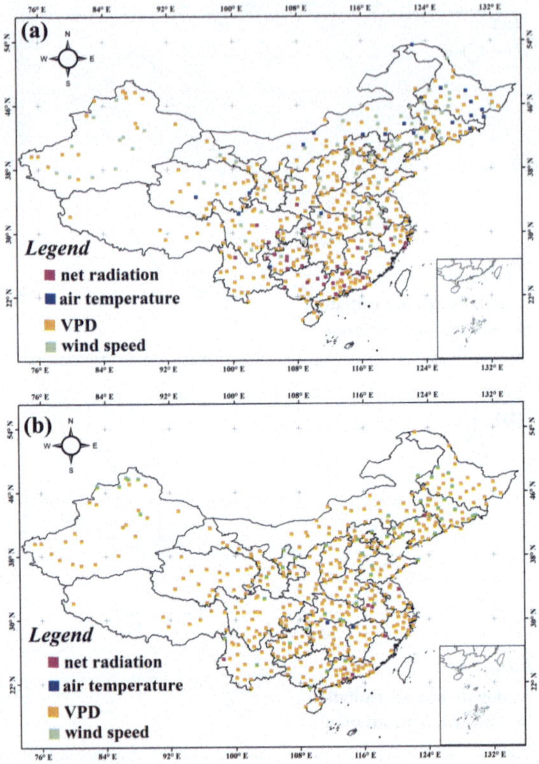

Fig. 7. The major influencing factors of E_{pan} (**a**) represents 1963–1996 and (**b**) represents 1997–2017.

4.2 Limitations of the Research

This study used a dataset from 462 meteorological stations, but the stations are not uniformly distributed across regions, especially in the northwest of the Northern region, where fewer stations are located, potentially affecting the representativeness of the data. In addition, the mechanism about how meteorological factors influence E_{pan} are complex, and interactions between factors are difficult to quantify. While partial differential derivation attribution method was employed, the individual contributions of each factor remain difficult to calculated Future research could use higher-resolution remote sensing models to further analysed the trends and influences factors on E_{pan}.

5 Conclusion

This study used long-term data reconstructed using the PenPan model to analyze the trends in E_{pan} at 462 meteorological stations across China, dividing the Northern region and Southern region, from 1963 to 2017. The study also examined the trends of meteorological factors during this period and their contributions to E_{pan} changes. The main conclusions are as follows:

From 1963 to 2017, E_{pan} at most stations across China showed a decline before 1996, followed by an increase afterward, with the Southern region exhibiting more significant changes than the Northern region. The combined effect of meteorological factors contributed to the changes in E_{pan}, with air temperature and VPD generally increasing, leading to higher E_{pan}, while wind speed and net radiation showed declining trends, contributing to a decrease in E_{pan}. VPD was the dominant factor at most stations, and the decline in wind speed prior to 1996 was significant for the downward trend in E_{pan}.

Acknowledgements. This research was funded by the Beijing Municipal Natural Science Foundation (grant no. 8222036), the National Natural Science Foundation of China (grants no. U2240223, 51979285), the Major Science and Technology Programs of the Ministry of Water Resources (grants no. SKS-2022126, WR110145B0042023), and the Research Fund of the State Key Laboratory of Simulation and Regulation of Water Cycle in River Basin of China Institute of Water Resources and Hydropower Research (grant no. SKL2022TS11).

References

1. McVicar, T.R., Van Niel, T.G., Li, L., Hutchinson, M.F., Mu, X., Liu, Z.: Spatially distributing monthly reference evapotranspiration and pan evaporation considering topographic influences. J. Hydrol. **338**(3), 196–220 (2007)
2. Wang, H., Sun, F., Liu, F., Wang, T., Liu, W., Feng, T.: Reconstruction of the pan evaporation based on meteorological factors with machine learning method over China. Agric. Water Manag. **287**, 108416 (2023)
3. Sun, F., Liu, W.: Assessing estimates of evaporative demand in climate models using observed pan evaporation over China. J. Geophys. Res. D Atmos. **121**(14), 8329–8349 (2016)
4. Fan, D., Yu, H., Liu, G., Yang, R., Li, X., Wang, L.: Effect of urbanization on the long-term change in pan evaporation: a case study of the Nanpan River Basin in China. Ecol. Indic. **145**, 109631 (2022)
5. Jin, Y., Wang, D., Feng, Y., Wu, J., Cui, W., He, X., Chen, A., Zeng, Z.: Decreasing relative humidity dominates a reversal of decreasing pan evaporation in mainland China after 1989. J. Hydrol. **608**, 127641 (2022)
6. Malik, A., Jamei, M., Ali, M., Prasad, R., Karbasi, M., Yaseen, Z.M.: Multi-step daily forecasting of reference evapotranspiration for different climates of India: a modern multivariate complementary technique reinforced with ridge regression feature selection. Agric. Water Manag. **272**, 107812 (2022)
7. Gelete, G., Yaseen, Z.M.: Hybridization of deep learning, nonlinear system identification and ensemble tree intelligence algorithms for pan evaporation estimation. J. Hydrol. **640**, 131704 (2024)
8. Rotstayn, L.D., Roderick, M.L., Farquhar, G.D.: A simple pan-evaporation model for analysis of climate simulations: evaluation over Australia. Geophys. Res. Lett. **33**(17) (2006)
9. Yang, H., Yang, D.: Climatic factors influencing changing pan evaporation across China from 1961 to 2001. J. Hydrol. **414-415**, 184–193 (2012)
10. Li, J., Gao, B.: How urbanization affects pan evaporation in China? Urban Clim. **49**, 101536 (2023)
11. Roderick, M.L., Rotstayn, L.D., Farquhar, G.D., Hobbins, M.T.: On the attribution of changing pan evaporation. Geophys. Res. Lett. **34**(17) (2007)

Calculation of Carbon Emission from Prefabricated Girders of Bridge Engineering and Low Carbon Strategy Research

Zhiqing Chen[1], Shu Yang[1], Zheshi Wang[1], and Mengyue Li[2(✉)]

[1] Guangzhou Expressway Company Ltd., Guangzhou, China
[2] Guangdong Provincial Transportation Planning and Research Center, Guangzhou, China
mengyuelee@163.com

Abstract. Prefabricated girders bridge is widely used in bridge engineering. In order to promote the low-carbon construction of bridge, in this study the carbon emission calculation method of prefabricated girders of bridge superstructure is established by using engineering budget data, and the carbon emission factor method is used to calculate the carbon emission of prefabricated girders. Based on the actual survey data of a highway girder yard in Guangzhou, the carbon emission calculation results are verified. Through carbon emission traceability analysis, the low carbon control strategy of bridge prefabricated girders yard is proposed.

Keywords: bridge · prefabricated girders · girder yard · carbon emission

1 Introduction

Bridges are the structural form with the largest amount of material per unit and the highest carbon emission index per unit among all types of structures in civil engineering [1], which is an important source of carbon emissions from road construction. In the existing studies, Xiao Lei [2] used the process estimation method, combined with the BIM model, to calculate the carbon emissions of railway bridges. Liu Muyu et al. [3] proposed a life cycle carbon emission calculation method for bridge engineering. Ma Jiaxing [4] established a carbon emission calculation model by using the carbon emission factor method, and conducted a quantitative analysis of the Ningbo Hangzhou cable-stayed Bridge. Hao Guojun [5] established the carbon emission calculation method during the construction period of the bridge, and analyzed the Xiongxin expressway bridge as a case. The carbon emission of bridge construction is calculated by combining the project budget document with the actual manual labour, material and machine consumption of on-site construction. However, there are many difficulties in the construction materials and mechanical equipment, and the data collection is difficult or incomplete. Prefabricated girder type bridge is usually centrally produced in prefabricated girder yards, and it is relatively easy to obtain construction information and data.

This study calculates the carbon emission of prefabricated girders in the girder yard, researches the main sources of carbon emission, the boundary and method of carbon

V. Achal et al. (Eds.): EPPC 2024, ESE, pp. 138–147, 2025.
https://doi.org/10.1007/978-3-031-92123-0_14

emission accounting. It also calculates the carbon emission and contributes to the analysis of the precast girders of a highway in Guangzhou as a case study. The carbon emission intensity index of the precast girders of bridges is obtained, puts forward the carbon reduction strategy for the main carbon emission contributing sources, and explores the path of low-carbon construction of bridges. The carbon reduction strategies are proposed for the main carbon emission contribution sources, and low-carbon bridge construction path is also discussed.

2 Sources of Carbon Emissions from Prefabricated Girders

2.1 Carbon Emission Calculation Boundary

Life cycle theory is currently an internationally recognised environmental impact and evaluation tool, which is based on the life cycle framework and environmental statistics, and can adequately and systematically describe the environmental impact degree of products to be tested [6].Many studies have applied the life cycle theory to the calculation of carbon emissions in the field of highway infrastructure construction. The whole life cycle of highway is divided into three stages: construction stage, maintenance stage and operation stage. According to the LCA definition, the construction stage can be divided into material production, transportation, construction stage and use stage. It has been found in the existing research that many influences on the measurement of carbon emissions related to manual labour are not easy to obtain and account for a small proportion of carbon emissions in the highway construction, which can be ignored [7].

Carbon emissions from prefabricated girders during the bridge construction phase are therefore defined as mainly reflected in the use of materials and mechanical equipment related to construction activities. Carbon emissions related to materials mainly refer to the carbon emissions from the entire materialisation process from raw material extraction, production, processing to transportation away from the production plant, etc. According to the type of construction machinery and construction site, carbon emissions related to construction machinery and transportation vehicles are usually classified into three types, which are related to off-site mixing, material transportation and construction site, etc. [8]These are essentially derived from the production and use of fuels, energy and electricity.

2.2 Carbon Emission Source Analysis

Carbon emissions from prefabricated girders during the bridge construction phase are related to the construction materials and mechanical equipment used in the production process of the girders, the girders transport and erection process. The production process includes the production of materials and transportation to the girder yard, while the transport and erection process includes the girders transport to the construction site after fabrication and their installation on site. The materials used in the prefabrication of girders include rebar, steel strand, steel mould plate, strand group anchors, cement and anchor gear, etc. The source of emissions is the carbon emissions generated by the electricity and fuel consumption in the material production process. Minor and sporadic materials were rounded off, and materials for which emission data were difficult to obtain and which accounted for less than 5% of the total material weight were ignored. The mechanical equipment used include transport equipment, processing equipment and installation equipment, such as flatbed trailer sets, trucks, concrete trucks, loaders, CNC steel processing machinery, pre-stressing tensioners, bridge erecting machines, etc., mostly using diesel, petrol, natural gas and electricity as energy consumption. Carbon emission sources are shown in Fig. 1 below.

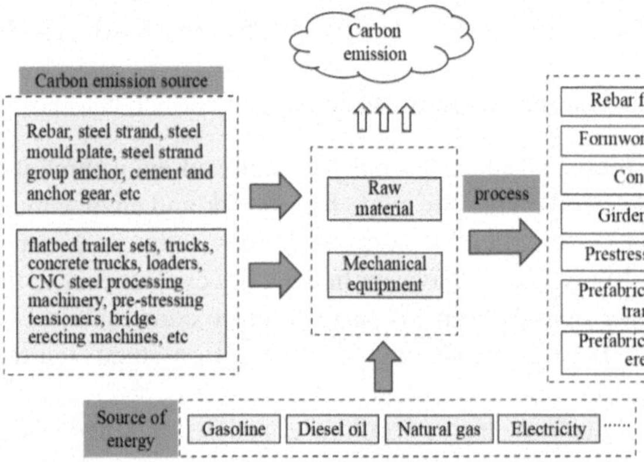

Fig. 1. Schematic diagram of the sources of carbon emissions from prefabricated girders

3 Methodology for Calculating Carbon Emissions

The emission factor method was used to establish the carbon emission calculation equations for materials, on-site machinery and equipment use and transport vehicle use of prefabricated girders production, transport and erection process respectively:

$$\mathrm{EC} = \mathrm{E_p} + \mathrm{E_t} + \mathrm{E_i} \tag{1}$$

Where: EC is the carbon emission from prefabricated girders; $\mathrm{E_p}$ is the carbon emission of prefabricated girders production process, and $\mathrm{E_t}$, $\mathrm{E_i}$ is the carbon emission of prefabricated girder transport and erection process.

1. Carbon emissions calculation at each process

$$E_p = \sum\nolimits_{j=1}^{n} \left(M_j \times MF_j + Y_j \times Q_j \times Ef_j \right) \tag{2}$$

where: M_j is the amount of j-th material used, and MF_j is the carbon emission factor of the jth material, and Y_j is the mileage of the jth material transport (including road, railway, waterway and other modes), and Q_j is the unit mileage energy consumption of the jth material transport, and Ef_j is the carbon emission factor of fuel or electricity consumption for the jth material transport.

$$E_t = \sum\nolimits_{i=1}^{n} Y \times Q_i \times Ef_i \tag{3}$$

Where: Y is the distance from the prefabricated girders transport to the construction site. Q_i is the energy consumption per mile of the ith transportation equipment, and Ef_i is the carbon emission factor of fuel or electricity consumed by the ith transport equipment.

$$E_i = \sum_{k=1}^{n} P_k \times S_k \times Ef_k \tag{4}$$

where: P_k is the quantity of work shifts of the kth type of construction machinery, and S_k is the energy consumption of the kth type of construction machinery work shift, and Ef_k is the carbon emission factor of fuel or electricity consumed by the kth type of construction machinery.

2. Establishment of carbon emission factor inventory

   Energy carbon emission factors includes fossil fuels such as diesel, gasoline, heavy oil, natural gas and the carbon emission factor of electricity. The carbon emission factor of materials includes existing road construction materials such as steel, cement, sand and gravel, as well as some self-developed materials, such as concrete, mortar and asphalt, which are calculated according to mixing ratios of raw material. The carbon emission factor of field operation equipment is the energy consumption per unit shift multiplied by the corresponding energy emission factor. The carbon emission factor of transport equipment is the energy consumption per kilometre multiplied by the corresponding energy emission factor. The carbon emission factor of prefabricated girders for bridge construction is selected by taking into account the existing domestic carbon emission database and the "Class Cost Quota for Highway Construction Machinery" and "Fuel Consumption Limitation Standard for Heavy Commercial Vehicles".

4 Analysis of the Results of Carbon Emission Calculations

4.1 Calculation of Carbon Emissions Based on Project Budgets

Taking a prefabricated girders yard of a highway in Guangzhou as an example, the girder yard concentrates on prefabricating 10,987 pieces of prestressed concrete box girders for 82 bridges along the highway, including various types such as 20 m, 25 m, 30 m, 35 m and 40 m. According to the bridge project prefabricated girder budget information, the main materials used in the process of prefabricated girders are obtained, as shown in Table 1 below.

Most of the construction machinery used in the project uses diesel and electricity as the energy, and the main construction machinery used in the process of prefabricating girders is shown in Table 2 below.

Table 1. Summary of quantities of building materials for prefabricated girder construction

Materials	Quantities	Unit (of measure)
Steel plate	98.067	t
Medium (coarse) sand	118574.014	m^3
Wire rope	109.311	t
Plastic bellows SBG-60Y	1736444.556	m
Ironware	202622.865	kg
20–22 iron wire	211806.081	kg
Section steel	126.303	t
Impact crusher stone	223242.471	m^3
Steel strand group anchor (6 holes)	16158.798	Suit
Electrode	246676.898	kg
Steel strand group anchor (4 holes)	93103.038	Suit
Steel strand group anchor (5 holes)	90454.461	Suit
Compacting material	4955.637	t
Steel mould plate	2214.714	t
Steel strand	13927.464	t
52.5 grade cement	135278.872	t
HRB400 rebar	57620.322	t

Table 2. Summary of prefabricated girder construction machinery

Construction machinery	Quantities	Unit (of measure)	Type of energy consumption
Self-propelled motor grader up to 120 kW	1.757	Work shift	Diesel oil
Truck-mounted cranes up to 25 t	6.480	Work shift	Diesel oil
18–21 t light wheel roller	10.022	Work shift	Diesel oil
AC butt welder up to 150 kV-A	16.052	Work shift	Electrical
Truck-mounted cranes up to 16 t	45.419	Work shift	Diesel oil
Cargo up to 20 t	20.689	Work shift	Diesel oil
Electric multi-stage water pump within Φ100mm (≤120 m)	122.405	Work shift	Electrical
Intelligent grouting system	535.739	Work shift	Electrical
Truck-mounted cranes up to 30 t	278.537	Work shift	Diesel oil
Truck-mounted cranes up to 8 t	418.253	Work shift	Diesel oil
Truck-mounted cranes up to 20 t	520.179	Work shift	Diesel oil
Tyre loader up to 1.0 m^3	791.333	Work shift	Diesel oil

(*continued*)

Table 2. (*continued*)

Construction machinery	Quantities	Unit (of measure)	Type of energy consumption
CNC steel bending machine	542.538	Work shift	Heavy oil/electricity
Crawler dozers up to 75 kW	791.333	Work shift	Diesel oil
Single drum slow moving electric winches up to 30 kN	10108.360	Work shift	Electrical
Flatbed trailer set up to 100 t	676.381	Work shift	Diesel oil
Concrete mixing plant up to 60 m^3/h	847.241	Work shift	Electrical
Intelligent tensioning system	16911.965	Work shift	Electrical
Concrete mixer trucks up to 6 m^3	3388.963	Work shift	Diesel oil
AC butt welder up to 100 kV-A	5203.591	Work shift	Electrical
AC arc welding machines up to 32 kV-A	42095.839	Work shift	Electrical
Single drum slow moving electric winches up to 50 kN	86126.530	Work shift	Electrical
CNC vertical steel bending Centre	7366.351	Work shift	Heavy oil/electricity

Calculate the total carbon emission and intensity of prefabricated girders as shown in Table 3, the total carbon emission is 426,000 tonnes, and the carbon emission per m3 prefabricated girder is 1.534 t. The vast majority of the carbon emission comes from the process of material production and use, which accounts for 97.44% of the total carbon emission, as shown in Table 4 below.

Table 3. Carbon emissions and intensity values of prefabricated girder pieces

Quantity of prefabricated girders (m^3)	Total carbon emissions (t)	Carbon intensity (t/m^3)
275317.595	426240.798	1.534

Table 4. Composition and share of carbon emissions from prefabricated girders

Material production (t)	Material processing (t)	Transport process (t)	Erection process (t)
414674.975 (97.3%)	5964.829 (1.4%)	1362.218 (0.3%)	4238.774 (1.0%)

According to the analysis of carbon emission contribution of materials, the carbon emission of HRB400 rebar, 52.2 grade cement, steel strand and steel mould plate accounted for most of the carbon emission of prefabricated girder materials, with a cumulative share of 96%, of which 49% was accounted for by HRB400 rebar, 33% was accounted for by 52.2 grade cement, 12% was accounted for by steel strand, and only 6% was accounted for by carbon emission of steel mould plate and other materials. Seen as Table 5 for details.

An analysis of the contribution of carbon emissions from construction machinery shows that 68% of the energy consumed by machinery is generated by electricity, while carbon emissions generated by diesel and heavy oil are closer, accounting for 15%and

Table 5. Summary of carbon emissions from key materials

Materials	Carbon emissions (t)	Percentage
Steel plate	351.963	0.1%
Medium (coarse) sand	355.722	0.1%
Wire rope	392.317	0.1%
Plastic bellows SBG-60Y	442.099	0.1%
Ironware	506.557	0.1%
20–22 iron wire	529.515	0.1%
Section steel	548.028	0.1%
Impact crusher stone	669.727	0.2%
Steel strand group anchor (6 holes)	695.927	0.2%
Electrode	885.323	0.2%
Steel strand group anchor (4 holes)	2673.066	0.6%
Steel strand group anchor (5 holes)	3246.200	0.8%
Compacting material	3979.377	0.9%
Steel mould plate	7948.607	1.9%
Steel strand	49985.667	11.9%
52.5 grade cement	140900.656	33.4%
HRB400 rebar	206799.336	49.1%

17% respectively. Emissions from CNC vertical rebar bending centers, single drum slow moving electric winches up to 50kN, and AC arc welding machines up to 32 kV-A accounted for 66%. Seen as Table 6 for details.

Table 6. Summary of carbon emissions from key materials

Construction machinery	Carbon emissions (t)	Emission share
Crawler dozers up to 75 kW	200.098	1.7%
Single drum slow moving electric winches up to 30 kN	213.952	1.8%
Flatbed trailer set up to 100 t	313.624	2.7%
Concrete mixing plant up to 60 m^3/h	345.538	3.0%
Intelligent tensioning system	550.248	4.7%
Concrete mixer trucks up to 6 m^3	862.396	7.4%
AC butt welder up to 100 kV-A	871.765	7.5%
AC arc welding machines up to 32 kV-A	2094.067	18.0%
Single drum slow moving electric winches up to 50 kN	2734.659	23.5%
CNC vertical steel bending centre	2808.442	24.2%

4.2 Result Verification Analysis

The research collects the production progress of the girder yard, monthly details of electricity and oil consumption, machinery and material entry ledger, concrete mix ratio testing ledger and other data and information, obtains the actual production progress of the girder piece, the actual materials and machinery energy consumption of prefabricated girders, converted to the level of per m3 unit, as shown in Table 7 below.

Table 7. Comparison of energy consumption of major materials and machinery for prefabricated girders

Source of carbon emissions	Unit (of measure)	Budgeted consumption	Actual consumption
HRB400 rebar	t/m^3	0.209	0.208
Steel stranded wire	t/m^3	0.051	0.041
52.5 grade cement	t/m-	0.491	0.477
Electricity used for machinery and equipment	kwh/m^3	49.63	25
Diesel oil used for machinery and equipment	kg/m^3	1.402	1.654
Heavy oil used for machinery and equipment	kg/m^3	1.702	1.82

The main carbon emission source was taken as the error influencing factor of the carbon emission intensity index of the prefabricated girders. Comparative analysis shows that the carbon emission per m3 calculated by budget is higher than the actual amount, and the difference is −5.3%. Among them, the main materials are closer to the actual consumption than the budgeted amount, while the actual consumption of electricity has a larger deviation, much less than the budgeted amount. Therefore, it is feasible and applicable to evaluate the carbon emission of prefabricated bridge girders by using the materials, machinery and equipment list data of the project budget when it is difficult to obtain the actual project material, machinery and equipment consumption.

5 Analysis of Carbon Reduction Strategies

Through the above analysis of carbon emission contribution, the main sources of prefabricated girder carbon emissions are rebar, cement, steel strand wire, steel mould plate and rebar fabrication equipment, girder transport equipment and concrete mixing equipment. In order to promote the low-carbon construction of bridges and reduce the carbon emission of the whole life cycle of production of prefabricated bridge girder, control measures should be taken in the girder construction and management.

The highway construction department should strictly manage the main materials, specify the brand, manufacturer and quality requirements of the materials used in the project in the bidding, select low-carbon steel and cement, and the contractor shall purchase the materials according to the requirements of the construction department after contracting the project, and make detailed material entry ledger records.

The highway management department should establish a elimination catalogue of high-emission construction machinery and equipment, and recommend energy-saving and environment-friendly machinery and equipment. The contractor should specify in the bidding that the machinery and equipment used in the project should comply with the selection requirements, and the contractor should select the machinery and equipment according to the requirements of construction department after contracting the project, and make detailed records of equipment entry and exit ledger and fuel consumption ledger.

The prefabricated girder yard applies photovoltaic energy storage technology to provide green electricity energy supply through photovoltaic power generation for the prefabricated girder yard and during the highway operation period.

Green and intelligent technologies such as ring production line and intelligent production management system are applied in the prefabricated girders yard to facilitate the construction management, improve the efficiency of production management, and reduce the energy consumption and carbon emissions in the whole bridge construction process.

6 Conclusion

Establishment of a carbon emission calculation method for materials, on-site mechanical equipment use and transport vehicle use in the production, transport and installation of prefabricated girders, which can be used for calculate and evaluate the carbon emissions level of prefabricated girders in bridge construction projects.

Based on the project budget, calculate the carbon emission of prefabricated girder yard of a highway in Guangzhou city, the carbon emission of prefabricated girder is 1.534 t per m^3.Through the verification and analysis of the actual project consumption data, the deviation between the consumption of main materials and budgeted amount is small, mainly due to the deviation of actual electricity consumption, and the error rate of carbon emission intensity calculation (t/m^3) of prefabricated girder caused by this is −5.3%.

According to the traceability analysis of carbon emissions, it is found that rebar, cement, steel strand wire, steel mould plate and rebar fabrication equipment, girders transport equipment and concrete mixing equipment are the main sources of carbon emissions, and strengthening the management of these types of materials and equipments in highway construction projects can greatly enhance the level of low-carbon construction of bridges.

References

1. Gervásio, H., da Silva, L.S.: Comparative life cycle analysis of steel-concrete composite bridges[J]. Struct. Infrastruct. Eng. **4**(4), 251–269 (2008)

2. Lei, X., Zhiyong, L., Cong, W., et al.: Research on carbon emission measurement of railway materialisation stage based on BIM[J]. Constr. Econ. **43**(S2), 295–230 (2022)
3. Muyu, L., Dan, O.: Calculation method of life cycle carbon emission of bridge engineering[J]. Civ. Constr. Environ. Eng. **33**(S1), 125–129 (2011)
4. Jiaxing, M.A., Jiannan, J., Hanjun, X.I.E., et al.: Calculation model of carbon emission in the whole life cycle of cable-stayed bridge[J]. J. Tianjin Univ. (Nat. Sci. Eng. Technol. Ed.). **57**(1), 31–41 (2024)
5. Guojun, H., Mengyue, L., Jiang L.: Research on carbon emission accounting for highway bridges in Lingnan Region[C]. In: ICEPG 2023
6. Shanshan, L., Liang, Y.: Carbon emission accounting and influencing factors of the coal industry in the whole life cycle[J]. J. Coal. **48**(7), 2925–2935 (2023)
7. Liu, Y., Wang, Y., Li, D., et al.: Life cycle assessment for carbon dioxide emissions from freeway construction in mountainous area: primary source, cut-off determination of system boundary[J]. Resour. Conserv. Recycl. **140**, 36–44 (2019)
8. Liu, Y., Wang, Y., An, D.: Life-cycle CO_2 emissions and influential factors for asphalt highway construction and maintenance activities in China[J]. Int. J. Sustain. Transp. **12**(7), 497–509 (2018)

Progress and Future Perspectives of Biochar on Environmental Sustainability

Hongduo Jia(✉)

Hebei University of Environmental Engineering, 8 Jingang Avenue, Beidaihe District, Qinhuangdao, Hebei, China
jiahongduo0128@qq.com

Abstract. The production of biochar is achieved through the pyrolysis of biological waste. Due to its distinctive chemical composition and pore structure, biochar exhibits a high adsorption capacity in environmental remediation, effectively removing organic pollutants and heavy metals. This paper presents a review of the characteristics, preparation, and application of biochar in the environmental field, with a particular focus on its use in pollution control. This paper reviews the functional mechanisms of biochar, including adsorption, catalytic conversion, and the promotion of biodegradation. The potential and value of biochar in the field of environmental protection are emphasized, and new ideas are provided for the advancement of technology, research and development, and the resolution of environmental issues. This paper also puts forward the shortcomings of biochar in the application process and the future improvement direction and looks forward to the practical application of biochar.

Keywords: Biochar · Sustainable development · Water environment · Air pollution · Soil environment

1 Introduction

It is very important to explore effective ways to solve environmental pollution problems, with the increasing emphasis on environmental protection and resource recycling in the world. As an emerging environmentally friendly material, biochar could solve the above problems. Studies indicate that biochar can remain in the soil for hundreds to thousands of years, enhancing soil quality and supporting agriculture with promising management applications. It is a recalcitrant carbon (C) and can absorb up to one billion tons of greenhouse gases per year. In addition, biochar can adsorb and fix organic pollutants in the water environment, reducing the migration and diffusion of pollutants. This paper presents a literature review of the characteristics, preparation methods, and application examples of biochar in the atmosphere, soil, and water environment. Through a comparative analysis, the remarkable effects and principles of biochar in environmental remediation and protection are elucidated, including adsorption, catalytic conversion and microbial carrier functions. The paper provides a basis for the optimization of subsequent biochar. At the same time, this paper also prospects the disadvantages of biochar in the practical application.

V. Achal et al. (Eds.): EPPC 2024, ESE, pp. 148–160, 2025.
https://doi.org/10.1007/978-3-031-92123-0_15

2 Characteristics and Preparation Technology of Biochar

Biochar can be divided into 3 types: plants, livestock and poultry manure, and municipal solid waste [4, 5]. Biochar mainly includes C, Hydrogen, Oxygen, Nitrogen and other elements. C is the most important component and its content usually is more than 50%. Compared with the C content of biochar, plants (60–90%) >municipal solid waste (60%–70%) >livestock and poultry manure (40–70%). Therefore, biochar has good stability of C. When it is applied to the soil, it may be in the soil for hundreds of years or longer. It will not cause potential harm to the soil due to its own decomposition, and has a soil improvement effect [22]. In addition, biochar also contains a small amount of Phosphorus, Sulfur, Potassium, Calcium, Magnesium and other elements, which has fertility effects on soil.

Biochar generally has a rich pore structure. The pore size and distribution will vary with the biomass feedstock and preparation technology. The pore system of biochar is characterised by the presence of three distinct pore types: large, medium and micropores. These pores are arranged in a complex network structure, which exerts a significant influence on the diffusion and transport speed of the material. Compared with macropores, micropores and mesopores have stronger adsorption capacity in terms of adsorption performance. The plant-based biochar exhibits a large specific surface area, allowing for an increased number of adsorption sites and enhanced interaction with the surrounding environment. For example, Shaheen et al. showed that the adsorption properties of date palm biochar treated by 500 °C pyrolysis can be comparable to those of activated carbon, showing great potential as an economical and efficient adsorption material [24]. Cow dung biochar has the potential to become a novel material for environmental purification due to its porous structure and strong adsorption capacity. The management of municipal solid waste is a crucial aspect of environmental protection, and cow dung biochar represents a promising avenue for demonstrating the dual potential of environmental restoration and resource utilisation. Xue et al. demonstrated that fruit peel biochar exhibited a greater specific surface area and total pore volume, while meat and bone biochar displayed a higher ash content [31]. The biochar characteristics from different biomass feedstock were listed in Table 1.

Table 1. Biochar characteristics from different biomass feedstock.

Biochar types	Carbon content (%)	Specific surface area (m^2/g)	Reference
Plant	60–90	Wood biochar: 200–500 Crop straw biochar: 50–150	(Afshar and Mofatteh, 2024, Ayaz, et al., 2021) [1]
Livestock and poultry manure	40–70	100–150	(Parthasarathy, et al., 2022) [21]
Municipal solid waste	60–70	180–320	(Afshar and Mofatteh, 2024, Raz-zak, 2024) [22, 23]

2.1 Preparation Technology of Biochar

The methods for producing biochar primarily encompass pyrolysis, hydrothermal processing, and gasification. Therefore, the biochar characteristics will also change.

2.1.1 Pyrolysis Technology

Pyrolysis technology refers to the process of heating biomass under anaerobic or oxygen-limited conditions, and biomass was decomposed into gas, liquid and solid product. The pyrolysis technology consists of 3 stages: water evaporation, organic matter degradation and carbon-containing product formation. Water evaporation stage shows that the water in the biomass is evaporated. The organic matter degradation stage shows that the cellulose and hemicellulose are degraded. The carbonization stage shows that the C structure is reorganized. Based on temperature, pyrolysis rate, and residence time, pyrolysis technology can be categorized into rapid pyrolysis (800–1300 °C, pyrolysis rate more than 200 °C/min, residence time less than 10 s) and slow pyrolysis (300–500 °C, pyrolysis rate 5–7 °C/min, residence time more than 60 min). In contrast to the two, the primary outputs of the fast pyrolysis process are bio-oil and pyrolysis gas, while the biochar yield is relatively small. Feedstock with higher C content and lower ash content were usually selected. Moore et al. prepared biochar from coconut shell with a high cellulose, hemicellulosic, lignin and carbon content. They found that increasing the pyrolysis temperature could effectively enhance the carbon fixation amount, specific surface area and adsorption efficiency of coconut shell biochar, and improve its pore structure. These findings provide a solid basis for the application of biochar in environmental purification [19].

2.1.2 Gasification Technology

The biomass gasification process refers to the process of generating biochar at higher temperature (usually 800–1000 °C) and under the condition of oxygen or steam. In this process, biomass is converted into gas, liquid and solid products by reacting with oxygen or steam. The main gaseous products include CO, H_2 and CO_2, and the solid product is biochar. Biochar generated through the gasification process exhibits an increased specific surface area and a refined pore structure, enabling it to effectively regulate heavy metals and adsorb organic contaminants [29]. It has been demonstrated that the gasification temperature and atmosphere exert a significant influence on the surface and pores of biochar. An increase in temperature results in the surface becoming rougher and the pores becoming more complex, while also enhancing the adsorption force. Reactions occurring in different atmospheres give rise to alterations in the pore structure, thereby providing a rationale for the customisation of biochar and its deployment in environmental energy applications. For example, for the production of Miscanthus biochar in a tubular furnace, biochar yield reduced with the gasification temperature increasing under any atmosphere. Biochar yield was highest under N_2 atmosphere, while the yield was at its minimum in a CO_2 atmosphere. Conversely, the specific surface area of Miscanthus biochar was greatest in a CO atmosphere at 800 °C. It decreased sharply at 1000 °C due to the effects of sintering and crushing.

2.1.3 Hydrothermal Carbonization Technology

Hydrothermal carbonization process refers to the process of utilizing water as the reaction medium, sealing treatment at higher pressure and 180–250 °C to hydrolyze biomass feedstocks, and then decarboxylation and aromatization to produce biochar. The application of hydrothermal carbonisation technology enables the precise transformation of biomass into a material with a high water content, which is characteristic of biochar. The feedstock do not need to be dried in advance, and the technology greatly shortens the dehydration and drying time. Higher yield and better hydrophobicity was formed for the technology. For example, Mueanpun et al. showed that fern biochar produced by hydrothermal carbonization can adsorb organic pollutants such as paraquat [20]. Hugo et al. selected the biochar produced by hydrothermal carbonization process to adsorb mercury (Hg) in acidic wastewater, and the removal rate was up to 89.4% [13].

2.1.4 Comparison of Preparation Technologies

The pyrolysis process is simple to operate. It has advantages of disposing a large amount of biomass, mass producing, and requiring any type of feedstock. Li et al. have shown that pyrolysis biochar, as a C sequestration material, has been recalcitrance in soil for hundreds or thousands of years. However, pyrolysis usually needs to be performed at higher temperatures and consumes more energy. When the slow pyrolysis will produce bio-oil and pyrolysis gas by-products. if not treated properly, the technology will cause pollution to the environment [5]. For example, Liu et al. have shown that the pyrolysis process is complicated and the product is unstable, which ultimately limits its industrial application [17, 18].

The biochar produced by the gasification process owns higher specific surface area, which makes it have a better effect in the adsorption and reaction process, such as the adsorption capacity of organic pollutants. Dissanayake et al. compared the adsorption capacity between biochar prepared from single biomass (sawdust) and mixed biomass (sawdust and chicken manure) by gasification process. The material displays favourable CO_2 adsorption characteristics, exhibiting an adsorption capacity of 2.92 mol/kg, a specific surface area of 1408 m^2/g, and a high abundance of micropores. Following numerous cycles, the adsorption remains stable, and the regeneration is robust, rendering the material suitable for CO_2 capture and separation. However, the process has the disadvantages of complicated operation and regulation process and high requirements for the equipment. It is essential to precisely control the reaction parameters in order to ensure the gasification effect and improve the biochar quality [10]. For example, Tian et al. compared a variety of temperatures and atmospheres when producing biochar by this technology, and performed gasification kinetic analysis. The experimental conditions are more strict [28].

The hydrothermal carbonization technology does not require the pretreatment and biomass drying in advance. It is energy-saving and economic. Kambo et al. showed that hydrothermal carbon is usually wet and in the form of slurry. The surface of the material is characterised by a high concentration of oxygen-containing functional groups, a high proportion of aromatics, and strong hydrophilicity. These properties contribute to enhanced soil water retention, promotion of plant growth, optimised water extraction, and the potential for improvement in soil quality and water management. However, compared with pyrolysis and other technologies (Table 2), biochar yield is relatively low according hydrothermal carbon method. It required more feedstock to keep the same yield of biochar, which will produce higher costs [14].

Table 2. Comparison of different pyrolysis technologies

Technology	Advantages	Deficiency	Reference
Pyrolysis	Convenient operation Large-scale production	Bio-oil and pyrolysis gas by-products will be produced during slow pyrolysis.	(Liu, et al., 2023) [17]
Gasification	It has a large surface area and effectively adsorbs organic contaminants.	The operation and control process are complex, and the requirements for equipment are high.	(Dissanayake, et al., 2020, Tian, et al., 2021) [10]
Hydrothermal carbonization	There is no need to pretreat and dry the biomass in advance, energy saving and economic	The yield of biochar is relatively low.	(Kambo and Dutta, 2015, Zhang, et al., 2019) [14, 34]

3 Biochar Application in the Air

According to the World Health Organization (WHO), air pollution refers to "If there are man-made pollutants in the outdoor atmosphere, its content, concentration and duration can cause discomfort to most residents, endanger public health in a large range, and make human, animal and plant life in a state of obstruction." In this paper, air pollution refers to the process of discharging waste gas into the air in the process of industrial and agricultural production. Air pollutants can be classified into two distinct categories: primary pollutants and secondary pollutants. Primary pollutants are substances that are released directly into the atmosphere from sources of pollution, such as CO_2, sulfur dioxide (SO_2), etc. Secondary pollutants refer to pollutants formed by chemical or photochemical reactions of primary pollutants, such as ozone, sulfate, nitrate (NO_3^-), organic particulate matter, etc.

3.1 Adsorption Fouling

Kumar et al. created particulate pollutants by burning aromatherapy sticks, mainly including PM_{10}, $PM_{2.5}$ and PM_1. The results showed that the removal efficiencies of PM_{10}, $PM_{2.5}$ and PM_1 by microalgae biochar (300 °C, 500 °C and 700 °C) were 69.6%, 82.7% and 80.8%, respectively. This is due to the fact that at a high temperature of

700 °C, the specific surface area of biochar is markedly enhanced following pyrolysis, and the pore size structure is optimised. The number of pores increases and the distribution becomes more uniform [15]. Choi et al. pyrolyzed cedar wood biochar at 500 °C, 600 °C and 700 °C in a fixed bed reactor. Under the condition of single gas NO adsorption, highest adsorption efficiency of about 90% had appeared the biochar produced at 600 °C. When NO and SO_2 were simultaneously adsorbed, the adsorption efficiency of NO and SO_2 by biochar produced was about 90% and 95% at each temperature, respectively [6]. Zhang et al. produced the biochar (wheat straw: granular activated carbon catalyst of 1:3) on the condition of microwave irradiation time of 10 min and power of 500 W. The results demonstrated that the specific surface area of biochar was 312.6 m^2/g, the total pore volume was 0.2218 cm^3/g, the micropore volume was 0.1380 cm^3/g, and the adsorption capacity of benzene VOCs was considerable. In comparison with conventional biochar, the adsorption capacities of p-benzene, toluene and xylene were 53.9, 75.8 and 63.0 mg/g, respectively, demonstrating an excellent potential for removal [35].

The adsorption capacity of biochar is influenced by three key characteristics: a large specific surface area provides an increased number of adsorption points, a reasonable pore size facilitates molecular diffusion and adsorption, and polar functional groups enhance adsorption through a range of mechanisms. The mechanism mainly lies in 3 aspects. (1) Due to its elevated specific surface area and porous structure, biochar exhibits remarkable adsorption capabilities for atmospheric particles and toxic gases. Its surface is characterised by a multitude of adsorption sites, while the internal pores facilitate diffusion and enhance the overall adsorption efficiency. (2) The surface functional groups of biochar enhance the adsorption capacity, stimulate the reaction, and augment the efficiency, which are the defining characteristics of effective atmospheric pollutant adsorption. (3) Size exclusion has also exerted important roles in biochar adsorption.

3.2 Reduce Pollutant Emissions

The 3 treatments were set up in this experiment: control, single application of biochar (application rate was 1% of soil weight) and organic fertilizer (application rate was 1% of soil weight). Compared with the control, biochar reduced the CO_2 emission flux by 20.5% and reduced the cumulative emission by 19.9%. Compared with single application of organic fertilizer, biochar decreased the CO_2 emission flux and cumulative emission. Lu et al. set up 3 treatments: control (no straw and biochar), straw returning (7.5 t/hm^2 per year), straw biochar returning (about 2.6 t/hm^2 per year). In comparison to the return of straw to the field, the application of biochar to soil can result in a notable reduction in the annual and seasonal emissions of nitrogen oxides. This has the potential to facilitate the protection of soil environments and the advancement of sustainable agricultural practices. Compared with CK, biochar reduced the total seasonal N_2O emissions by 17.5% on average [27]. Dai et al. pyrolyzed biochar at 500 °C for 2 h to conduct a split-plot design. The implementation of continuous flood irrigation and alternate dry and wet irrigation strategies, in conjunction with the addition of 20 t/hm^2 biochar, resulted in a significant reduction in CH_4 emissions, amounting to 25.8% and 38.9%, respectively, over the two experimental years. These findings highlight the efficacy and potential for broad application of biochar in the control of greenhouse gases in agricultural settings. The highest CH_4 emission reduction was achieved by applying biochar (20 t/hm^2) under alternating wet and dry irrigation conditions, which was 86.8% and 91.1% lower than that of the treatment without biochar during two experimental years under continuous flooding irrigation, respectively [7].

The primary processes through which biochar diminishes greenhouse gas emissions from soil are outlined below: (1) Biochar can increase soil carbon storage and reduce greenhouse gases such as CO_2 and N_2O in soil to the atmosphere. (2) Biochar changes the pH and redox potential of the soil, so that the soil mirco-environment is not conducive to the growth and reproduction of methanogens.

4 Biochar Application in the Water

The word "water environment" appeared in the 1970 s. The water environment refers to "all kinds of water bodies distributed on the earth and various environmental factors closely related to them, such as riverbed, coast, vegetation, soil and so on". In this paper, water environment mainly refers to tail water. Tail water treatment refers to the process of purifying water by removing harmful substances in water through physical, chemical and biological methods. Pollutants in water include heavy metal ions, organic pollutants and nutrients. Among them, heavy metal ions include copper, zinc, nickel, etc.; organic pollutants include antibiotics (such as penicillin, cephalosporins, etc.), polycyclic aromatic hydrocarbons, phenolic compounds (such as phenol, o-cresol, p-cresol, etc.), synthetic dyes (including acid dyes, alkaline dyes), etc.; nutrients include nitrogen, phosphorus and potassium.

4.1 Immobilizing Heavy Metal

Wang Yi et al. applied biochar (0 t/hm^2, 5.0 t/hm^2, 7.5 t/hm^2 and 10.0 t/hm^2) into the paddy field. The results showed Cd concentration in paddy soil decreased with the increase of biochar application. The incorporation of 10.0 t/hm^2 biochar has been demonstrated to markedly diminish the concentration of Cd in soil. This outcome can be attributed to the porous structure of biochar and the effective adsorption capacity of functional groups on Cd [30]. Biswal et al. conducted a comparison of the economic viability and practicality of biochar and carbon-based catalytic materials(graphene, carbon nanotubes, etc.) as adsorbents. Research has demonstrated that the functional groups present on the surface of more affordable biochar can participate in ion exchange reactions with heavy metal ions [2].

4.2 Adsorption of Organic Pollutants

Feng et al. compared the conventional Fenton process and the Fenton process with the addition of biochar (1.0–5.0 g/L). A number of studies have demonstrated that the degradation efficiency of two organic pollutants, 2,4-dichlorophenoxyacetic acid (2,4-D) and 2,4-dichlorophenol (2,4-DCP), can be significantly enhanced when the traditional Fenton process is combined with biocharcoal. The degradation degree increased to 94.8% [11]. Dalahmeh et al. compared the traditional sand filter and sand filter with biochar (3.0 g/L), and found that the latter could effectively remove carbamazepine, and the effect was more stable and effective. After 22 weeks of operation, biochar increased the adsorption rate of carbamazepine to 99%, while that of the control treatment was only 7% [8]. The findings of Boehm et al. further indicated that biochar has the potential to improve the effectiveness of biofilters and has a better removal effect on trace organic pollutants (such as herbicides atrazine, bisphenol-A, plastic agents, pesticides, etc.). Studies have also shown that the synergistic effect of biochar and biofilm can remove toxic organic pollutants in industrial wastewater [3]. Moore et al. colonized coconut shell biochar with natural microorganisms at different pyrolysis temperatures (450 °C, 550 °C and 700 °C) for 11 months to enhance the efficacy of biochar disposing microcystin. After 48 h, the concentration of microcystins in water was reduced to below the detection limit (0.1 ng/mL) [19]. Gupta et al. found that biochar removed harmful substances from dyes (acidic and alkaline) in the textile industry. Due to the higher pH of biochar, it is beneficial to the adsorption of alkaline dyes. But It is not helpful for adsorption of acid dyes [12].

4.3 Adsorption of Nutrients

Deng et al. set up a control (pure gravel) and 3 constructed wetlands (biochar addition volume at 10%, 20% and 30%). Biochar removed $NH_4{}^+$-N content by 52.2%, 49.7% and 63.5%, respectively, when the biochar addition at 10%, 20% and 30%. While, $NH_4{}^+$-N removal was only 47.4% by pure gravel [9]. Under neutral or weak acid conditions, the material has better adsorption effect on tetracycline hydrochloride. Following five cycles of adsorption and desorption, the material demonstrated consistent adsorption efficiency, ranging from 60.1% to 68.4%. Additionally, the adsorption capacity reached 525.6 mg/g, indicating favourable reusability. This reusability is advantageous in terms of cost control and potential improvement in efficiency.

The mechanism may be due to (1) biochar has more pores and larger specific surface area; (2) biochar is used as a microbial membrane scaffold is predicated on its exceptional physical and chemical properties. Proliferating microorganisms secrete a variety of polymers to form extracellular biofilms and attach to the surface of biochar.

5 Biochar Application in the Soil

The Soil Pollution Prevention and Control Law of the People's Republic of China provides a clear definition of "soil pollution". The definition of soil pollution as set forth in the legislation in question indicates that such contamination is a consequence of human activities and encompasses a range of substances, including chemicals, heavy metals, organic pollutants, and other materials, which enter the soil surface. These substances alter the chemical composition, physical structure, and biological characteristics of soil, thereby impairing its functionality, affecting the utilization of resources, and potentially endangering human health and the ecological environment. This definition underscores the gravity of soil pollution and provides a legal foundation and guidance for prevention and control efforts. Soil pollutants include heavy metals, organic pollutants, plastic and antibiotics. Heavy metal includes cadmium (Cd), mercury (Hg), arsenic (As), etc.; organic pollutants include pesticides (HCHs, DDT and 2,4-D, etc.), PAHs and pPCBs, etc. According to the National Soil Pollution Survey Bulletin published in 2014, the concentrations of inorganic substances such as cadmium, mercury, arsenic, copper, lead, chromium, zinc, and nickel, as well as organic compounds including BHC, DDT, and PAHs in Chinese soil, exceed established safety standards, highlighting the extensive and serious nature of soil contamination. This data serves as a foundation for developing effective strategies for the prevention and control of soil pollution and further underscores the urgent need to enhance soil environmental protection and management.

5.1 Resisting Heavy Metals

Ying et al. modified biochar by magnesium chloride to resist the effect of Cd (II). The enhanced biochar elevated the removal rate of Cd (II) by 23.7%, which is 11.2 times greater than that of the unaltered biochar [33]. Su et al. rubber biochar and coconut shell biochar were enhanced through treatment with a manganese chloride solution (mass ratio of Mn^+ to biochar was 1:25) and pyrolyzed at 500 °C to obtain 2 types of biochar: two distinct varieties of biochar are manganese-modified rubber biochar and coconut shell biochar. In order to evaluate their performance, the specific surface area was measured. The specific surface area of the manganese-modified coconut shell biochar was found to be as high as 604.1 m^2/g, which was significantly higher than that of ordinary biochar. This indicates that its pore structure and adsorption properties were excellent. Although the specific surface area of Mn-modified rubber biochar is relatively low at 56.5 m^2/g, it still exhibits certain porosity and adsorption characteristics. The observed differences between the two may be attributed to variations in the raw materials, preparation technology, and manganese modification conditions, which warrant further investigation. Modified coconut shell biochar reduced the concentrations of available As and available Cu by 5.4–49.4% and 5.7–54.6%; Modified rubber biochar reduced the concentrations of available As and available Cu by 33.1–37.3%, 15.5–39.3%. Both MG-modified rubber biochar and MG-modified coir biochar have been demonstrated to effectively reduce the availability and fluidity of heavy metals in soil, and have a significant effect on soil heavy metal pollution control. In particular, manganese-modified coir biochar, with its stronger resistance to heavy metals, shows greater application prospects, and introduces new materials and strategies for soil heavy metal pollution control(Su, et al.,2024). Chen et al. modified indoor soil column experiments by magnesium chloride to resist Cr in sewage sludge. After biochar treatment, the pH of leachate in the Cr-contaminated soil increased slightly from 7.92 to 8.14. The leachate pH of heavy Cr contaminated soil showed a trend of decreasing first (7.75–7.85) and then increasing (7.75–8.09). The incorporation of biochar has been demonstrated to effectively inhibit the dissolution and flow of heavy metal chromium in soil. This is achieved by increasing the pH of soil leachate, which in turn reduces the bioavailability and toxicity of the metal. This represents a novel and effective approach to the control of soil heavy metal pollution [4].

The mechanism is mainly as follows: (1) Following treatment, the specific surface area of modified biochar is increased, thereby significantly enhancing its physical adsorption efficiency for heavy metals. This results in an improved effect and durability of soil heavy metal pollution control, as well as providing an efficient and stable material choice for the remediation of heavy metal-contaminated soil; (2) modified biochar increased other ions such as Mg^{2+} by modification, and ion exchange with soil heavy metal ions; (3) biochar co-precipitated and functional group complexation with heavy metals by regulating soil pH; (4) the removal mechanism of modified biochar on soil heavy metals is not limited to one mechanism, but may be a combination of several mechanisms. Research suggests that modified biochar removes cadmium through: co-precipitation mechanism (73.4%) >ion exchange mechanism (22.7%) >electrostatic adsorption mechanism (3.9%), but the ability of each mechanism is different.

5.2 Degradation of Organic Pollutants

Shen et al. modified biochar by saturated ruthenium chloride solution (pyrolysis temperature 400 °C) to degrade 2,2′-azino-bis(3-ethylbenzothiazoline)-6-sulfonate(ABTS/BC) and ruthenium Ruthenium (Ru/BC) under heating for 3 h. The removal rate of phenanthrene in the control group was 79.9%, while 2 biochar increased the removal rates by 94.6% and 95.5%, respectively. The ruthenium element, when supported on the surface of biochar, has the potential to significantly enhance the adsorption efficiency of phenanthrene. This is achieved by strengthening the interaction between the organic pollutant phenanthrene and the biochar surface, thereby improving the surface structure and chemical properties of the biochar. This, in turn, offers an effective and durable treatment scheme for environmental pollution control, particularly in the removal of organic pollutants. Biochar provided a suitable soil microenvironment for microbial degradation of phenanthrene [25]. Silvan et al. introduced earthworms and PE passive samplers into contaminated (PCBs) soils and compared the degradation mechanisms of PCBs by rice husk biochar and woody biochar. The findings indicated that biochar exerted a pronounced adsorption impact on PCBs in soil, resulting in a notable reduction in the concentration of PCBs in soil pore water and in earthworms. This provided a viable solution for the remediation of soil PCB pollution. Both biochars reduced the absorption of PCBs by earthworms. Rice husk biochar degraded PCBs by 52–91%; wood biochar degraded PCBs by 19–63% [26].

5.3 Increase of Soil Organic Carbon

Chen et al. selected tobacco straw biochar (pyrolysis at 500 °C for 120 min) and applied biochar (0.563 g) to the culture flask. A control without biochar was also set up. The findings indicated that the application of biochar resulted in a significant increase in soil organic carbon content, ranging from 9.4% to 25% in comparison to the control group. This observation underscores the potential of biochar to enhance soil quality, improve fertility, and facilitate carbon storage, thereby offering novel strategies for sustainable agricultural advancement [4]. Linam et al. applied rice husk biochar at 6.8 t/hm^2 (pyrolyzed at 450 °C) and rice husk at 14.7 t/hm^2 to paddy fields (the same C input). Results found that biochar increased soil organic carbon by 0.47–0.76% [16]. Yang et al. selected corn straw biochar at 3 different temperature (300 °C, 450 °C and 600 °C), mixed with 80 g sandy clay loam (dry weight) and put it into a 1 L glass bottle. The amount of biochar added was 2% (wt) of the soil weight. The results of comparative experiments demonstrated that the application of biochar led to a significant increase in soil organic carbon content, with a range of 12–15%. This finding verified the impact of biochar in enhancing soil quality and promoting carbon fixation, which is of paramount importance for the sustainability of agricultural practices [32].

The main mechanisms are as follows. (1) The aromaticity and carbon content of biochar was increased with temperature rising. Adding biochar to the soil lowered its mineralization capacity. It is more difficult for the C in the soil to be decomposed, so that soil organic carbon could accumulate. (2) As a soil additive, biochar has been demonstrated to promote the growth, reproduction and diversity of microorganisms by providing rich carbon sources and energy for these organisms. This, in turn, contributes to the long-term accumulation and stability of soil organic carbon, enhances soil fertility, optimises soil structure, and improves soil adaptability and ecosystem functionality and stability.

6 Conclusion and Prospectives

This paper presents a review of the characteristics of biochar and its applications in the environment. Based on its adsorption, catalytic and physical and chemical properties, biochar has been demonstrated to have excellent performance in air purification, water treatment and soil improvement, which is of great significance to environmental protection and the sustainable development of agriculture. At present, some challenges were still present during the application, such as the standardization of production processes, the uncertainty of long-term environmental impacts, and the cooperation with local industrial systems. It is urgent to strengthen scientific research and practice and reduce its costs through methods such as government standards.

The performance of biochar in the air is mainly to adsorb pollutants and reduce pollutant emissions. The biochar adsorption mechanism is intricate and does not depend on a single set of characteristics; however, these characteristics can still be beneficial. Modification of biochar to enhance its adsorption capacity can optimise its structural properties, improve its efficiency, and provide cost-effective solutions for environmental governance. The performance of biochar in the water environment is mainly to immobilize heavy metals, adsorb solid organic pollutants and nutrients. Biochar is helpful to form a biofilm scaffold in the water to provide an environment for the proliferation of microorganisms. The organic contaminants present in aquatic ecosystems do not exist in isolation, and usually combined. Therefore, it is necessary to strengthen the exploration of the synergistic effect of biochar and other remediation technologies. The performance of biochar in the soil is mainly to regulate heavy metals, degrade organic pollutants and enhance soil C sequestration. Biochar application in different types of soil and different crop planting seasons should be demonstrated, and long-term experimental researches should be carried out. it is urgent to build targeted biochar application and implementation plans for different soils, such as saline-alkali soil, secondary salinization soil, coastal field, etc.

References

1. Ayaz, M., et al.: Biochar role in the sustainability of agriculture and environment [J]. Sustain. For. **13**(3), 1330 (2021)
2. Biswal, K.B., Balasubramanian, R.: Use of biochar as a low-cost adsorbent for removal of heavy metals from water and wastewater: a review [J]. J. Environ. Chem. Eng. **11**, 110986 (2023)
3. Boehm, A., et al.: Biochar-augmented biofilters to improve pollutant removal from stormwater–can they improve receiving water quality? [J]. Environ. Sci. Water Res. Technol. **6**(6), 1520–1537 (2020)
4. Chen, L., et al.: Straw and straw biochar differently affect fractions of soil organic carbon and microorganisms in farmland soil under different water regimes [J]. Environ. Technol. Innov. **32**, 103412 (2023)
5. Chen, L., et al.: Effect of metal-modified sewage sludge biochar tubule on immobilization of chromium in unsaturated soil: groundwater table fluctuations induced by rainfall [J]. Chemosphere. **365**, 143378 (2024)
6. Choi, H., Lee, Y.: Impact evaluation of NO and SO2 reduction by biochar characteristics [J]. Preprints. (2024)
7. Dai, W., et al.: Biochar incorporation increases grain yield, net ecosystem CO2 exchange, and decreases CH4 emissions in an alternate wetting and drying paddy ecosystem [J]. Environ. Technol. Innov. **34**, 103577 (2024)
8. Dalahmeh, S., et al.: Potential of biochar filters for onsite sewage treatment: adsorption and biological degradation of pharmaceuticals in laboratory filters with active, inactive and no biofilm [J]. Sci. Total Environ. **612**, 192–201 (2018)
9. Deng, C., et al.: Response of microbes to biochar strengthen nitrogen removal in subsurface flow constructed wetlands: microbial community structure and metabolite characteristics [J]. Sci. Total Environ. **694**, 133687 (2019)
10. Dissanayake, P., et al.: Sustainable gasification biochar as a high efficiency adsorbent for CO2 capture: a facile method to designer biochar fabrication [J]. Renew. Sust. Energ. Rev. **124**, 109785 (2020)
11. Feng, D., et al.: Biochar enhanced the degradation of organic pollutants through a Fenton process using trace aqueous iron [J]. J. Environ. Chem. Eng. **9**(1), 104677 (2021)
12. Gupta, R., et al.: Potential and future prospects of biochar-based materials and their applications in removal of organic contaminants from industrial wastewater [J]. J. Mater. Cycl. Waste Manage. **24**(3), 852–876 (2022)
13. Hugo, R., et al.: Production of hydrochar by low-temperature hydrothermal carbonization of residual biomass from cocoa production for mercury adsorption in acidic aqueous solutions [J]. Case Stud. Chem. Environ. Eng. **10**, 100938 (2024)
14. Kambo, H., Dutta, A.: A comparative review of biochar and hydrochar in terms of production, physico-chemical properties and applications [J]. Renew. Sust. Energ. Rev. **45**, 359–378 (2015)
15. Kumar, R., Dalvi, V., Pant, K., et al.: Microalgal biochar assisted simultaneous removal of particulate matter, formaldehyde, and total volatile organic compounds (TVOC's) from indoor air [J]. Chemosphere. **355**, 141866 (2024)
16. Linam, F., Limmer, M., Ebling, A., et al.: Rice husk and husk biochar soil amendments store soil carbon while water management controls dissolved organic matter chemistry in well-weathered soil [J]. J. Environ. Manag. **339**, 117936 (2023)
17. Liu, G., Zhang, R., Sun, Z., et al.: Simulation and process analysis of hydrogen-rich syngas production by chemical looping reforming of biomass pyrolysis volatiles based on decoupling strategy [J]. J. China Coal Soc. **48**(6), 2351–2358 (2023)

18. Lu, J., Zhang, W., Liu, X., et al.: Can converting raw straw into biochar incorporation achieve both higher maize yield and lower greenhouse gas emissions intensity in drought-prone environment? [J]. Environ. Technol. Innov. **35**, 103683 (2024)
19. Moore, J., Jayakumar, A., Soldatou, S., et al.: Nature-based solution to eliminate cyanotoxins in water using biologically enhanced biochar [J]. Environ. Sci. Technol. **57**(43), 16372–16385 (2023)
20. Mueanpun, N., Srisuk, N., Chaiammart, N., et al.: Nanoporous activated carbons derived from water ferns as an adsorbent for removal of paraquat from contaminated water [J]. Materialia. **15**, 100986 (2021)
21. Parthasarathy, P., et al.: A review on prominent animal and municipal wastes as potential feedstocks for solar pyrolysis for biochar production [J]. Fuel. **316**, 123378 (2022)
22. Rathnayake, D., et al.: Biochar from animal manure: a critical assessment on technical feasibility, economic viability, and ecological impact [J]. GCB Bioenergy. **15**(9), 1078–1104 (2023)
23. Razzak, S.: Municipal solid and plastic waste derived high-performance biochar production: a comprehensive review [J]. J. Anal. Appl. Pyrolysis, 106622 (2024)
24. Shaheen, J., Fseha, Y., Sizirici, B.: Performance, life cycle assessment, and economic comparison between date palm waste biochar and activated carbon derived from woody biomass [J]. Heliyon. **8**(12), e12388 (2022)
25. Shen, X., Zhang, J., Xie, H., et al.: Electron shuttles enhance phenanthrene removal in constructed wetlands filled with manganese oxides-coated sands [J]. Chem. Eng. J. **426**, 131755 (2021)
26. Silvani, L., Hjartardottir, S., Bielská, L., et al.: Can polyethylene passive samplers predict polychlorinated biphenyls (PCBs) uptake by earthworms and turnips in a biochar amended soil? [J]. Sci. Total Environ. **662**, 873–880 (2019)
27. Su, J., et al.: Remediation of multi-metal (loid) contaminated soils using Mn-modified biochar: mechanistic insights and influencing factors [J]. Process Saf. Environ. Prot. **192**, 36–48 (2024)
28. Tian, H., Hu, Q., Wang, J., et al.: Kinetic study on the CO2 gasification of biochar derived from Miscanthus at different processing conditions [J]. Energy. **217**, 119341 (2021)
29. Wang, L., Ok, Y., Tsang, D., et al.: New trends in biochar pyrolysis and modification strategies: feedstock, pyrolysis conditions, sustainability concerns and implications for soil amendment [J]. Soil Use Manag. **36**(3), 358–386 (2020a)
30. Wang Y, Wang B, Zhou W, et al. Effects of biochar application and water management coupling on cadmium transfer and accumulation in late-season rice [J]. 2020: 1479–1485
31. Xue, S., Zhang, X., Ngo, H., et al.: Food waste based biochars for ammonia nitrogen removal from aqueous solutions [J]. Bioresour. Technol. **292**, 121927 (2019)
32. Yang, Y., Sun, K., Han, L., et al.: Biochar stability and impact on soil organic carbon mineralization depend on biochar processing, aging and soil clay content [J]. Soil Biol. Biochem. **169**, 108657 (2022)
33. Ying, D., Hong, P., Jiali, F., et al.: Removal of uranium using MnO2/orange peel biochar composite prepared by activation and in-situ deposit in a single step [J]. Biomass Bioenergy. **142**, 105772 (2020)
34. Zhang, X., Lou, J., Su, J., et al.: Preparation of biochar from spent pleurotus geesteranus substrate and its physicochemical properties [J]. J. Zhejiang Univ. Sci. Technol. **31**(01), 50–57 (2019)
35. Zhang, X., et al.: Microwave biochars produced with activated carbon catalyst: characterization and sorption of volatile organic compounds (VOCs) [J]. Sci. Total Environ. **827**, 153996 (2022)

Research on Environmental Science and Water Treatment Technology

Metal-organic Framework Derived Co-based Porous Carbon for Water Purification

Ming Ma[1] and Xiang Li[2](✉)

[1] School of Materials, Beijing Institute of Technology, Beijing, China
[2] Advanced Research Institute of Multidisciplinary Science, Beijing Institute of Technology, Beijing, China
xiangli0369@bit.edu.cn

Abstract. Emerging Contaminants (ECs) in water environments brings huge potential threat to human health and even ecosystems. However, traditional wastewater treatment plants can't completely remove them through existing technologies, so developing a more efficient water purification technology was urgent. The activation of peroxymonosulfate (PMS) using heterogeneous Cobalt-based catalysts to generate reactive oxygen species (ROS) such as sulfate radicals ($SO_4{\cdot}^-$) with high oxidation potential and selectivity has been proven to be an effective advanced oxidation processes (AOPs). The Cobalt-based porous materials derived from Metal-Organic-Frameworks (MOFs) have uniformly dispersed metal active sites and a large number of mesoporous structures, which have great potential advantages in enriching and oxidizing organic pollutants. In this study, 10-CoHNPC-1000 was successfully synthesized. Research has found that the 10-Co-NHPC-1000 material has excellent saturation adsorption capacity and degradation rate constant for SMX, which were 81.67 mg/g and 0.056 min^{-1}, respectively. This study provides ideas for designing efficient integrated materials for adsorption and degradation.

Keywords: AOPs · ZIF-8@ZIF-67 · PMS

1 Introduction

The progress in technology, agriculture, and medicine has improved human life quality. However, with the development of science and technology, more serious secondary environmental problems have emerged. In 2008, the US environmental protection agency defined ECs as chemicals or materials that pose a perceived, potential, or real threat to human health or the environment, or chemicals or materials that lack publicly available health standards [1]. The prominent categories in ECs include personal care products and drugs (PPCPs), perfluorinated and perfluorinated compounds, flame retardants, pesticides, plasticizers, and surfactants [2]. In recent years, research has shown that emerging pollutants (ECs) exhibit characteristics such as difficulty in biodegradation and environmental persistence. In addition, most ECs are lipophilic, ultimately leading to their continuous accumulation in organisms. In fact, the immediate toxicity caused by the

V. Achal et al. (Eds.): EPPC 2024, ESE, pp. 163–170, 2025.
https://doi.org/10.1007/978-3-031-92123-0_16

continuous release of chemical substances in recent decades, as well as the bioaccumulation and continuous concentration of pollutants in the environment, can have long-term adverse effects on ecosystems [3, 4]. Among all kinds of ECs, PPCPs include all kinds of drugs and personal care products (shampoo, soap, perfume, sunscreen, skin anti-aging agents, etc.). At present, the efficiency of sewage treatment plants for removing emerging pollutants is low [5, 6].

Traditional methods for removing PPCPs have poor removal efficiency for PPCPs. Among these methods, adsorption has the characteristics of easy to be operated, without causing secondary pollution and low cost, so it is considered to be one of the most ideal technologies for removing PPCPs. Carbon based materials have become the most ideal PPCPs adsorbent due to their low preparation cost, large specific surface area, and high removal efficiency. Numerous studies have shown that porous carbon samples (MDCs) derived from metal organic frameworks (MOFs) have unique advantages in adsorbing PPCPs in water compared to most adsorbents [7].

Advanced oxidation processes (AOPs) can mineralize organic compounds (OCs) into harmless byproducts by generating reactive oxygen species (ROS) with high oxidation potential. According to the types of free radicals generated, AOPs are usually divided into HR-AOP based on hydroxyl radicals (·OH) and SR-AOPs based on sulfate radicals ($SO_4\cdot^-$). HR-AOP includes most advanced oxidation technologies, such as ozone oxidation, Fenton method, electrochemical oxidation technology, and photocatalytic technology. However, in recent years, advanced oxidation processes (SR AOPs) based on sulfate free radicals ($SO_4\cdot^-$) have received widespread attention due to their high selectivity and oxidation efficiency in water purification [8]. Among various methods for generating $SO_4\cdot^-$, the use of heterogeneous catalysts to activate peroxymonosulfate (PMS) or persulfate (PS) has been proven to be an effective strategy. PMS can directly degrade OCs through electron transfer processes, but its demonstrated oxidation ability is very limited. PMS has a unique asymmetric structure ($HO\text{-}OSO_3^-$), which can be activated by various transition metal catalysts and carbon-based material catalysts, followed by the production of highly reactive free radicals $SO_4\cdot^-$, ·OH with high oxidation potential, superoxide free radicals ($\cdot O_2^-$) [9, 10], as well as other non-radical reactive oxygen species such as singlet oxygen ($^1O_2$), thereby achieving efficient oxidative degradation of OCs.

In this study, MOF precursors (10-ZIF-8@ZIF-67) was prepared at room temperature. And 10-ZIF-8@ZIF-67 was then pyrolyzed at 1000 °C to obtain cobalt based porous material, 10-CoHNPC-1000. After systematically characterizing its morphology, crystal structure, and pore structure, the performance of material adsorption and degradation of typical high-risk trace drug sulfamethoxazole (SMX) was evaluated through pseudo first/second order kinetics fitting, respectively. Research has found that the 10-Co-NHPC-1000 material has higher saturation adsorption capacity and degradation rate constant for SMX compared to C-ZIF-67, with values of 81.67 mg/g and 0.056 min^{-1}, respectively. This study provides ideas for designing efficient integrated materials for adsorption and degradation.

2 Results and Discussion

The crystal structure of MOFs precursors was characterized through X-ray diffraction patterns (XRD). Figure 1 shows the XRD spectra of experimentally synthesized ZIF-8 and ZIF-67, as well as their standard XRD diffraction spectra. Their main diffraction peak positions and relative intensities were consistent, demonstrating their successful synthesis. At the same time, due to the similar cell parameters and the same spatial structure of ZIF-8 and ZIF-67, the positions of XRD peaks of ZIF-8@ZIF-67 were almost the same as ZIF-8 and ZIF-67. As displayed in Fig. 2, it could be concluded that ZIF-8 and ZIF-67 only had micropores and their surface area were high, which were 1782 m^2/g. However, there were mesopores structure founded in 10-ZIF-8@ZIF-67. So 10-ZIF-8@ZIF-67 had higher surface area, which was 1853 m^2/g. TEM image were used to characterize the morphology and further verify the core-shell structure of 10-ZIF-8@ZIF-67, as was shown in Fig. 3. The EDS mapping of 10-ZIF-8@ZIF-67 displayed that C and N elements were uniformly distributed among the sample, while Co element was mainly distributed at the edge of the sample and Zn element was accumulated in the center. It indicated the successful fabrication of 10-ZIF-8@ZIF-67 with core-shell structure.

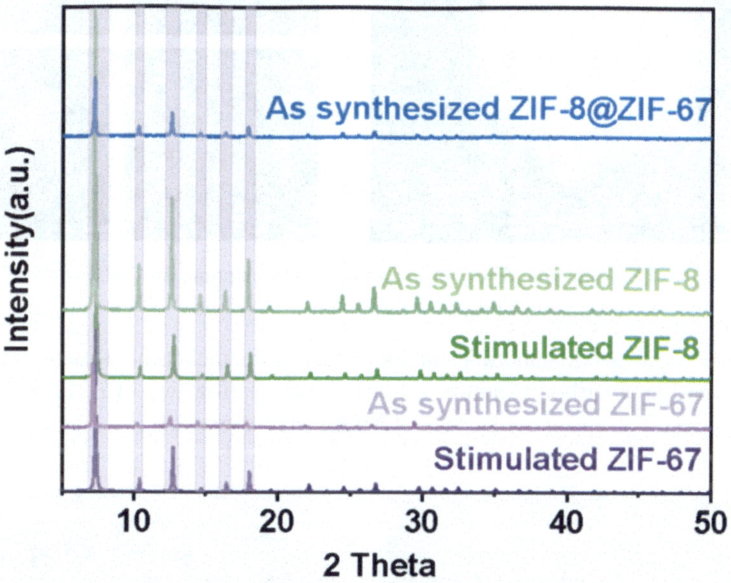

Fig. 1. XRD patterns of as-prepared ZIF-8, ZIF-67 and ZIF-8@ZIF-67

Figure 4a showed the scanning electron microscope image of sample C-ZIF-67 by pyrolyzing ZIF-67. It can be seen that after carbonization, the surface of the material appeared to be depressed, but the overall shape still maintained the shape of rhombic dodecahedron. Figure 4b exhibited the SEM image of 10-Co HNPC-1000 which was carbonized at 1000 °C by 10-ZIF-8@ZIF-67. As was shown in Fig. 4, 10-Co-HNPC-1000 had an uneven and twisted surface, but retained the original rhombic dodecahedral

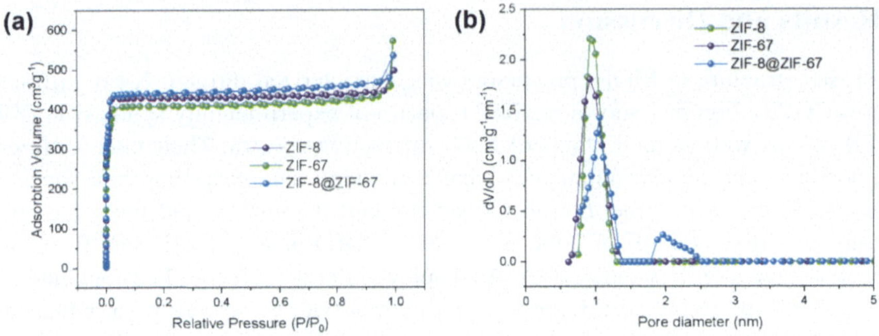

Fig. 2. (**a**) N_2 adsorption-desorption isotherms and (**b**) pore size distributions of ZIF-8, ZIF-67 and 10-ZIF-8@ZIF-67

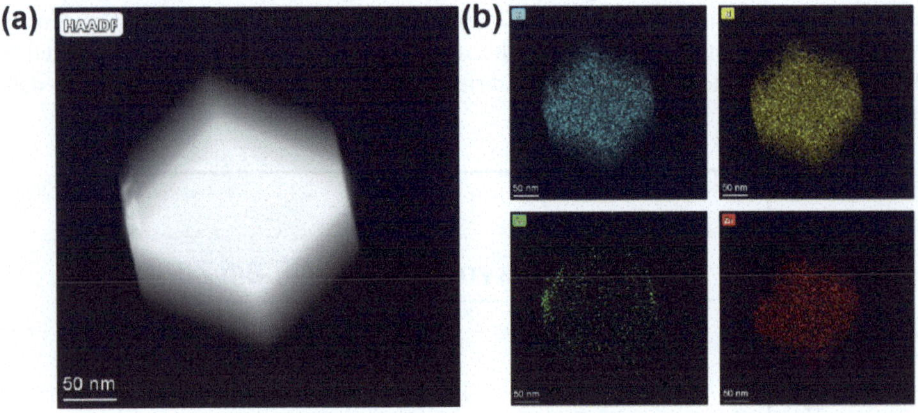

Fig. 3. (**a**) TEM image and (**b**) corresponding EDS mapping of 10-ZIF-8@ZIF-67

structure and grew carbon nanotubes on its surface in the meantime. Research has found that the formation of carbon nanotubes is due to the volatilization of Zn element and the catalysis of cobalt atoms at the surface of MOFs precursors.

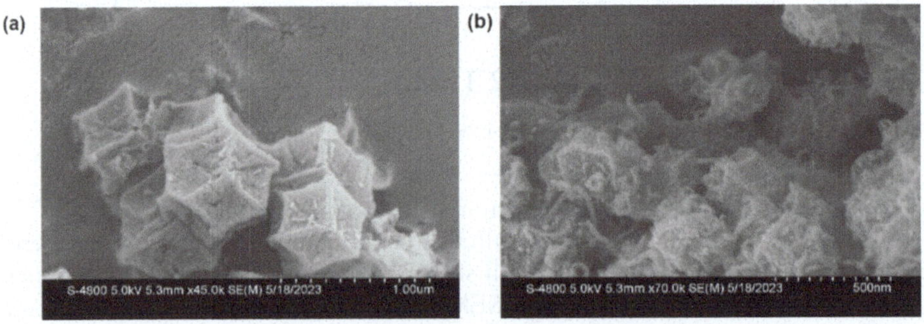

Fig. 4. SEM images of (**a**) C-ZIF-67 and (**b**) 10-Co-NHPC-1000

Figure 5 showed the XRD patterns of C-ZIF-67 and 10-Co-HNPC-1000. The diffraction peaks at 44.0° and 51.5° were attributed to the (200) and (220) crystal planes of metal Co (JCPDS No.15-0806), respectively, while the diffraction peak at 26.4° corresponded to the (002) crystal plane of graphite carbon, which was consistent with the results of HRTEM and EDS results. In the XRD diffraction pattern of 10-Co-HNPC-1000, a new graphite carbon peak (corresponding to the (100) crystal plane) appeared at 44.2°. BET tests were applied to characterize the specific surface area and pore structure of MOFs derived porous cobalt based materials, as shown in Fig. 6. Compared with C-ZIF-67, the nitrogen adsorption-desorption isotherm of the 10-Co-HNPC-1000 significantly increased at $P/P_0 > 0.8$. In addition, compared with C-ZIF-67, the specific surface area and total pore volume of 10-Co-HNPC-1000 were higher, which will be more conducive to exposing active sites and accelerating electron transfer.

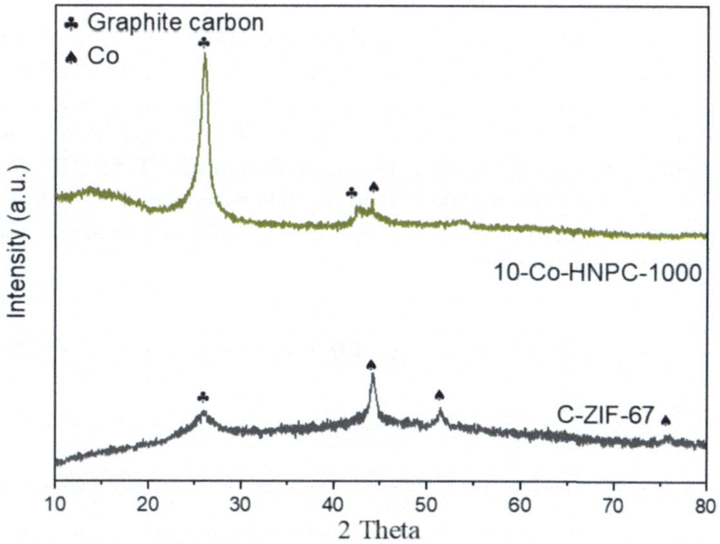

Fig. 5. XRD Patterns of 10-Co-HNPC-1000 and C-ZIF-67

To evaluate the adsorption performance of cobalt based porous samples on SMX, adsorption kinetics experiments were conducted. Figure 7a displayed the adsorption kinetics curves of C-ZIF-67 and 10-Co-HNPC-1000 for SMX, and Fig. 7b showed the fitting graphs of pseudo first/second order adsorption kinetics models. Both C-ZIF-67 and 10-Co HNPC-1000 can reach adsorption equilibrium for SMX within 60 min, and the adsorption curve clearly exhibited that the adsorption capacity of 10-Co-HNPC-1000 for SMX was higher than that of C-ZIF-67. And the saturated adsorption capacity (Q_e) and initial adsorption rate(v_0) of 10-Co-HNPC-1000 were 81.67 mg/g and 48.51 mg/(g · min) respectively. For the sake of comprehensively evaluating the degradation performance of cobalt based porous samples derived from MOFs to SMX, this study conducted degradation kinetics experiments. Figure 8a showed the degradation kinetics curves of C-ZIF-67 and 10-Co-HNPC-1000 degrading SMX by activating PMS, Fig. 8b exhibited

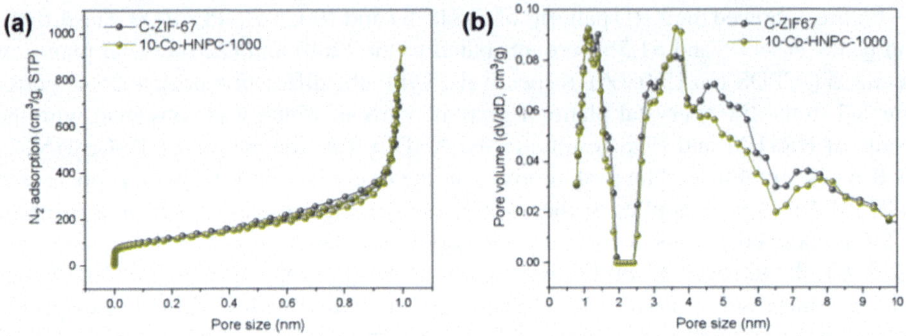

Fig. 6. (**a**) N_2 adsorption-desorption isotherms and (**b**) pore size distributions of 10-Co-HNPC-1000 and C-ZIF-67

the pseudo first order catalytic kinetics model fitting curves of C-ZIF-67 and 10-Co-HNPC-1000. Figure 8a indicated that the degradation effect of PMS alone on SMX was negligible, but the addition of carbonized samples synthesized in this experiment to activate PMS can efficiently degrade and remove the pollutant SMX in water. The degradation efficiency of 10-Co-HNPC-1000 derived from 10-ZIF-8@ZIF-67 was superior to C-ZIF-67, with which catalytic kinetic constant was 0.056 min^{-1}. As was shown in Table 1 the catalytic performance of 10-Co-HNPC-1000 was relatively outstanding compared to the catalysts reported.

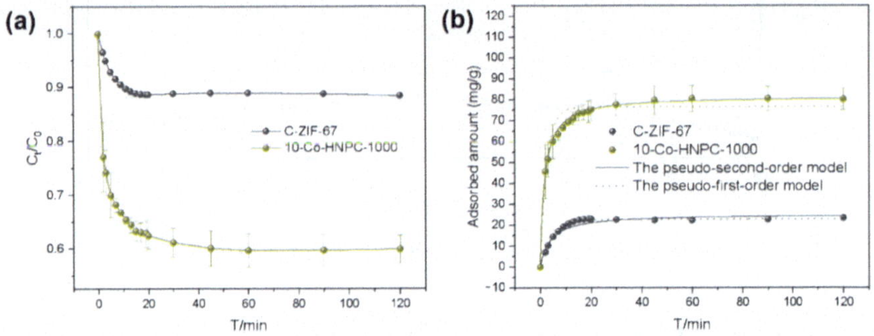

Fig. 7. (**a**) Adsorption curves and (**b**) corresponding fitting curves using pseudo-first/second order adsorption kinetics models of 10-Co-HNPC-1000 and C-ZIF-67 to SMX

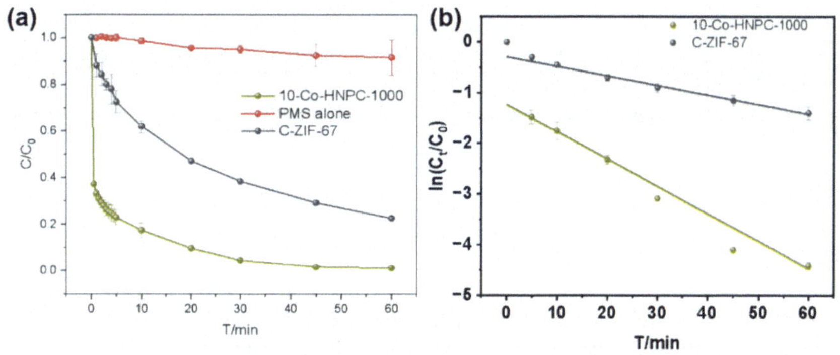

Fig. 8. (**a**) Degradation curves and (**b**) pseudo first order catalytic kinetics models fitting curves of 10-Co-HNPC-1000 and C-ZIF-67 to SMX

Table 1. comparison of catalytic performance between 10-Co-HNPC-1000 and other reported catalysts

Catalyst	Pollutants	Removal rate within 10 min	K_{app} (min^{-1})	Literature
10-Co-HNPC-1000	SMX	85%	0.056	This work
$CoFe_2O_4$	SMX	91%	–	11
Co-BC	SMX	80%	0.0108	12
Cu-N-C	SMX	92%	0.155	13

3 Conclusions

In summary, 10-CoHNPC-1000 was successfully synthesized by calcinating core-shell structured 10-ZIF-8@ZIF-67. It has found that the 10-Co-NHPC-1000 material has excellent saturation adsorption capacity and degradation rate constant for SMX, which were 81.67 mg/g and 0.056 min^{-1}, respectively. This study provides ideas for designing efficient integrated materials for adsorption and degradation.

References

1. Pingitore, E.V., Salvucci, E., Sesma, F., et al.: Different strategies for purification of antimicrobial peptides from lactic acid bacteria (LAB)[J]. Commun. Curr. Res. Educ. Topics Trends Appl. Microbiol. **1**, 557–568 (2007)
2. Balducci, C., Perilli, M., Cecinato, R.A.: New developments on emerging organic pollutants in the atmosphere[J]. Environ. Sci. Pollut. Res. **9**, 1875–1884 (2012)
3. Pal, A., He, Y., Jekel, M., et al.: Emerging contaminants of public health significance as water quality indicator compounds in the urban water cycle[J]. Environ. Int. **71**, 46–62 (2014)
4. Armin, Z., Cicero, M.R., et al.: Bioaccumulation and biomagnification potential of pharmaceuticals with a focus to the aquatic environment[J]. J. Environ. Manag. **133**(1), 378–387 (2014)

5. Ericson, J.F., et al.: Comment on 'pharmaceuticals, hormones, and other organic wastewater contaminants in U.S. streams, 1999-2000: a national reconnaissance'[J]. Environ. Sci. Technol. **36**(18), 4005–4006 (2002)
6. Bradle, P.M., Journey, C.A., Romanok, K.M., et al.: Expanded target-chemical analysis reveals extensive mixed-organic-contaminant exposure in U.S. streams[J]. Environ. Sci. Technol. **51**(9), 4792–4802 (2017)
7. Kim, S., Chu, K.H., Al-Hamadani, Y.A.J., et al.: Removal of contaminants of emerging concern by membranes in water and wastewater: a review[J]. Chem. Eng. J. **335**, 896–914 (2018)
8. Ma, Y., Wang, H., Lv, X., et al.: Three-dimensional ordered mesoporous Co3O4/peroxymonosulfate triggered nanoconfined heterogeneous catalysis for rapid removal of ranitidine in aqueous solution[J]. Chem. Eng. J. **443**, 136495 (2022)
9. Tan, J., Xu, C., Zhang, X., et al.: MOFs-derived defect carbon encapsulated magnetic metallic Co nanoparticles capable of efficiently activating PMS to rapidly degrade dyes[J]. Sep. Purif. Technol. **289**, 120812 (2022)
10. Pan, Y., Cao, J., Xing, M., et al.: Current mechanism of peroxymonosulfate activation by Cobalt-based heterogeneous catalysts in degrading organic compounds[J]. ACS Environ. Sci. Technol. Eng. **4**(1), 19–46 (2024)

Research Progress on the Treatment Technology of High-Concentration Organic Wastewater

Baitong Shi[1(✉)] and Huaduo Li[2]

[1] School of Chemistry, University of Edinburgh, Edinburgh EH8 9YL, UK
s2485023@ed.ac.uk

[2] School of Physics, University of Edinburgh, Edinburgh EH8 9YL, UK

Abstract. With the rapid development of industrialization and urbanization, the treatment of high-concentration organic wastewater has become an urgent problem to be solved. This type of wastewater has complex components and contains a large amount of organic matter, making its treatment challenging. This article comprehensively analyzes the characteristics and treatment methods of different types of organic wastewater such as food wastewater, pesticide wastewater, and chemical wastewater. Physical, chemical and biological methods are currently commonly used treatment technologies, but there are limitations to a single method. Therefore, it is necessary to use coupling technologies of different methods to treat high-concentration organic wastewater. Finally, this article looks forward to the possible development directions of organic wastewater treatment research in the future, including intelligent and efficient treatment technologies and the application of new biodegradants and green chemistry principles to achieve more environmentally friendly treatment of high-concentration organic wastewater.

Keywords: Food wastewater · pesticide wastewater · chemical wastewater · treatment methods · coupling technology

1 Introduction

With the continuous acceleration of industrialization and urbanization, a large amount of high-concentration organic wastewater is produced, and the treatment of organic wastewater has become one of the problems that humans need to solve urgently. The main characteristics of high-concentration organic wastewater are high organic content with a chemical oxygen demand (COD) of tens of thousands or even hundreds of thousands of mg/L, and complex composition, mostly containing heteroatoms (nitrogen, sulfur, oxygen, halogen, etc.) organic matter, heavy metal ions and other toxic substances [1]. High-concentration organic wastewater mainly includes food wastewater, pesticide wastewater, and chemical wastewater. If these wastewaters are discharged directly without treatment, they will not only cause pollution to water bodies, but also pose potential threats to the ecological environment and human health.

Food wastewater, as a by-product produced during food processing, contains a large amount of organic matter and microorganisms. Pesticide wastewater mainly comes from

V. Achal et al. (Eds.): EPPC 2024, ESE, pp. 171–178, 2025.
https://doi.org/10.1007/978-3-031-92123-0_17

the use of pesticides and farmland irrigation in the agricultural production process, and may contain various organic compounds and their metabolites. Chemical wastewater covers all types of organic wastewater discharged from the chemical production process, including organic solvents, organic acid and alkali wastewater, etc. Its composition is complex and diverse, and it is difficult to be treated. Therefore, it is urgent to carry out efficient, economical and environmentally friendly treatment for these different types of organic wastewater.

Traditional wastewater treatment technologies mainly include physical, chemical and biological methods, but the treatment effects of these technologies are very limited [2]. Therefore, it is particularly urgent to research and develop new organic wastewater treatment technologies. This article discusses different types of organic wastewater treatment technologies, focusing on their treatment principles, advantages and disadvantages, and applicable scenarios. Through comparative analysis of the characteristics and effects of different technologies, it aims to provide reference for research and application in the field of organic wastewater treatment.

2 Types and Characteristics of Organic Wastewater

2.1 Food Wastewater

Food wastewater refers to wastewater generated during food processing, production and consumption [3]. Its main feature is that it contains a large amount of organic matter and microorganisms. This type of wastewater usually has high concentrations of organic matter, and different types of food processing produce wastewater with different properties. Meat processing may produce wastewater containing large amounts of protein and fat, while fruit and vegetable processing may contain higher concentrations of fructose and fruit acids. For example, the total nitrogen (TN) content of cow slurry liquid fraction wastewater is about 2000 mg/L, and the COD of cheese dairy wastewater exceeds 23,000 mg/L [4], making its treatment difficult and complex. In addition, the treatment of food wastewater is also affected by seasonal changes and production scale. During certain holidays or specific production seasons, the amount of wastewater generated by food processing may increase dramatically, and corresponding treatment measures are required to deal with this change. To sum up, food wastewater has the characteristics of high organic matter content, complex composition and seasonal changes.

2.2 Pesticide Wastewater

Pesticide wastewater refers to the wastewater produced due to the use of pesticides in the agricultural production process, including production wastewater from pesticide manufacturing plants and washing wastewater from pesticide application equipment. Its COD content can reach more than 30,000 mg/L, and its biochemical oxygen demand (BOD) content can reach 10,000 mg/L or above. At the same time, pesticide wastewater is highly toxic and contains toxic heavy metals such as copper, lead, mercury and arsenic, as well as salt compounds such as NaCl, PO_4^{3-}, NH_4^+ and Ca^{2+} [5]. The total salt content can reach 40,000 mg/L [6]. These pesticide ingredients have certain toxicity and stability, and may have serious impacts on water bodies and the ecological environment. Traditional wastewater treatment methods are often difficult to completely remove pesticide residues, resulting in low treatment efficiency and difficulty in meeting discharge standards.

2.3 Chemical Wastewater

Chemical wastewater refers to process wastewater, waste gas washing water, flushing site and equipment water, etc. produced in chemical manufacturing. Chemical wastewater comes from petrochemical, coal chemical, papermaking, printing and dyeing and other industries, and has complex composition and high pollution [7]. For example, as illustrated in Table 1, the COD content of pulping wastewater exceeds 17,000 mg/L [8], the total phenol content in petrochemical wastewater is 150–250 mg/L [9], and the total nitrogen content in coal chemical wastewater is 70 mg/L [10].

Table 1. Characteristics of chemical wastewater (mg/L)

	COD	BOD	TN	Total phosphorus	Reference
Petrochemical wastewater	1900–2800	475–980	–	–	[9]
Coal chemical wastewater	500	300	70	4.5	[10]
Pulping wastewater	17,388	–	80–100	20–30	[8]

3 Sewage Treatment Process

3.1 Physical Methods

Physical method refers to a way to treat sewage using physical principles and technical means. In wastewater treatment, physical methods cover processes such as screening, sedimentation, filtration and flotation. Adsorption and membrane separation are commonly used physical treatment methods for wastewater. The adsorption method is a method that uses porous materials to adsorb pollutants in wastewater on the surface of the material, thereby purifying wastewater. It has the characteristics of low cost, high efficiency, and simple process. Commonly used adsorbents include activated carbon, resin, diatomaceous earth, bentonite, etc. (Table 2). When cetyltrimethylammonium bromide-modified calcium bentonite (0.9 g/L) is used as an adsorbent to treat petrochemical wastewater (concentration 120 mg/L, pH value 9), the removal rate of COD in the wastewater exceeds 80% [7]. Membrane separation refers to the process of separating pollutants from water by utilizing the selective permeability of membranes. It has the advantages of energy saving, flexible design, easy to be upgraded and high product quality. Nanofiltration membrane is used to treat phosphorus-containing wastewater from pesticide plants, and the removal rate of phosphorus in the sewage is close to 90% [5]. Physical methods play a role in removing impurities such as solid waste, suspended solids, and grease in sewage treatment. However, physical methods themselves still have certain limitations. For example, adsorption methods can only exert their advantages within a certain pH range when treating petroleum wastewater. When the membrane separation method is used to treat pesticide wastewater, since the pesticide components easily interact with the filter membrane, it is easy to cause membrane fouling and thus reduce the treatment efficiency.

Table 2. Advantages and disadvantages of different sewage treatment methods

Methods	Classification	Materials	Advantages	Disadvantages
Physical methods	Adsorption method	Activated carbon, resin, bentonite clay, etc.	High efficiency and simple process	Limited adsorption capacity, and high cost
	Membrane separation	Nanofiltration, ultrafiltration, reverse osmosis membrane, etc.	Energy saving, flexible design, and high product quality	Easy to cause membrane fouling
Chemical methods	Chemical oxidation method	Air, ozone, chloride, etc.	Good organic pollutant removal effect	Some oxidants have limited oxidation capacity and high cost
	Advanced oxidation method	Hydrogen peroxide	High oxidation efficiency, wide application range, simple operation	High cost
Biological methods	Aerobic treatment	Activated sludge, biofilm	High COD removal rate	Need oxygen
	Anaerobic treatment	Anaerobic sludge	Suitable for treating high-concentration organic wastewater	There is a blockage problem

3.2 Chemical Methods

Chemical method refers to a method that uses the principle of chemical reaction to change the composition and properties of sewage by adding an appropriate amount of chemical agents in sewage treatment, thereby achieving water purification. Chemical methods mainly include chemical oxidation and advanced oxidation methods (Table 2). Among them, chemical oxidation methods include air oxidation, ozone oxidation, chlorine oxidation and wet oxidation, etc., which can remove most organic pollutants in water. For example, activated carbon catalyzed ozone oxidation technology is used to treat yeast fermentation wastewater. The COD and chroma removal rates in the wastewater can exceed 61% and 96% respectively [11]. However, the chemical properties of ozone are unstable and require on-site production and preparation, which increases the cost of wastewater treatment. Commonly used advanced chemical oxidation technologies include hydrogen peroxide oxidation technology, photocatalytic technology, Fenton reagent method, electrochemical oxidation method, ultrasonic oxidation method, supercritical water oxidation method, etc. It can decompose organic pollutants in wastewater that are difficult to biodegrade or are toxic to organisms. For example, Fenton oxidation technology can achieve a removal rate of more than 97% of organophosphorus and more than 90% of organochlorine in pesticide wastewater. However, this method is suitable for a narrow pH range (2.5–4), and it is easy to form in the treated wastewater. Ferrous sludge leads to increased operating costs [5].

3.3 Biological Methods

Biological treatment is a method that uses organisms such as microorganisms, plants or animals to degrade, transform or adsorb organic matter, nitrogen, phosphorus and other pollutants in sewage. Biological treatment methods include aerobic treatment and anaerobic treatment (Table 2). Aerobic treatment methods mainly include activated sludge method and biofilm method. The strong oxidation and adsorption properties of the activated sludge method can effectively degrade COD in wastewater. For example, aerobic sludge particles are used to treat noodle production wastewater, and the removal efficiency of COD and TN reaches 92% and 83% respectively [11]. The biofilm method degrades organic matter, nitrogen, phosphorus, etc. in wastewater by forming a biofilm on the solid surface. It is commonly used in treatment units such as biological filters and biological contact oxidation tanks. It can effectively remove organic pollutants in wastewater, but the treatment cost is relatively high. In addition, the anaerobic treatment method can be divided into three stages: hydrolysis and acidification, hydrogen and acetic acid production, and methane production. There is no need to introduce oxygen during the treatment process, and it is suitable for treating high-concentration organic wastewater [12]. Using the upflow anaerobic sludge bed process to treat sugar manufacturing wastewater can achieve a removal rate of 92% of volatile acids and COD in the sewage [11].

3.4 Comprehensive Processing Method

Single physical, chemical and biological treatment methods each have some disadvantages in the treatment of high-concentration organic wastewater, such as limited treatment effects, high treatment costs and secondary pollution. Therefore, these methods should be combined and matched through synergy and complementation between methods. The realization of advanced treatment of high-concentration organic wastewater is the future development direction [2]. At present, some research results have been achieved through different method combinations to treat high-concentration organic wastewater.

Pesticide wastewater contains bactericides, which have an inhibitory effect on microorganisms, so purely biochemical treatment methods are not ideal. Tian et al.

used a combination of air flotation/iron carbon/Fenton+ hydrolysis+ A/O+ contact oxidation process to treat pesticide wastewater. The COD and toluene removal rates in the wastewater exceeded 94%. The treated wastewater can meet China's Class B discharge standards, and the combined process is economical and reasonable. Air flotation can remove grease and organic solvents from wastewater, thereby reducing COD, iron-carbon micro-electrolysis and Fenton can destroy the structure of macromolecular organic matter and remove some toxic and harmful substances, laying the foundation for subsequent biological treatment. At the same time, the combination of iron-carbon micro-electrolysis and Fenton oxidation as a pretreatment can reduce the amount of acid added. In addition, the Fe2+ produced by iron-carbon micro-electrolysis can be used as a catalyst for the Fenton reaction, thereby reducing the amount of ferrous sulfate and reducing operating costs. The biochemical system adopts the process of hydrolysis+ A/O+ contact oxidation. Hydrolysis and acidification improve the biodegradability of wastewater. A/O and contact oxidation can respectively enrich high-load microbial flora and low-load microbial flora, thus improving biochemical treatment efficiency to achieve discharge standards for effluent [13].

Grease wastewater has complex components, high oil content, and low carbon-to-nitrogen ratio. Floating oil, dispersed oil, emulsified oil, and a small amount of dissolved oil are difficult to remove. A single method cannot achieve advanced treatment of grease wastewater. Bai et al. used a combined treatment process of inclined plate oil separation-vortex air flotation-ABR-contact oxidation. After treatment, the COD of the effluent was less than 100 mg/L, the mass concentration of animal and vegetable oil was 7.0 mg/L. Inclined plate oil separation can remove floating oil, dispersed oil and a small amount of emulsified oil, reducing the oil load of subsequent process units, vortex air flotation effectively removes emulsified oil and emulsified oil in wastewater by adding coagulant poly aluminum chloride and coagulant aid polyacrylamide. ABR converts difficult-to-degrade macromolecules into small molecules through anaerobic hydrolysis, significantly reducing COD concentration. The biofilm of the catalytic oxidation tank is very rich in microorganisms, and there are nitrifying bacteria and denitrifying bacteria, which further removes BOD and TN in the wastewater, and finally allows the wastewater to meet the discharge standards [14].

4 Conclusion

There are many types of high-concentration organic wastewater, including food wastewater, pesticide wastewater, and chemical wastewater. The main characteristics of these wastewaters are high organic matter content and complex composition. At present, these wastewater treatment methods include physical, chemical and biological methods. Commonly used physical methods include adsorption, membrane separation and air flotation. The adsorption method has a simple process and high efficiency, but has high requirements for the pH value of the sewage and limited adsorption capacity; membrane separation method is energy saving, but the membrane is easily contaminated. Chemical methods include chemical oxidation and advanced oxidation. The chemical oxidation method has good organic pollutant removal effect but limited oxidation ability. The advanced oxidation method has high oxidation efficiency and wide application range but high cost. Biological methods include aerobic treatment and anaerobic treatment. It is more effective in treating medium and low concentration organic wastewater. The treatment cost is low but the treatment efficiency is also low. Overall, single physical, chemical and biological treatment methods still have some drawbacks in the treatment of high-concentration organic wastewater. Combining different methods to treat high-concentration organic wastewater is the future development direction. For example, the combined process of air flotation/iron-carbon/Fenton+ hydrolytic acidification + A/O+ contact oxidation is used to treat pesticide wastewater, and the combined process of inclined plate oil separation-vortex air flotation-ABR-contact oxidation is used to treat grease wastewater, both of which have achieved good results. This study can provide reference for research and application in the field of organic wastewater treatment in the future.

5 Prospects

In the future, organic wastewater treatment technology will continue to face new challenges and opportunities. Coupling technologies of different treatment processes will continue to be innovated and improved to meet increasingly stringent environmental regulations and social needs. In terms of food wastewater treatment, more intelligent and efficient treatment technologies may emerge, such as the use of bioreactors combined with artificial intelligence to achieve real-time monitoring and optimal control of the wastewater treatment process to improve treatment efficiency and save resources. For pesticide wastewater, future research may focus on the development and application of new biodegradants to improve the removal rate and degradation efficiency of pesticide residues and protect the ecological environment of farmland and water sources. In the field of chemical wastewater treatment, more new treatment technologies based on green chemistry principles may appear in the future, such as the use of environmentally friendly oxidants and catalysts to achieve efficient degradation and conversion of organic matter and reduce the generation of secondary pollution. Generally speaking, the future development of organic wastewater treatment technology will pay more attention to environmental protection, efficiency and sustainability, provide more solid technical support for the realization of clean production and circular economy, and create a cleaner and more beautiful living environment for mankind.

References

1. Li Zhuwan, M., Dandan, L.Y., et al.: Catalytic technologies for treatment of high concentration organic wastewater for resource recovery. Ind. Cataly. **29**(10), 25–31 (2021)
2. Beibei, Z., Yi, C.: Research progress of high-concentration organic wastewater treatment technology. Environ. Protect. Energy Saving. **3**, 40–41 (2021)
3. Huiyan, H.: Debugging of AO biochemical treatment process for meat wastewater. Guangdong Chem. Ind. **50**(20), 129–132 (2023)
4. Ramanaiah, S.V., Chandrasekhar, K., Cordas, C.M., et al.: Bioelectrochemical systems (BESs) for agro-food waste and wastewater treatment, and sustainable bioenergy-a review [J]. Environ. Pollut. **325**, 121432 (2023)

5. Wenxuan, M., Yanling, Q.: Research progress of pesticide wastewater treatment technology. Chem. Eng. **3**, 72–75 (2024)
6. Hongyan, S.: Application of recycling process for high-salt pesticide wastewater treatment. Guangzhou Chem. **47**(05), 73–78 (2022)
7. Zhongqiang, X.: Research and application of chemical wastewater treatment. Refin. Chem. Ind. **34**(02), 9–12 (2023)
8. Jiawei, L., Wenning, M., Jia, W., et al.: Performance and microbial communities of a novel integrated industrial-scale pulp and paper wastewater treatment plant. J. Clean. Prod. **278**, 123896 (2021)
9. Jothinathan, L., Cai, Q.Q., Ong, S.L., et al.: Organics removal in high strength petrochemical wastewater with combined microbubble-catalytic ozonation process. Chemosphere. **263**, 127980 (2021)
10. Mengbo, L.: Typical coal chemical park production wastewater treatment project case. Shandong Chem. Ind. **52**(12), 244–246 (2023)
11. Liuzhou, C., Quanlin, Z., Zhengfang, Y.: Research progress wastewater treatment technology in food industry. Appl. Chem. Eng. **51**(8), 2332–2336 (2022)
12. Lei, L.: Research progress of integrated food wastewater treatment technology. Leath. Manufact. Environ. Protect. Technol. **3**(17), 13–15 (2022)
13. Qiping, T., Meng, X., Lai, L., et al.: Pesticide wastewater treatment by combination process of air flotation/iron carbon/Fenton+ hydrolytic acidification+ A/O+ contact oxidation. Ind. Water Treat. **44**(2), 190–193 (2024)
14. Wangjun, B., Jun, N., Jianzhong, Z.: Treatment of oil and grease wastewater by combined process of inclined plate deoiling-cavitation air floatation-ABR-contact oxidation. Ind. Water Wastewater. **51**(3), 61–64 (2020)

Changes of Hydrochemical Characteristics of Reservoir Water Under the Influence of Saline Groundwater

Lina Yi[1,2], Haiming Li[1,2](✉), and Sihui Su[1,2]

[1] Tianjin University of Science and Technology, College of Marine and Environmental Sciences, Laboratory of Coastal Groundwater Utilization and Protection, Tianjin 300457, China
lhm@tust.edu.cn

[2] Groundwater Utilization and Protection Laboratory, Tianjin 300457, China

Abstract. Salinization of reservoir water will affect the utilization of water resources. In this paper, Beidagang Reservoir in Tianjin Binhai New Area is taken as the research object, and the Majuan intake gate is taken as the research point. Through column experiment, the influence of saline groundwater on reservoir water was studied, and the hydrogeochemical changes of reservoir water were studied, and the mechanism of reservoir water salinity is revealed. The change of hydrochemical index concentration will go through three stages: static period, rising period, dynamic balance period. There was a concentration gradient in the vertical direction (10–30 cm). The maximum increase of TDS concentration at the depth of 10 cm is 5.05 g/L, and the increase of TDS concentration at 30 cm is 1.19 g/L.

Keywords: Beidagang reservoir · Salnization of reservoir · Saline groundwater · Salt release

1 Introduction

The construction of coastal reservoir can effectively solve the water supply problem in coastal areas. However, in many coastal areas, the problem of salinization of water quality generally exists in coastal reservoirs. Salinization of reservoirs leads to low utilization rate of water resources, which affects the safety of water supply, aggravates the shortage of water resources, wastes limited fresh water resources and seriously threatens the safety of water supply [1–5].

After many scholars' research, reservoir salinization is the result of many factors, among which the role of saline groundwater has not been studied and its influence cannot be ignored.

V. Achal et al. (Eds.): EPPC 2024, ESE, pp. 179–185, 2025.
https://doi.org/10.1007/978-3-031-92123-0_18

2 Materials and Methods

2.1 Water Samples

The Reservoir water was taken from the Majuan intake gate. The saline groundwater was taken from the vicinity of Beidagang Reservoir. The hydrochemical indexes are shown in Table 1.

Table 1. Hydrochemical indexes of the reservoir water and saline groundwater

Index	pH	TDS (g/L)	Mass concentration of main ion (mg/L)							
			Na^+	Ca^{2+}	Mg^{2+}	K^+	Cl^-	SO_4^{2-}	HCO_3^-	CO_3^{2-}
Majuan intake gate	8.08	4.45	777.82	135.22	249.88	16.10	2205.20	917.76	305.10	–
Saline ground water	8.31	18.90	3927	18.43	1074	100.9	12,240	1239	610.2	–

2.2 Experimental Setup and Procedures

Prepare two columns with a diameter of 10 cm and a height of 65 cm, and their bottoms are closed. It has seven rounds holes: one is filled into the reservoir water of the Majuan intake gate; The other is filled with saline groundwater. The test flow is as shown in the Fig. 1.

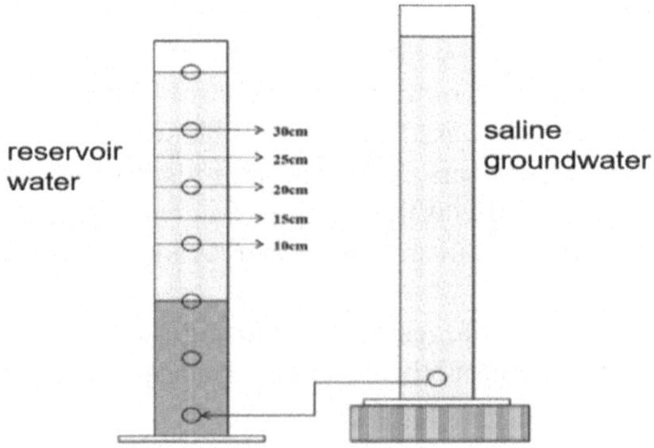

Fig. 1. Injection of saline groundwater

Fill the column with sediment with a thickness of about 20 cm, and spread quartz sand with a thickness of about 2 cm on the upper and lower layers. Seal the opening with plastic wrap to avoid evaporation. Pour the reservoir water into the column and let it stand for a week. After 1 week, the saline groundwater is injected from the bottom round hole, and water samples are taken at the positions as shown in the picture. The pH, TDS and plasma concentrations of Na^+, Mg^{2+}, Ca^{2+}, K^+, Cl^-, $SO_4{}^{2-}$, $CO_3{}^{2-}$ and $HCO_3{}^-$ in the overlying water were measured.

3 Result and Discussion

3.1 TDS

The Fig. 2 shows the variation of TDS with time. TDS was an important index to measure the total salt content of water. The initial TDS concentration (10–30 cm) of Majuan intake gate is 2.77 g/L, 2.72 g/L, 2.75 g/L, 2.76 g/L and 2.73 g/L in turn, with an average of 2.74 g/L. The maximum concentration increment at 10 cm is 5.05 g/L, and the maximum concentration increment at 30 cm is 1.19 g/L.

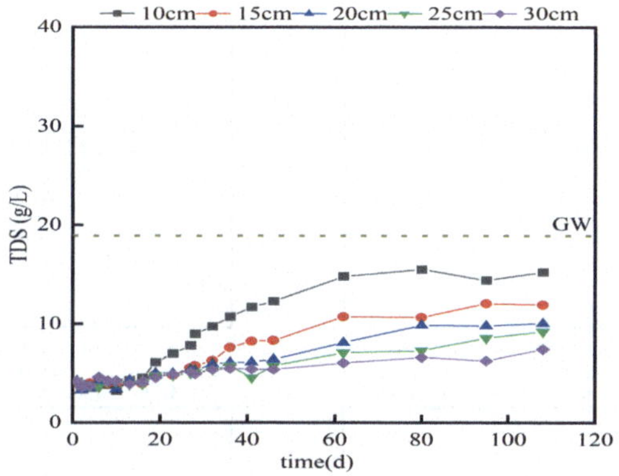

Fig. 2. Variation of TDS with time of Majuan intake gate

During the week before the injection of saline groundwater, the TDS concentration hardly changed under the interaction between reservoir water and sediments. It shows that during this period, sediments have little effect on the salinization of overlying water. The variation law of TDS concentration of Majuan intake gate goes through three stages: static stage, rising stage and dynamic balance stage: (1) static stage, the TDS concentration of reservoir water has not changed significantly within about 1 week after the injection of saline groundwater, and the TDS concentration remains stable, indicating that the influence of saline groundwater on reservoir water needs a certain period of time. From the figure, the static period of the horse circle entering the sluice is about 1 week. (2) The rising stage, the TDS concentration at different depths gradually increased with the passage of time. The TDS concentration at the depth of 10 cm was the highest, and the TDS concentration at the depth of 30 cm was the lowest, forming a concentration gradient in the vertical direction, and the TDS concentration showed a state of up, down and up. (3) During the dynamic balance period, the TDS concentration no longer increased significantly, but fluctuated up and down within a certain range.

The experimental results show that the TDS of overlying water will increase with the input of saline groundwater, but it is not infinite. After the input of saline groundwater, the original salt dynamic balance will be broken, and after a period of increase, it will tend to be stable and reach a new dynamic balance. The salt change of the overlying water body is not a simple linear process, but a complex dynamic process.

3.2 Cations

The dynamic change of cations concentration with time is shown in the Fig. 3. The change law of Na^+, Mg^{2+} and Ca^{2+} concentration with time is roughly the same as that of TDS. It first goes through a period of static period, then grows rapidly and gradually reaches a certain concentration, reaching a dynamic balance stage.

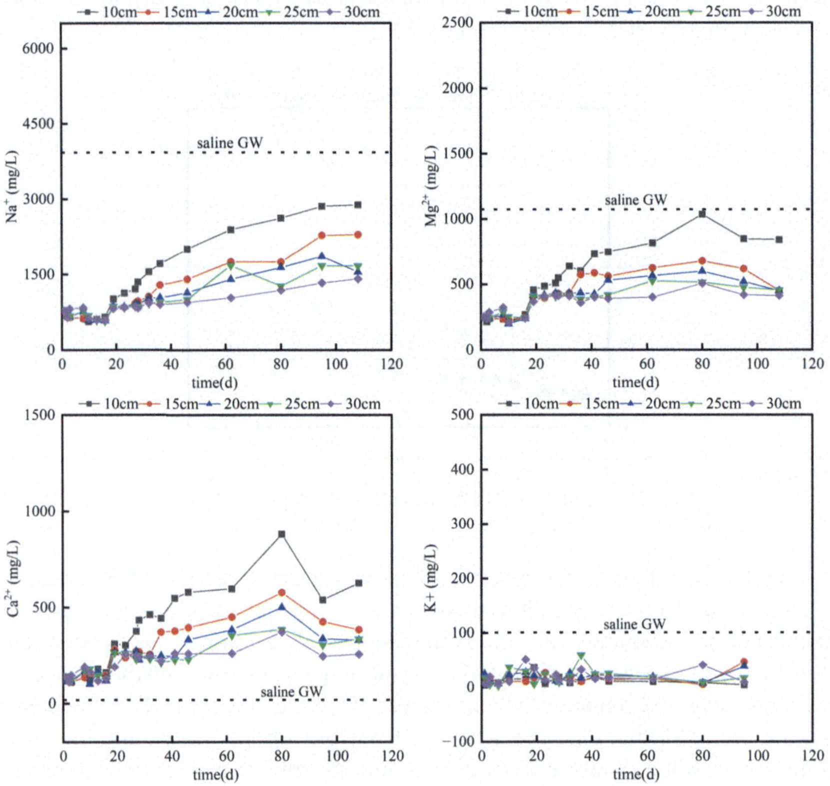

Fig. 3. Variation of cations with time of Majuan intake gate

The average concentration of Na^+ in Majuan intake gate increased from 777.82 mg/L to 1967.6 mg/L, and the maximum increase of Na^+ concentration in 10–30 cm was 2101.6 mg/L, 1539.7 mg/L, 1092.3 mg/L, 894.4 mg/L and 641.9 mg/L respectively. The average concentration of Mg^{2+} in Majuan intake gate increased from 249.88 mg/L to 524.7 mg/L, and the maximum increase of Mg^{2+} concentration in 10–30 cm was

787 mg/L, 435.2 mg/L, 350.7 mg/L, 380.9 mg/L and 417.6 mg/L in turn. The average concentration of Ca^{2+} in Majuan intake gate increased from 135.22 mg/L to 387.94 mg/L, and the maximum increment of Ca^{2+} concentration in 10–30 cm was 739.4 mg/L, 438.7 mg/L, 385.9 mg/L, 245 mg/L and 236.3 mg/L in turn. After experimental observation, the concentration of K^+ did not increase and there was no obvious change law. It is inferred that the influence of saline groundwater on K^+ in overlying water is not significant, and the relationship between the salinization of reservoir water and the concentration of K^+ ions is not significant. This result suggests that we need to further explore other factors that may affect the salinization of reservoir water in order to understand and solve this problem more comprehensively.

The results show that saline groundwater will lead to the increase of Na^+, Mg^{2+} and Ca^{2+} concentration. The increase of cations concentration in overlying water is not infinite, and it will keep a dynamic balance when it reaches a certain level, and the cation concentration, such as Na^+ and Mg^{2+}, in water may not necessarily reach that in saline groundwater. Although the Na^+ and Mg^{2+} concentration at the depth of 30 cm increased to some extent with the input of saline groundwater, it was far from the Na^+ and Mg^{2+} concentration in saline groundwater. It shows that in the natural environment, the higher the water level of the reservoir will inhibit the upward diffusion of ions and form a concentration gradient in the vertical direction, so the salinity of the surface water body will be lower than that of the bottom water body. However, saline groundwater will still increase the ion concentration in the surface water, and its salinity will also deepen.

3.3 Anions

The dynamic change of anion concentration with time is shown in the Fig. 4. The dynamic change of Cl^- and $SO_4{}^{2-}$ concentration with time is the same as TDS, and it goes through the static period, rising period and dynamic balance period. Cl^- and $SO_4{}^{2-}$ concentration forms a concentration gradient in the vertical direction. The average concentration of Cl^- in Majuan intake gate increased from 2205.2 mg/L to 6714.4 mg/L, and the maximum increase of Cl^- concentration in 10–30 cm was 7374 mg/L, 5277 mg/L, 4322 mg/L, 3471 mg/L and 2162 mg/L in turn. The average $SO_4{}^{2-}$ concentration in the Majuan intake gate increased from 917.76 mg/L to 1383.6 mg/L, and the maximum increase of $SO_4{}^{2-}$ concentration in 10–30 cm was 692.7 mg/L, 605.1 mg/L, 418.9 mg/L, 385.9 mg/L and 318.6 mg/L respectively. The concentration of $HCO_3{}^-$ has no obvious regularity, fluctuates greatly, and the overall change is small, unlike the obvious increase or decrease of anions such as Cl^- and $SO_4{}^{2-}$. It is inferred that saline groundwater has no great influence on the concentration of $HCO_3{}^-$ in reservoir water.

3.4 Hydrochemical Characterization

Through Piper's three-line diagram, the chemical composition and evolution characteristics of water can be observed intuitively. According to the percentage normalized concentration of main ions, Piper's three-line diagram is drawn in the unit of meq/l% [6]. The hydrochemical evolution of Majuan intake gate at various depths is shown in the Fig. 5. The proportion of Na^+ in 10 cm is from 55.67% to 55.56%, and the proportion of Na^+ in 30 cm is from 55.40% to 56.97%. The proportion of Mg^{2+} did not change

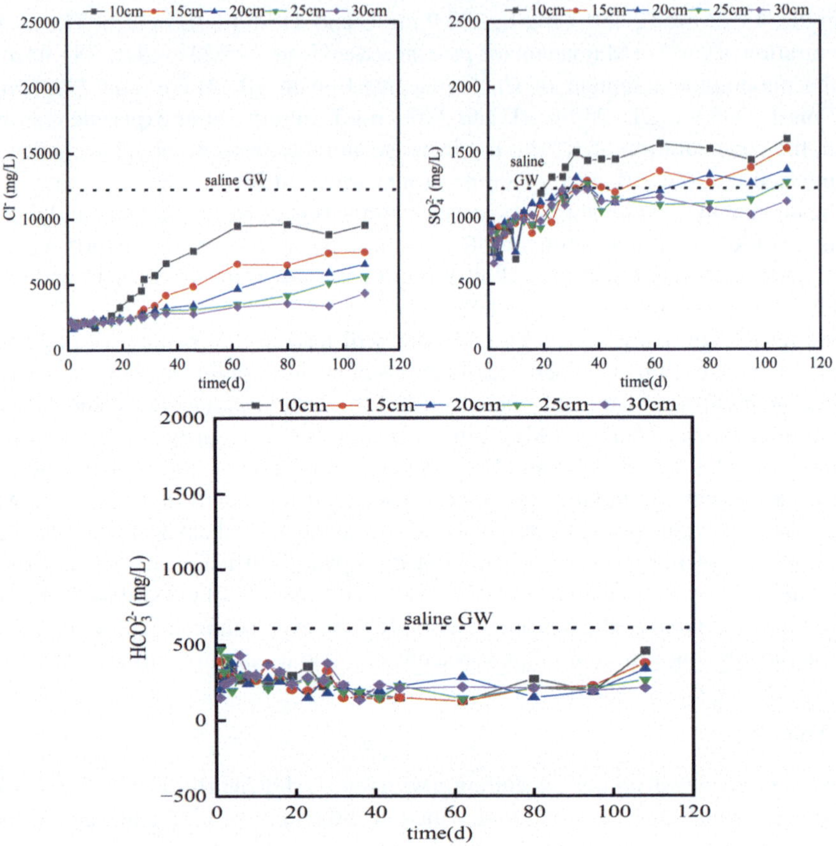

Fig. 4. Variation of anions with time of Majuan intake gate

much on the whole, and the average proportion of Mg^{2+} in 10 ~ 30 cm decreased from 33.24% to 29.17%. The average proportion of Ca^{2+} increased from 10.91% to 12.93%. On the whole, Na^{+} is dominant in cations, and the proportions of Na^{+}, Mg^{2+} and Ca^{2+} have little change. Anions showed the trend of Cl^{-} increasing and $SO_4{}^{2-}$ decreasing. The equivalent percentage of Cl^{-} in 10 cm depth increased from 72.32% to 86.81%, and Cl^{-} in 30 cm increased from 74.37% to 81.97%. The proportion of $SO_4{}^{2-}$ in 10 cm and 30 cm decreased from 22.02% and 2.63% to 10.77% and 5.69% respectively. According to Shukalev's hydrochemical classification, the hydrochemical types at the depths of 10 cm and 30 cm in Majian sluice have not changed, and the hydrochemical types at the depth of 10 cm are Na·Mg-Cl and Na·Mg-Cl·SO_4 at the depth of 30 cm.

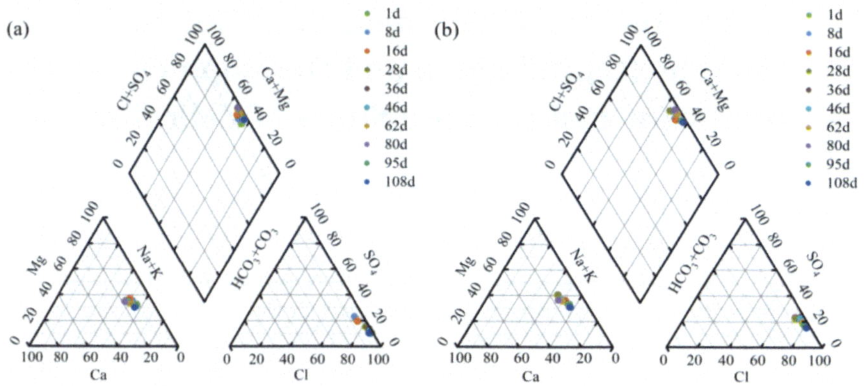

Fig. 5. Piper diagram of different depth at Majuan intake gate. (a) 10 cm; (b) 30 cm

4 Conclusion

The change of indexes of hydrochemistry in overlying water is not a simple linear process, but a complex dynamic process. The dynamic changes of TDS, cations and anions with time go through three stages: static stage, rapid rising stage and dynamic balance stage. The saline groundwater had no significant effect on the concentration changes of K^+ and HCO_3^-. The order of cation proportion is $Na^+ > Mg^{2+} > Ca^{2+}$; The order of anion proportion: $Cl^- > SO_4^{2-} > HCO_3^-$. Na^+ and Cl^- are the cations and anions with the highest proportion, both exceeding 50%, and they are also the main ions that cause salinization of reservoir.

References

1. Hua, L., Zengwen, G., Jing, L.: Inclusion of slow infiltration in determining the influence time of saline sediments on reservoir water [J]. J. Hydrol. **603**(PA), 126853 (2021)
2. Jalali, L., Zarei, M., Gutiérrez, F.: Salinization of reservoirs in regions with exposed evaporites. The unique case of upper Gotvand dam, Iran [J]. Water Res. **157**, 587–599 (2019)
3. do Patrocinio Hora, A.J., Rodrigues, L.M.H., Raimunda, D.J., et al.: Hydrogeochemical characterization of reservoir waters undergoing salinization processes in Northeastern Brazil [J]. RBRH. **26**, e24 (2021)
4. Mo, Y., Jin, G., Zhang, C., et al.: Effects of the deep pool on groundwater flow and salinization in coastal reservoir and adjacent aquifer [J]. Water Resour. Manag. **35**(8), 1–18 (2021)
5. Oz, I., Shalev, E., Yechieli, Y., et al.: Saltwater circulation patterns within the freshwater–saltwater interface in coastal aquifers: laboratory experiments and numerical modeling [J]. J. Hydrol. **530**, 734–741 (2015)
6. Ijioma, U.D.: Delineating the impact of urbanization on the hydrochemistry and quality of groundwater wells in Aba, Nigeria [J]. J. Contam. Hydrol. **240**, 103792 (2021)

The Impact of Microplastic Pollution on the Reproduction and Development of Marine Fish: A Case Study of Zebrafish

Yuhan Jin(✉)

School of Science, Xi'an Jiaotong-Liverpool University, Suzhou 215123, China
jinyuhan2004@icloud.com

Abstract. In recent years, microplastic pollution has become a global environmental concern, significantly impacting marine ecosystems, particularly fish species. This study investigates the potential effects of microplastic exposure on the reproductive systems and offspring development of zebrafish, serving as a model organism. The findings reveal that microplastic exposure disrupts the reproductive systems of both male and female zebrafish, leading to hormonal imbalances and impaired development of reproductive cells. The adverse effects may also extend to subsequent generations, affecting the embryonic and larval development of zebrafish. Additionally, microplastics can adsorb heavy metals and organic pollutants, creating more complex pollution scenarios and intensifying their detrimental effects on the reproductive health of zebrafish. Consequently, ecological risk assessments of microplastic pollution should consider the combined effects of microplastics and other contaminants. The results of this study underscore the importance of further investigating the mechanisms of microplastics' impact on aquatic life and contribute to the formulation of effective pollution control and prevention strategies.

Keywords: Microplastic Pollution · Zebrafish · Reproductive System · Offspring Development

1 Introduction

Microplastic pollution has emerged as one of the major challenges facing the global marine environment. Microplastics refer to plastic particles smaller than 5 mm in diameter, mainly originating from the fragmentation of plastic waste, plastic microbeads in cosmetics, and the washing of synthetic fiber clothing. Due to their small size, buoyancy, and resistance to degradation, microplastics can be transported over long distances by ocean currents and are widely distributed in the marine environment [1]. The harm of microplastics to marine ecosystems is mainly reflected in two aspects: First, microplastics may physically damage the digestive tracts of marine organisms, causing inflammation and malnutrition; Second, the surface of microplastics can easily adsorb toxic chemicals such as heavy metals and organic pollutants, which can accumulate and transfer through the food chain, ultimately endangering the health of marine life and even

V. Achal et al. (Eds.): EPPC 2024, ESE, pp. 186–193, 2025.
https://doi.org/10.1007/978-3-031-92123-0_19

humans. Marine fish are important receptors of microplastic pollution [2]. On the one hand, fish can be directly exposed to microplastic pollution by consuming prey containing microplastics or mistakenly ingesting microplastic particles; on the other hand, microplastics in the water can also enter fish bodies through gill respiration, skin penetration, and other pathways [3–5]. The accumulated microplastics and their associated toxic substances may interfere with the physiological metabolism, immune function, and reproductive development of fish.

Zebrafish is a small freshwater fish species that has been widely used as a model organism in toxicology and environmental pollution research due to its complete genome information, transparent embryonic development, and short reproductive cycle. The reproductive development process of zebrafish is highly sensitive to environmental pollutants, and microplastic exposure is likely to have significant impacts on key reproductive events such as gonadal development, gamete formation, fertilization, and embryonic development. Moreover, experimental results from zebrafish can be extrapolated to some extent to other fish species for assessing the potential risks of microplastic pollution to marine fish populations and ecosystems. Therefore, conducting zebrafish microplastic exposure experiments to systematically investigate the effects and mechanisms of microplastic pollution on fish reproductive development is of great significance for comprehensively understanding the ecotoxicological effects of microplastics, scientifically assessing the ecological risks of microplastic pollution, and formulating effective pollution prevention and control strategies.

2 The Impact of Microplastic Exposure on the Reproductive System of Zebrafish

2.1 Potential Impact of Microplastic Exposure on the Female Zebrafish Reproductive System

Microplastic exposure may disrupt the reproductive endocrine function of female zebrafish through multiple mechanisms. Environmental pollutants such as bisphenol A and polychlorinated biphenyls adsorbed on the surface of microplastics can directly bind to estrogen receptors (ERs) and interfere with endogenous estrogen signalling pathways, leading to disorders in ovarian steroid hormone synthesis. Microplastic particles can also be ingested by zebrafish and accumulate in the digestive tract, stimulating intestinal inflammatory responses and indirectly affecting ovarian function through the gut-ovary axis. In addition, microplastic exposure may induce oxidative stress and immunotoxicity, damaging oocyte DNA and accelerating oocyte apoptosis [6–8]. Combining the above mechanisms, microplastic exposure may lead to reproductive toxic effects such as imbalanced sex steroid hormone levels, impaired oocyte development, and ovulation disorders in female zebrafish (Table 1). Common types of microplastics in the environment, such as polyethylene (PE), polystyrene (PS), and polyvinyl chloride (PVC), may all interfere with the female zebrafish reproductive system to varying degrees, and this effect may be dose- and time-dependent. For example, low-dose microplastic exposure may primarily affect ovarian function through endocrine disruption effects, while high-dose exposure may exacerbate reproductive damage through toxic effects such as inducing inflammation and oxidative stress.

Table 1. Potential effects of microplastic exposure on the female zebrafish reproductive system

Reproductive indicator	Potential change	Main mechanism
Sex steroid hormone levels	Disordered	Endocrine disruption
Oocyte development	Impaired	Oxidative stress, DNA damage
Ovulation	Disordered	Endocrine disruption
Fertility	Decreased	Germ cell apoptosis

2.2 Potential Impact of Microplastic Exposure on the Male Zebrafish Reproductive System

Microplastic exposure can also interfere with the reproductive endocrine function of male zebrafish through similar mechanisms. Endocrine disruptors carried by microplastics can bind to androgen receptors (ARs) and inhibit the normal signal transduction of male hormones such as testosterone, leading to spermatogenesis disorders. Microplastic particles can also accumulate in testicular tissues, inducing local inflammatory and oxidative stress responses, damaging seminiferous tubules and spermatogenic cells, and accelerating sperm DNA damage and apoptosis [9, 10]. In addition, microplastic exposure may affect the hypothalamic–pituitary-gonadal axis, inhibiting the synthesis and release of pituitary gonadotropins and indirectly suppressing the synthesis of testicular steroid hormones. Combining the above mechanisms, microplastic exposure can lead to reproductive toxic effects such as sex hormone disorders, decreased sperm quality, and impaired fertility in male zebrafish (Table 2). Microplastics of different types, particle sizes, and concentrations may all interfere with the male zebrafish reproductive system to varying degrees. For example, smaller microplastic particles are more likely to penetrate the blood-testis barrier and directly damage spermatogenic cells; more hydrophobic polymer microplastics are more likely to accumulate environmental pollutants and produce stronger endocrine disruption effects; high-concentration microplastic exposure is more likely to induce inflammatory and oxidative stress responses in reproductive tissues.

Table 2. Potential effects of microplastic exposure on the male zebrafish reproductive system

Reproductive indicator	Potential change	Main mechanism
Sex steroid hormone levels	Disordered	Endocrine disruption
Spermatogenesis	Impaired	Oxidative stress, DNA damage
Sperm quality	Decreased	Endocrine disruption
Fertility	Decreased	Germ cell apoptosis

2.3 Mechanisms of Toxic Effects of Microplastic Exposure on Zebrafish Reproductive System

Microplastics can interfere with the reproductive endocrine function of zebrafish through multiple molecular mechanisms. First, environmental pollutants adsorbed on the surface of microplastics can directly bind to ERs and ARs, interfering with endogenous sex hormone signalling pathways and leading to disorders in gonadal steroid hormone synthesis. Second, ingested microplastic particles can accumulate in the digestive tract and gonadal tissues, inducing local inflammatory and oxidative stress responses, activating signalling pathways such as NF-κB and MAPK, upregulating the expression of pro-inflammatory cytokines and reactive oxygen species, damaging reproductive cells, and accelerating gamete DNA damage and apoptosis. In addition, microplastics can also affect the hypothalamic-pituitary-gonadal axis, inhibiting the synthesis and release of key signalling molecules such as GnRH, FSH, and LH, and indirectly suppressing the synthesis of gonadal steroid hormones. The reproductive toxicity of microplastics may be closely related to their physicochemical properties. For example, smaller microplastic particles are more likely to penetrate biological barriers and directly act on reproductive tissues; more hydrophobic polymers are more likely to accumulate environmental pollutants and produce stronger endocrine disruption effects; irregularly shaped microplastic particles may cause greater physical damage to tissues. Therefore, to comprehensively assess the reproductive health risks of microplastic pollution, it is necessary to systematically consider the inherent relationships between the physicochemical properties, environmental behavior, bioaccumulation, and toxicological effects of microplastics, and to deeply explore the combined pollution effects and molecular mechanisms of microplastics and endocrine disruptors.

3 The Impact of Parental Microplastic Exposure on the Development of Zebrafish Offspring

3.1 Potential Impact of Parental Microplastic Exposure on Zebrafish Embryonic Development

Parental zebrafish exposed to microplastics during their reproductive development stage may affect the quality of their germ cells, which in turn can have potential transgenerational effects on the development of their offspring embryos. Microplastics and their adsorbed environmental pollutants can be vertically transmitted to the offspring through the parental germ cells, interfering with key signalling pathways and gene expression regulatory networks during embryonic development. For example, endocrine disruptors carried by microplastics can affect early embryonic sex differentiation and gonadal development; oxidative stress induced by microplastics can cause DNA damage and apoptosis in embryonic cells; microplastic particles may also physically hinder the normal morphogenesis of embryos. The combined effects of these factors may lead to delayed embryonic development, increased malformation rates, and decreased hatching rates (Table 3). In addition, parental microplastic exposure may also affect the gene expression patterns of offspring embryos through epigenetic mechanisms (such as DNA methylation, histone modifications, etc.), resulting in transgenerational epigenetic effects.

Table 3. Potential effects of parental microplastic exposure on zebrafish embryonic development

Indicator	Potential change	Cause
Development rate	Delayed	Microplastics interfere with the expression of key developmental genes
Malformation rate	Increased	Microplastics induce oxidative stress and DNA damage
Hatching rate	Decreased	Microplastics physically hinder the hatching process of embryos
Sex differentiation	Disordered	Microplastics carry endocrine disruptors that interfere with the expression of sex-determining genes

3.2 Potential Impact of Parental Microplastic Exposure on the Growth and Development of Zebrafish Larvae

The microplastic exposure history of parental zebrafish may have persistent negative effects on the growth and development of their offspring larvae. As microplastics and their adsorbed pollutants accumulated in the parental germ cells can be passed on to the offspring through maternal effects, the larvae may be exposed to the potential toxic environment of microplastics from an early developmental stage. Microplastic exposure may interfere with important physiological processes in larvae, such as the development of the nervous system, maturation of immune function, and skeletal formation. For example, chemical additives released from microplastics may affect neurotransmitter transmission and synaptic plasticity in larvae, leading to abnormal neurobehavioral development; inflammatory responses induced by microplastics may inhibit the development and maturation of immune organs in larvae, increasing their sensitivity to pathogen infections; microplastic particles may also interfere with the mineralization process of larval skeleton, resulting in reduced bone density and skeletal malformations. The combined effects of these factors may lead to slow growth, poor development, and decreased survival rates in larvae (Table 4). In addition, the transgenerational epigenetic effects caused by parental microplastic exposure may also affect gene expression regulation during larval development, further exacerbating the toxic effects of microplastics.

3.3 Potential Impact of Parental Microplastic Exposure on the Reproductive Health of Adult Zebrafish

Parental zebrafish exposed to microplastics during critical periods of reproductive development may have long-term effects on the reproductive health of their adult offspring. On the one hand, parental microplastic exposure may alter the gene expression patterns of germ cells in the offspring through transgenerational epigenetic effects, affecting their gonadal development and gamete quality. For example, microplastic exposure may change the methylation levels of genes related to sex differentiation and gonadal development in parental germ cells, leading to sex ratio imbalance and abnormal gonadal development in the offspring. On the other hand, parental microplastic exposure may also

Table 4. Potential effects of parental microplastic exposure on the growth and development of zebrafish larvae

Indicator	Potential change	Cause
Growth rate	Decreased	Microplastics interfere with the expression of growth-related genes and endocrine regulation
Survival rate	Decreased	Microplastics cause oxidative stress and immune suppression, increasing the risk of pathogen infection
Skeletal development	Abnormal	Microplastics interfere with skeletal mineralization and calcium-phosphorus metabolism
Neurobehavioral development	Abnormal	Microplastics interfere with neurotransmitter transmission and synaptic plasticity

affect the endocrine homeostasis and reproductive endocrine axis function of adult offspring through maternal effects. For example, endocrine disruptors carried by microplastics may continue to be released in the bodies of adult offspring, interfering with the synthesis and secretion of their sex steroid hormones, leading to reproductive cycle disorders and decreased gamete quality. In addition, the oxidative stress response caused by parental microplastic exposure may also have persistent DNA damage effects on the germ cells of adult offspring, increasing the apoptosis and mutation rates of germ cells, thereby affecting their fertility. Combining the above factors, parental microplastic exposure may lead to poor gonadal development, reduced fertility, and decreased quantity and quality of offspring in the adult progeny (Table 5), potentially negatively impacting the sustainable development of zebrafish populations.

Table 5. Potential effects of parental microplastic exposure on the reproductive health of adult zebrafish

Indicator	Potential change	Cause
Gonadal development	Abnormal	Microplastics interfere with the expression of genes related to gonadal development through transgenerational epigenetic effects
Fertility	Decreased	Microplastics interfere with the synthesis and secretion of sex steroid hormones, causing apoptosis and mutation of germ cells
Offspring quantity	Reduced	Microplastics reduce the gamete quality and fertilization rate of parents, leading to a reduced number of offspring
Offspring quality	Decreased	Microplastics affect offspring development through maternal effects, leading to abnormal development and decreased survival rates of offspring individuals

4 Conclusion

The comprehensive analysis of studies on the impact of microplastics on the reproduction and development of zebrafish highlights that microplastic pollution is an environmental issue that cannot be ignored. Microplastics severely affect the reproductive health of zebrafish through mechanisms such as endocrine disruption, induction of oxidative stress, and triggering of inflammatory responses, and may have long-term negative effects on their offspring. Given the ubiquity of microplastics and their potential ecological harms, further research on their impact mechanisms on aquatic organisms, ecological risk assessments, and the development of effective pollution control and management strategies are necessary. Moreover, the interaction between microplastics and other pollutants should also become a focus of future studies to fully understand and address the challenges posed by microplastic pollution.

References

1. Postel, S.L., Daily, G.C., Ehrlich, P.R.: Human appropriation of renewable fresh water[J]. Science. **271**(5250), 785–788 (1996)
2. Savenije, H.H.G.: Water scarcity indicators; the deception of the numbers[J]. Phys. Chem. Earth Part B Hydrol. Oceans Atmosp. **25**(3), 199–204 (2000)
3. Yan, Y., Wang, R., Chen, S., et al.: Three-dimensional agricultural water scarcity assessment based on water footprint: a study from a humid agricultural area in China[J]. Sci. Total Environ. **857**, 159407 (2023)
4. Al-Busaidi, A., Yamamoto, T., Inoue, M., et al.: Effects of zeolite on soil nutrients and growth of barley following irrigation with saline water[J]. J. Plant Nutr. **31**(7), 1159–1173 (2008)
5. Mekonnen, M.M., Hoekstra, A.Y.: Four billion people facing severe water scarcity[J]. Sci. Adv. **2**(2), e1500323 (2016)

6. Bozorg-Haddad, O., Zolghadr-Asli, B., Sarzaeim, P., et al.: Evaluation of water shortage crisis in the Middle East and possible remedies[J]. J. Water Supply Res. Technol. AQUA. **69**(1), 85–98 (2019)
7. Akhtar, M., Hassan, F.-U., Ahmed, M., et al.: Is rainwater harvesting an option for designing sustainable cropping patterns for Rainfed agriculture?[J]. Land Degrad. Dev. **27**(3), 630–640 (2016)
8. Lee, H.-J., Kim, K.Y., Hamm, S.-Y., et al.: Occurrence and distribution of pharmaceutical and personal care products, artificial sweeteners, and pesticides in groundwater from an agricultural area in Korea[J]. Sci. Total Environ. **659**, 168–176 (2019)
9. Moreau, M., Hadfield, J., Hughey, J., et al.: A baseline assessment of emerging organic contaminants in New Zealand groundwater[J]. Sci. Total Environ. **686**, 425–439 (2019)
10. Sjerps, R.M.A., Kooij, P.J.F., Van Loon, A., et al.: Occurrence of pesticides in Dutch drinking water sources[J]. Chemosphere. **235**, 510–518 (2019)

Study on Enhanced Biological Filtration of Sedimentation Effluent

Shuo Zhang[1(✉)], Jiajiong Xu[1], Min Rui[2], and Xuejun Tan[1]

[1] Shanghai Municipal Engineering Design Institute (Group) Co., Ltd., Shanghai, China
zhangshuo@smedi.com

[2] Shanghai Water Industry Design Engineering Co., Ltd., Shanghai, China

Abstract. At low temperature, the treatment of raw water with high ammonia nitrogen and organic matter has always been one of the difficulties in the plant. The removal of ammonia nitrogen and COD_{Mn} by ceramide biofiltration was studied in pilot and the effects of assisted filtration and backwashing were analyzed. The results showed that the mean turbidity of the filtered water was stable below 0.3NTU if the particle size of the biofilter column was 2 mm to 4 mm. When ammonia nitrogen in raw water was lower than 2.5 mg/L, the mean value of ammonia nitrogen in filtered water could be stable below 0.3 mg/L and the maximum would be no more than 0.5 mg/L. The average removal rate was about 82%. When ammonia nitrogen in raw water was 2.5 mg/L to 3.0 mg/L, the removal rate of ammonia nitrogen was still higher than 50%, and the effluent would rise to 0.8 mg/L to 1.0 mg/L. The removal rate of COD_{Mn} by ceramic filter was 24% to 32%, and the effluent was stable below 3 mg/L. The average UV_{254} effluent was about 0.05 cm^{-1}, and the removal rate was 20% to 30%. The filtration cycle would be up to 72 h in the low temperature period, and would be appropriately reduced in the high temperature period from 24 h to 36 h.

Keywords: Biological ceramide filtration · Ammonia nitrogen · COD_{Mn} · Backwash · Low temperature

1 Introduction

Although the river network in southern China is dense and the water system is developed. Almost of them have been polluted. The odor, ammonia nitrogen and organic indicators of raw water obviously exceed the dose that the human body can tolerate, so the drinking water safety risk is very prominent. High ammonia in water poses taste and odor problem, and has adverse health impacts [1]. The effectiveness of some treatment processes could be reduced. In some areas of the Yangtze River, only micro-polluted surface water has to be used as raw water. It is difficult to meet the quality of plant water by using the existing conventional treatment process due to the contaminated water. Although the zeolite was used to remove ammonia nitrogen [2], the disposal of the regenerated liquid had been a problem. Membrane processes could effectively remove ammonia, but high capital and operational costs would hinder the application [3]. Biological ammonia removal in biofilters is being increasingly applied as an economical and environment-friendly

V. Achal et al. (Eds.): EPPC 2024, ESE, pp. 194–200, 2025.
https://doi.org/10.1007/978-3-031-92123-0_20

process [4–6]. In addition to the traditional sand filter, the biological activated carbon filter used has been widely introduced to enhance ammonia and organic matter removal [7, 8]. When one or more of the COD_{Mn}, ammonia nitrogen, iron, manganese and odor substances in raw water would be high, it is necessary to adopt biological pretreatment, enhanced conventional treatment and advanced treatment to improve the water quality. However, the quality of raw water in low temperature period would often deteriorate significantly. Some whole-process water plants have excessive ammonia nitrogen and COD_{Mn}, causing public complaints, even affecting social stability, which has become a long-term problem.

The removal effect and influencing factors of ceramide biofiltration were analyzed for raw water with high ammonia nitrogen and organic matter during low temperature. The experimental results could be used to guide the production practice of biological filtration in water plants.

2 Experimental Method

2.1 Pilot Equipment

The process was enhanced coagulation precipitation and biological filtration., as shown in Fig. 1. The 2–4 mm broken ceramic particles were used as filter material for biofilter columns. The filter layer was 1.5 m and the supporting layer was 0.3 m. Filter velocity was 8 m/h. Short handle filters were used for air flushing and water flushing. The liquid polyaluminum chloride was 10 mg/L as the filter aid. The dosage of liquid aluminum ferric chloride was 40 mg/L. Two points of 0.05 mg/L PAM were added to the sedimentation tank and the return sludge respectively. The ratio of returned sludge was 10%. The filter column diameter was 900 mm. The inlet flow rate was 5 m^3/h.

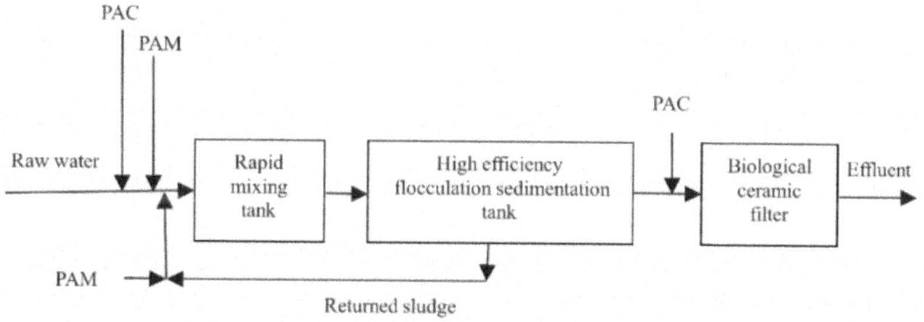

Fig. 1. The pilot equipment.

2.2 Raw Water

Raw water from the plant was used for the pilot. The water temperature was 6 °C to 12 °C. Ammonia nitrogen increased from 1.6 mg/L to 2.7 mg/L.

2.3 Starting Film

The biological ceramic filter column was cultured by continuous water feeding and dynamic natural membrane hanging. After 3 weeks, the removal rate of ammonia nitrogen was stable above 80%, which indicated that the biofilm had been acclimated.

3 Results and Discussion

3.1 Ammonia Nitrogen

The removal effect of biological ceramic filter column on ammonia nitrogen was shown in Fig. 2. At low temperature, the effect of ceramic biofiltration on removing ammonia nitrogen was obvious. If raw water ammonia nitrogen was from 1.0 mg/L to 1.5 mg/L, the average ammonia nitrogen concentration in filtered water was stable below 0.3 mg/L, and the maximum was no more than 0.5 mg/L. The removal amount of ammonia nitrogen was from 1.1 mg/L to 1.4 mg/L. The removal rate was about 82% on average, and the maximum was 90%.

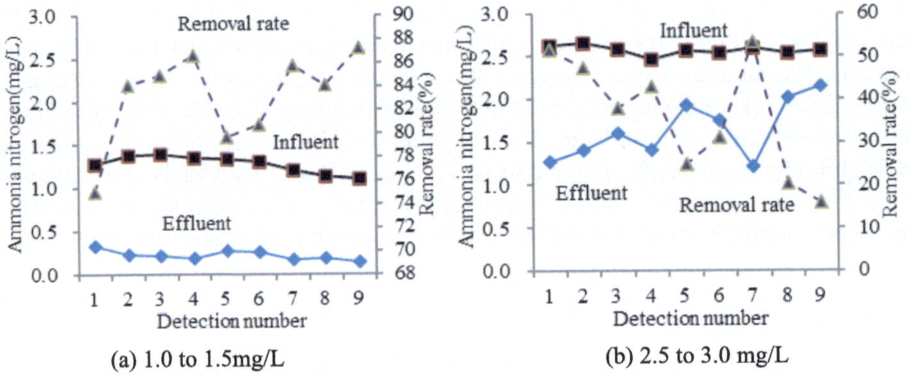

Fig. 2. Removal of ammonia nitrogen by biological filtration process. (**a**) 1.0–1.5 mg/L. (**b**) 2.5–3.0 mg/L.

When raw water ammonia nitrogen was among 2.5–3.0 mg/L, the removal rate of ammonia nitrogen by biological filtration was still higher than the 50% while removal amount was about 1 mg/L. But the effluent ammonia nitrogen concentration rises to 0.8–1.0 mg/L due to the high inlet concentration.

When inlet ammonia nitrogen increased to more than 3 mg/L, the removal rate decreased significantly to 16% and the removal rate in the combined process decreased to less than 30%. The nitroobacteria in the filter were shown in Fig. 3. AOB, AOA, and comammox often appear at biofilters simultaneously. Most AOB had difficulty thriving while AOA were more suitable to grow which could incorporate inorganic carbon more energy-efficiently [9]. AOA would not increase with increasing influent ammonia concentration and decreasing temperature [10].

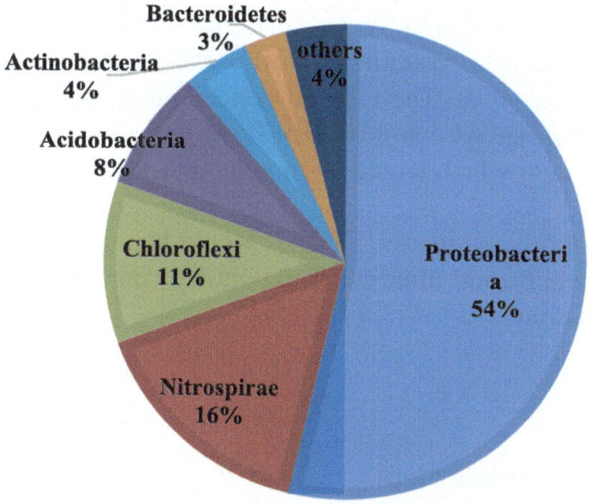

Fig. 3. Distribution of nitrobacteria.

3.2 COD_{Mn}

The removal COD_{Mn} of biological ceramic filter column was shown in Fig. 4. The removal rate was about from 15% to 32% and the removal amount was generally 1 mg/L. The effluent concentration was stable below 3 mg/L. The removal rate and removal amount can be increased with the increase of inlet concentration of filter column. The reason may be that the microbial generation cycle of decomposing organic matter was short, and there was still room for improvement of microbial proliferation and vitality under low temperature conditions.

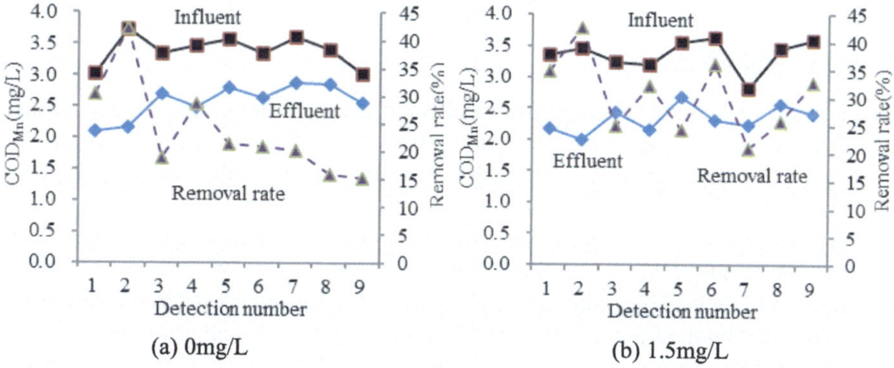

Fig. 4. Removal of COD_{Mn} by biological filtration process under potassium permanganate. (**a**) 0 mg/L. (**b**) 1.5 mg/L.

It was reported that permanganate pre-oxidation could enhance ammonia removal efficiency in biofilters [9]. The combination of permanganate composite chemicals and biological aerated filter treatment also presented a high removal efficiency of ammonia [11]. The removal effect of COD_{Mn} by biofilter was improved by adding more potassium permanganate before sedimentation. When the dosage was 1.5 mg/L, the average COD_{Mn} removal rate of ceramic filter column was 30% and the removal amount was about 0.8 mg/L accounting for 40%. The effluent concentration was about 2.3 mg/L. This indicated that chemical agents could oxidize large molecular organic matter to small molecular organic matter, which would be more conducive to microbial utilization and degradation.

As an indicator of organic matter, COD_{Mn} would be mainly removed by biodegradation. The residence time and filter material volume were substituted into the organic reaction kinetics equation, as shown in Eq. (1).

$$ln\frac{S_{out}}{S_{in}} = -K_V T = -K_V\frac{AH}{Q} = -K_A\frac{A}{Q} = \frac{-K_A}{q} \tag{1}$$

Where S_{out} = effluent concentration
S_{in} = influent concentration
K_V = volume reaction rate constant
T = hydraulic retention time
A = section area of the filter layer of the reactor
H = height of filter layer
Q = inlet water flow rate
K_A = area reaction rate constant
q = hydraulic load per unit area

If COD_{Mn} average inlet concentration, average effluent concentration, filter diameter and flow rate were substituted into Eq. (1), the K_A under the two working conditions could be 2.2 and 2.9 respectively. Therefore, when the influent concentration increased, the reaction rate would be correspondingly increased by 30%. The effluent concentration should still be lower than the national standard 3 mg/L.

3.3 UV_{254} and UV_{410}

The removal UV_{254} and UV_{410} of biological ceramic filter column was shown in Figs. 5 and 6. The average UV_{254} effluent was about 0.05 cm^{-1} and the removal amount were 0.019 cm^{-1}. The removal rate was about 20–30%. The average effluent UV_{410} was about 0.004 cm^{-1} and the removal amount were about 0.08 cm^{-1}.

Under adding powdered activated carbon before sedimentation, the removal efficiency of UV_{254} and UV_{410} was improved to a certain extent by biological filtration. The removal rate of UV_{410} could be as high as 80%. The residual powdered activated carbon was trapped by the filter column after precipitation and played a certain adsorption role on the ceramic particles.

3.4 Backwash

Backwashing was carried out periodically every few days in intermittent mode. After each backwash, the biofilm had a short recovery period. There was a certain residence time of biofilm named membrane age. Single gas impact strength was 50 $m^3/h/m^2$ and the duration was 3 min. When the gas and water flush together, the water flush intensity

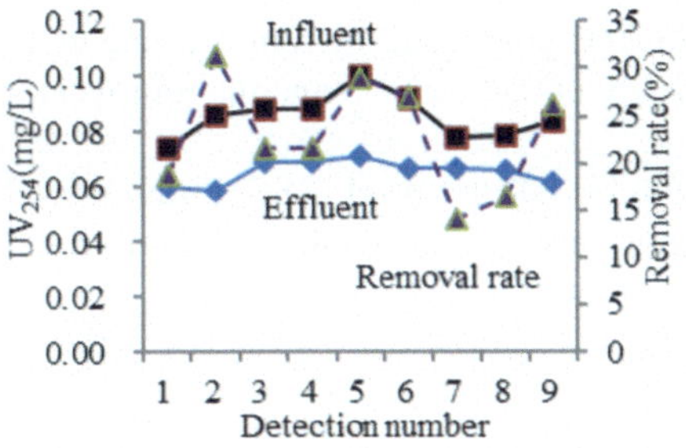

Fig. 5 Removal effect of UV_{254}.

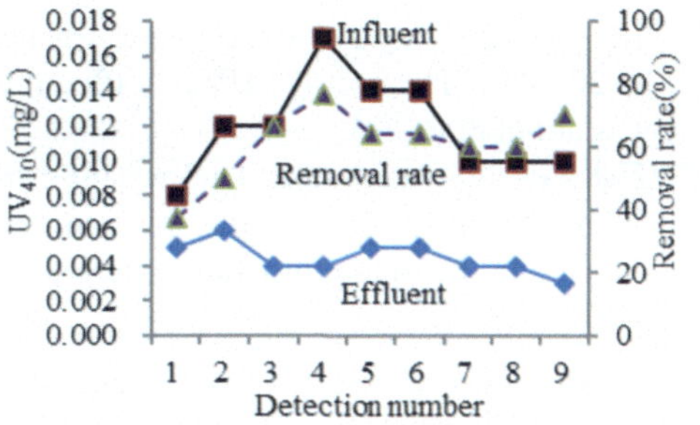

Fig. 6 Removal effect of UV_{410}.

was 10 $m^3/h/m^2$ and the duration was 4 min. When single water flushing, the flushing strength was 17 $m^3/h/m^2$ and the duration was 5 min. In order to minimize the biological impact of backwashing, the filtration cycle in the low temperature period could be up to 72 h. In the high temperature period, the filtration cycle would be appropriately reduced in 24–36 h because of the large number of organisms, high activity and fast reproduction.

3.5 Application in Water Plant

In the water plant, the filter speed of the biological filter was 8 m/h. After precipitation and biological sand filtration, the effluent turbidity was below 0.4NTU and the average ammonia nitrogen was 0.5 mg/L. The removal rate of COD_{Mn} was 50%.

After the subsequent ozone activated carbon and sand filter, the factory water generally reached the Chinese national standard. The turbidity was under 0.1 NTU. Chroma was not more than 5. COD_{Mn} was about 2.5 mg/L.

4 Conclusions

The effect of biological ceramic filtration on the removal of organic matter and ammonia nitrogen in submerged water had been studied, which was significant to solve the adverse effects of turbidity on organisms and improve the treatment efficiency of the whole water purification process.

1. The biological ceramic filter column was cultured by continuous water feeding and dynamic natural membrane hanging. In the water temperature of about 10 °C, after 3 weeks, the removal rate of ammonia nitrogen was stable above 80%, and the biofilm was mature.
2. At low temperature, the removal effect of ammonia nitrogen was obvious. When the raw water ammonia nitrogen was lower than 2.5 mg/L, the average ammonia nitrogen concentration in filtered water was stable below 0.3 mg/L and the average removal rate was about 82%. The removal rate of ammonia nitrogen could be increased by 11% after reducing the filtration rate.
3. The removal rate of COD_{Mn} was 24–32% and the effluent concentration was stable below 3 mg/L. The addition of high concentration of permanganate was beneficial to improve the removal effect of COD_{Mn} by biofilter.

4. The average UV_{254} of the effluent was about 0.05 cm^{-1} and the removal rate would be 20–30%. The average UV_{410} effluent was about 0.003 cm^{-1} and the removal rate was about 80%.
5. The gas-gas-water-water backwashing method could be used. The filtration cycle would be up to 72 h in the low temperature period, and would be appropriately reduced in the high temperature period from 24 h to 36 h.

Acknowledgments. This work was financially supported by the National Key Research and Development Program (No. 2023YFC3804700).

References

1. Talukder, A., Rabby, S.H., Ashraf, A.M.: Removal of ammonia from drinking water using synthetic zeolites. AIP Conf. Proc. **060004** (2023)
2. Ansari, W.R., Harahap, H., Husin, A.: Fixed-bed column adsorption performance for ammonia removal using adsorbent from zeolite. IOP Conf. Ser. Mater. Sci. Eng. **1122**, 012076 (2021)
3. Aung, H.Z., Spitsov, D., Pervov, A.: Improvement of drinking water quality and reduction of concentrate discharges using nanofiltration membranes. E3S Web Conf. **457**, 02002 (2023)
4. Luo, X.N., Shen, T., Guan, C.T., Li, N., Jiang, J.: Ammonia-oxidizing microbes and biological ammonia removal in drinking water treatment. Water Res. Technol. **8**(6), 1152–1172 (2022)
5. Zeng, J., Chabi, K., Hu, Y., Zhang, S., Yu, X.: Ammonium removal of biological roughing filter for rural drinking water pretreatment. Water Supply. **20**(7), 2768–2778 (2020)
6. Liu, H.L., Li, X., Li, N.: Application of bio-slow sand filters for drinking water production: linking purification performance to bacterial community and metabolic functions. J. Water Process Eng. **53**, 103622 (2023)
7. Rui, M., Chen, H., Ye, Y., Deng, H., Wang, H.: Effect of flow configuration on nitrifiers in biological activated carbon filters for potable water production. Environ. Sci. Technol. **54**(22), 14646–14655 (2020)
8. Jantarakasem, C., Kasuga, I., Kurisu, F., Furumai, H.: Temperature-dependent ammonium removal capacity of biological activated carbon used in a full-scale drinking water treatment plant. Environ. Sci. Technol. **54**(20), 13257–13263 (2020)
9. Xiaonan, L., Tong, S., Chaoting, G., Li, N., Jin, J.: Ammonia-oxidizing microbes and biological ammonia removal in drinking water treatment. Water Res. Technol. **8**(6), 1152–1172 (2022)
10. Ma, B., LaPara, T.M., Kim, T., Hozalski, R.M.: Multi-scale investigation of ammonia-oxidizing microorganisms in biofilters used for drinking water treatment. Environ. Sci. Technol. **57**(9), 3833–3842 (2023)
11. Han, M., Zhao, Z., Gao, W., Tian, Y., Cui, F.: Effective combination of permanganate composite chemicals (PPC) and biological aerated filter (BAF) to pre-treat polluted drinking water source. Desalin. Water Treat. **57**(58), 28240–28249 (2000)

Research on Waste Treatment Method for Cutting Fluid and Its Application

Shihua Yuan[1], Bing Dai[1], Qiang Xu[1], Yonggao Li[2], and Penghui Wang[2(✉)]

[1] AVIC Chengdu Aircraft Industry (Group) Co., Ltd., Chengdu, Sichuan, China
[2] Jihua Laboratory, Foshan, Guangdong, China
wangph@jihualab.ac.cn

Abstract. A large amount of cutting fluid waste was generated during the mechanical processing of metal and non-metal material parts, which increased the cost of cutting fluid waste treatment and part manufacturing. This work studied the cutting fluid waste treatment method mainly based on low-temperature evaporation of cutting fluid waste and filtration method of distilled water. A cutting fluid waste treatment system with a water production rate of 200 L/h and a reuse rate of 90% was constructed. After the cutting fluid waste was treated by the system, the distilled water obtained satisfied the requirements for direct emission and reusing, significantly reducing the cost of cutting fluid waste treatment for manufacturing enterprises. In detail, the conductivity, turbidity, COD value, total ammonia nitrogen, total nitrogen content, total phosphorus content, and microbial content of distilled water significantly decreased, with a decline rate of 99.5%, 100.0%, 100.0%, 99.8%, 98.8%, 99.9%, and 99.9%, respectively. The pH value of distilled water decreased obviously, which was very close to the pH value of pure water at room temperature and normal pressure. This work provides a strong theoretical and technical support for the construction of green manufacturing capability.

Keywords: Low-temperature evaporation · Cutting fluid · Waste treatment

1 Introduction

Currently, the majority of metal and non-metallic parts are manufactured by mechanical processing, which requires a large amount of cutting fluid to cool, lubricate, clean, and protect cutting tools and parts [1, 2]. During the service life of cutting fluid, a large amount of solid impurity, oil impurity, microbial impurity and ionic impurity are inevitably mixed in, directly reducing the performance of the cutting fluid or indirectly accelerating the degradation of its performance [3–5]. This phenomenon makes the cutting fluid no longer meeting the requirements of mechanical processing of parts, which generates a large amount of cutting fluid waste. Cutting fluid waste has complex ingredient, high organic content, high turbidity, and pungent odor. It poses certain safety hazards in the ecological environment and human health, and is one of the most difficult industrial waste liquids to treat. In addition, most manufacturing companies entrust environmentally friendly enterprises with hazardous waste liquid treatment qualification

V. Achal et al. (Eds.): EPPC 2024, ESE, pp. 201–209, 2025.
https://doi.org/10.1007/978-3-031-92123-0_21

to handle cutting fluid waste, which poses problems such as high treatment costs and strict treatment standards. Taking commonly cutting fluids as an example, the price of fresh concentrated cutting fluid is about 20–60 RMB/kg, the price of industrial water is about 3–6 RMB/ton. Then the economic cost of preparing fresh diluted cutting fluid is about 4000 RMB/ton, and the economic cost of treating cutting fluid waste is about 3000–5000 RMB/ton. In a word, the economic cost of fresh diluted cutting fluid and cutting fluid waste are very close, which is ridiculous. With the popularization of green manufacturing concepts and the implementation of environmental protection policies, the emission allowance of cutting fluid waste is strictly limited [6, 7]. Thus, it difficult to meet the requirements of mechanical processing of metal and non-metallic parts through continuous replacement of fresh cutting fluid, resulting in a certain technical contradiction between the emission of cutting fluid waste and the functional performance of cutting fluid. Unfortunately, it is still difficult to treat cutting fluid waste through physical, chemical and biological methods. Most machining enterprises lack relevant equipment, personnel, and qualification to treat cutting fluid waste by themselves.

Based on above analysis, researching an efficient treatment technology for cutting fluid waste suitable for manufacturing enterprises, achieving near zero emission of cutting fluid waste, will significantly reduce the economic costs and environmental pressure of manufacturing enterprises, and provide a strong theoretical and technical support for the construction of green manufacturing capability.

2 Methodology

2.1 Waste Concentrated Treatment by Low-Temperature Evaporation

2.1.1 Principle of Low-Temperature Evaporation

In the cutting fluid waste, the mass proportion of organic component and suspended solid impurity was about 6–10%, and the mass proportion of water was approximately 90%. The concentration of cutting fluid waste was achieved by recycling water in cutting fluid waste through a certain process, which could reduce the amount of cutting fluid waste by nearly 90%. Cutting fluid waste had a high boiling point, heat capacity, and evaporation enthalpy. Conventional distillation methods required a large amount of thermal energy to complete the evaporation of water, resulting in very high treatment costs for cutting fluid waste. Therefore, it was necessary to study a technical method for evaporating cutting fluid waste at room temperature/near room temperature, in order to achieve efficient and low-cost concentration and water reuse of cutting fluid waste.

As shown in Fig. 1, the boiling point of water decreased with decreasing ambient pressure. When the ambient pressure was 100 kPa, the boiling point of water was about 100 °C. When the e ambient pressure was 10 kPa, the boiling point of water was about 46 °C. When the ambient pressure was 1 kPa, the boiling point of water was about 7 °C. Considering the working environment of the cutting fluid waste treatment process, this work recovered the water from the cutting fluid waste at an ambient pressure of 8–9 kPa and an evaporation temperature of 35–40 °C. Under the ambient pressure and evaporation temperature, the volatile amount of organic compounds and other impurities in the cutting fluid waste was relatively low. By this way, it was easy to achieve efficient separation of water from main ingredients of cutting fluid and other impurities, maximizing the concentration ratio of the cutting fluid waste and the purity of distilled water.

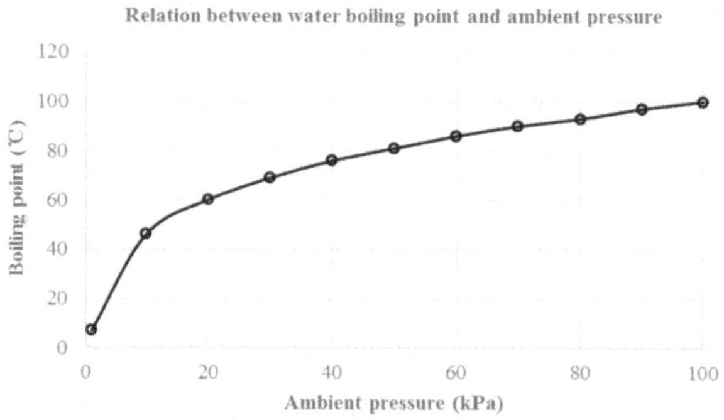

Fig. 1. Relation between water boiling point and ambient pressure

2.1.2 Low-Temperature Evaporation Equipment of Cutting Fluid

The low-temperature evaporation equipment made the rapid evaporation of cutting fluid waste at lower temperatures (30–40 °C), and its core components included an evaporation tank, heater, vacuum pump, etc. The evaporation tank was a well sealed chamber used to store the cutting fluid waste that was being evaporated. The heater was used to provide the energy required for the evaporation of cutting fluid waste, allowing the cutting fluid waste to be in a boiling state. The vacuum pump was used to maintain the vacuum degree inside the evaporation tank, so that the ambient pressure of the cutting fluid waste was between 8–9 kPa. Actually, most electrical energy was used to maintain the vacuum of evaporator tank. As shown in Fig. 2, the inlet of cutting fluid waste and the outlet of concentrated waste in the evaporation tank kept closed for a long time to prevent external air from entering the evaporation tank. When it was necessary to supplement the cutting fluid waste, the cutting fluid waste inlet of the evaporation tank was opened. When it was necessary to remove the concentrated waste from the evaporation tank, the outlet of the concentrated waste in the evaporation tank was opened. The heater and vacuum pump kept on for a long time, allowing the water in the cutting fluid waste to continuously evaporate and then condense to form distilled water. During the entire low-temperature evaporation, the cutting fluid waste did not undergo violent chemical and physical reactions. Thus, low-temperature evaporation was friendly to surrounding environment and safe to working personnel. In this work, the production rate of distilled water and deionized water was about 200 L/h, the power consumption was about 30 kW, the ambient pressure was between 8–9 kPa, the evaporation temperature was between 35–40 °C.

2.2 High-Precision Filtration Equipment of Distilled Water

After being processed by low-temperature evaporation equipment, nearly 90% water in cutting fluid was separated by evaporation and recovered by condensation. However,

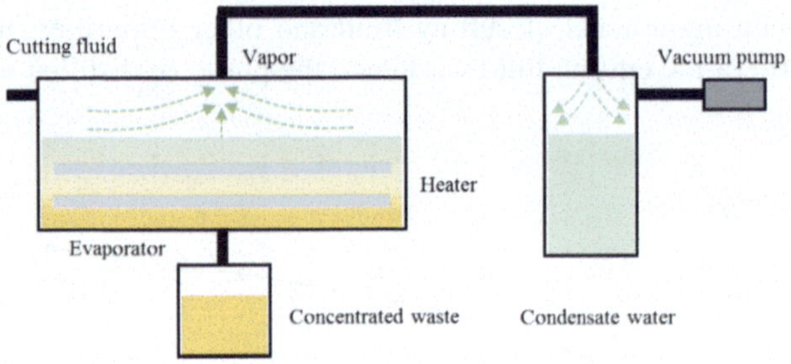

Fig. 2. Low-temperature evaporation equipment

distilled water often contained small amounts of organic impurities, ionic impurities, etc., and could not be directly discharged or reused. It need to be further treated through certain filtration processes. As shown in Fig. 3, distilled water was firstly filtered through a filter with 100 nm precision to remove impurities with high content and large particle size, reducing their impacts on subsequent filtration processes. The filtered water was filtered again through a reverse osmosis membrane with 1 nm precision to remove soluble, small particle impurities and some ionic impurities, resulting in a deionized water. The mechanism of nanoscale filtration and reverse osmosis membrane filtration were consistent. The filter element was a hollow tubular porous material, where unfiltered liquid flows in from one end of the filter element, filtered liquid flows out from the side wall of filter element, and liquid containing concentrated impurities flowed back into the unfiltered liquid from the other end of the filter element. During the process of nanoscale filtration and reverse osmosis membrane filtration, liquid waste containing concentrated impurities flowed into the unfiltered liquid multiple times, achieving high concentration of impurities. The liquid waste with multiple concentrations of impurities was subjected to low-temperature evaporation treatment again to minimize the production of waste liquid. In this work, the filtration speed of nanoscale filtration and reverse osmosis membrane filtration was about 250 L/h, to satisfy distilled water production rate of low-temperature evaporation equipment.

3 Results

3.1 Results of Low-Temperature Evaporation

As shown in Fig. 4, the untreated cutting fluid waste had a high impurity content and a turbidity of 9844 NTU. After low-temperature evaporation, the recovered distilled water was clear and transparent, with a turbidity of 0 NTU. From the appearance, micron/submicron level solid impurity and oil impurity in the cutting fluid waste had been removed mostly. As shown in Table 1, after low-temperature evaporation, the conductivity, turbidity, COD value, total ammonia nitrogen, total nitrogen content, total phosphorus content, and microbial content of distilled water significantly decreased, with a decline rate of 97.7%, 100.0%, 99.5%, 99.3%, 98.6%, 99.4%, and 90%, respectively. The pH value of distilled water has decreased obviously, approaching the pH value of pure water at room temperature and normal pressure. From a microscopic perspective, the impurity in the cutting fluid waste had been removed mostly.

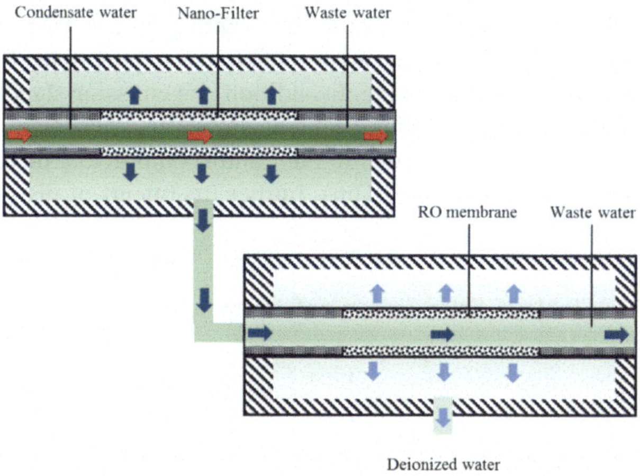

Fig. 3. High-precision filtration equipment of distilled water

Fig. 4. Result of low-temperature evaporation (Left: Before, Right: After)

Table 1. Parameter of cutting fluid before and after low-temperature evaporation

Parameter	Before	After	Decline rate (%)
pH value	8.74	7.94	9.2
Conductivity (μS/cm)	3658	86	97.7
Turbidity (NTU)	9844	1	100.0
COD value (mg/L)	128,517	626	99.5
Total ammonia nitrogen (mg/L)	1624	12	99.3
Total nitrogen content (mg/L)	2106	29	98.6
Total phosphorus content (mg/L)	17	0.1	99.4
Microbial content (CFU/mL)	10^5	10^4	90

The emission standards of cutting fluid waste strictly limited its multiple parameters such as pH value, turbidity, COD value, total ammonia nitrogen, total phosphorus content, etc. The specific requirements were show in Table 2. After low-temperature evaporation, the COD value of distilled water did not satisfy the requirements for the first, second, and third level emission standards of cutting fluid waste. The total amount of ammonia nitrogen in distilled water did not satisfy the requirements for the first level emissions standard of cutting fluid waste. In addition, the emission standards for cutting fluid waste did not strictly limit microbial content. However, when the microbial content in distilled water was high, it was prone to odor and should not be directly discharged or used to prepare fresh cutting fluid. Thus, further treatments of distilled water were essential.

Table 2. Emission standards of cutting fluid waste

Parameter	First	Second	Third
pH value	6–9	6–9	6–9
Turbidity (NTU)	70	100	100
COD value (mg/L)	100	130	500
Total ammonia nitrogen (mg/L)	10	20	–
Total phosphorus content (mg/L)	0.5	1	–

3.2 Results of High-Precision Filtration

The distilled water recovered by low-temperature evaporation equipment had some impurities. It need to be deeply filtrated and purified by nanoscale filtration and reverse osmosis membrane filtration to remove residual cutting fluid impurities. After these treatments, the distilled water could satisfy cutting fluid waste emission standards and reuse requirements. In detail, the precision of nanoscale filtration was 100 nm, and the precision of reverse osmosis membrane filtration was 1 nm. As shown in Table 3, after nanofiltration and reverse osmosis membrane filtration, the conductivity, turbidity, COD value, total ammonia nitrogen, total nitrogen content, total phosphorus content and microbial content of distilled water significantly decreased, with a decline rate of 99.5%, 100.0%, 100.0%, 99.8%, 98.8%, 99.9%, and 99.9%, respectively. The pH of distilled water had decreased and was closer to the pH value of pure water at room temperature and normal pressure. From a microscopic perspective, the cutting fluid ingredients and other impurities in distilled water had been deeply purified and removed, meeting the requirements for cutting fluid waste emission standard. As shown in Table 4, after nanofiltration and reverse osmosis membrane filtration, the pH value, conductivity, turbidity, microbial content of distilled water were very consistent with those of normal industrial water and deionized water used in cutting fluid. The total ammonia nitrogen content, total nitrogen content, and total phosphorus content of distilled water were very close to those of normal industrial water and water used in cutting fluid. The total ammonia nitrogen content of distilled water accounted for 0.2% of the total ammonia nitrogen content of fresh cutting fluid, and the total nitrogen content of distilled water accounted for 0.2% of the total nitrogen content of fresh cutting fluid. Meanwhile, the total phosphorus content of distilled water accounted for 0.3% of the total phosphorus content of fresh cutting fluid; The COD value of distilled water was slightly higher than that of conventional industrial water and deionized water used in cutting fluid, and the COD value of distilled water accounted for 0.07% of the COD value of fresh cutting fluid. The total amount of ammonia nitrogen, total nitrogen content, total phosphorus content, and COD value reflected the main ingredients of cutting fluid, which were far less than the normal range of relevant ingredient content in fresh cutting fluid. The impact of residual COD value on the physical and chemical properties and usability of the cutting fluid could be ignored. From a microscopic perspective, after nanofiltration and reverse osmosis membrane filtration, the physical and chemical parameters of distilled water were similar to those of deionized water used in cutting fluid. These results indicated that the distilled water could be used to prepare fresh cutting fluid and achieve direct water recovery and reuse in cutting fluid waste.

Table 3. Distilled water before and after high-precision filtration

Parameter	After evaporation	After reverse osmosis membrane filtration	Decline rate (%)
pH value	7.69	7.52	14.0
Conductivity (μS/cm)	67	19	99.5
Turbidity (NTU)	2	0	100.0
COD value (mg/L)	447	76	100.0
Total ammonia nitrogen (mg/L)	12	3	99.8
Total nitrogen content (mg/L)	16	4	98.8
Total phosphorus content (mg/L)	0.1	0.02	99.9
Microbial content (CFU/mL)	$\sim 10^4$	$<10^2$	99.9

4 Discussion

Based on the sketch map of low-temperature evaporation equipment and filtration purification equipment (Figs. 1 and 2), a cutting fluid waste treatment system was constructed and used, for actual production line cutting fluid waste treatment and water recovery. In detail, the production of distilled water and deionized water were about 200 L/h, the power consumption was about 30 kW, and the water recovery rate of cutting fluid waste was about 100%. By using this cutting fluid waste treatment system, only 10% cutting fluid waste need to be treated by complex waste treatments, which concluded main ingredient of cutting fluid, solid impurity, oil impurity and microbial impurity.

Assuming that the normal treatment cost of cutting fluid waste was 4000 RMB/ton and the electric cost was 1 RMB/kWh, the economic benefit generated by the equipment

Table 4. Reuse performance of distilled water

Parameter	After filtration	Normal industrial water	Deionized water	Fresh cutting fluid
pH value	7.52	7.54	7.51	8.2
Conductivity (μS/cm)	19	636	15	3436
Turbidity (NTU)	0	0	0	12
COD value (mg/L)	76	37	11	110,572
Total ammonia nitrogen (mg/L)	3	1	0	1583
Total nitrogen content (mg/L)	4	2	0	1757
Total phosphorus content (mg/L)	0.02	0	0	8
Microbial content (CFU/mL)	$<10^2$	$<10^2$	$<10^2$	$<10^2$

for processing 1 ton of cutting fluid waste was approximately 3450 RMB. The economic benefit was calculated by Formula 1.

$$4000 - 30 \times 5 \times 1 - 4000 \times 10\% = 3450 \text{ RMB} \tag{1}$$

For large mechanical processing enterprises, the annual production of cutting fluid its cutting fluid waste exceeded 300 tons. The cutting fluid waste treatment method proposed in this work could significantly reduce the cost of cutting fluid waste treatment for manufacturing enterprises, and it had good application value. In addition, by combining the low-temperature evaporation and high-precision filtration, the waste liquid treatment technology proposed in this work had no special requirements for the type of liquid waste and could be widely applied to the treatment and reuse of other waste liquids.

5 Conclusion

This work introduced a waste treatment method for cutting fluid, which aimed to reduce cutting fluid waste by recycling water. The key points are as follows.

1. After low-temperature evaporation treatment, the conductivity, turbidity, COD value, total ammonia nitrogen, total nitrogen content, total phosphorus content and microbial content of distilled water significantly decreased, with a decline rate of 97.7%, 100.0%, 99.5%, 99.3%, 98.6%, 99.4%, and 90%, respectively; The pH value of distilled water had decreased, approaching the pH value of pure water at room temperature and normal pressure.

2. After low-temperature evaporation and high-precision filtration treatment, the conductivity, turbidity, COD value, total ammonia nitrogen, total nitrogen content, total phosphorus content and microbial content of distilled water significantly decreased, with a decline rate of 99.5%, 100.0%, 100.0%, 99.8%, 98.8%, 99.9%, and 99.9%, respectively; The pH value of distilled water had decreased, very close to the pH value of pure water at room temperature and normal pressure.
3. Based on the low-temperature evaporation equipment and high-precision filtration equipment, a cutting fluid waste treatment system with a water production rate of 200 L/h and a recovery rate of 90% was constructed, achieving high concentration of cutting fluid waste and water reuse.
4. The deionized water still contained a small amount of impurities, and relative research on purification method need to be continue.

References

1. Li, H., Zhang, Y., Li, C., et al.: Cutting fluid corrosion inhibitors from inorganic to organic: progress and applications. Korean J. Chem. Eng. **39**, 1107–1134 (2022)
2. Wu, X., Li, C., Zhou, Z., et al.: Circulating purification of cutting fluid: an overview. Int. J. Adv. Manuf. Technol. **117**, 2565–2600 (2021)
3. Sankaranarayanan, R., Krolczyk, G.M.: A comprehensive review on research developments of vegetable-oil based cutting fluids for sustainable machining challenges. J. Manuf. Process. **67**, 286–313 (2021)
4. Jurina, F., Peterka, J., Vozar, M., et al.: Design of real-time monitoring system for cutting fluids. Designs. **7** (2023)
5. Korkmaz, M.E., Gupta, M.K., Ross, N.S., Sivalingam, V.: Implementation of green cooling/lubrication strategies in metal cutting industries: a state of the art towards sustainable future and challenges. Sustain. Mater. Technol. **36**, e00641 (2023)
6. D'Angelo, V., Cappa, F., Peruffo, E.: Green manufacturing for sustainable development: the positive effects of green activities, green investments, and non-green products on economic performance. Bus. Strateg. Environ. **32**, 1900–1913 (2023)
7. Thomas, J., Patil, R.: Enabling green manufacture of polymer products via vegetable oil epoxides. Ind. Eng. Chem. Res. **62**, 1725–1735 (2023)

Chemical Characteristics and Genesis Analysis of Groundwater in Dongting Lake Plain

Dongqin Chen[1], Fulong Li[2](✉), and Denghui Lu[1]

[1] Changjiang Polytechnic, Wuhan 430205, China
[2] Hainan Vocational University, Haikou 570216, China
18689672896@163.com

Abstract. This study found out hydrochemical characteristics and genesis mechanism of groundwater in Dongting Lake Plain of the groundwater. The study collected 693 groups of groundwater and surface water samples collected for 6 years, and adopted several mathematical statistics methods (i.e., Piper diagram, Gibbs diagram, chlor-alkali indices and ion ratio analysis) to analyze the hydrochemical evolution of groundwater based on the chemical data. The shallow groundwater and deep groundwater in the study area were mainly SO_4 type, HCO_3^-·SO_4 type and HCO_3 type water. The groundwater composition is mainly determined by water-rock interaction. Gypsum and mirabilite are the main sources of sulfate in groundwater, and dolomite and calcite are the main sources of carbonate. Ca^{2+} and Mg^{2+} mainly come from dissolution of dolomite, calcite and gypsum. Na^+ is not only derived from rock salt, but also from the dissolution of albite and mirabilite and cation exchange in water and soil.

Keywords: Dongting Lake plain · Hydrochemical characteristics of groundwater · Genesis analysis

1 Introduction

Dongting Lake Plain is located in the southern bank of Ching River of the middle reaches of the Yangtze River, straddling Hunan and Hubei Provinces and connecting Jianghan Plain in the north [1]. It is an important area of favorable economic conditions. Groundwater is an important source of water for local residents. Studying the hydrochemical characteristics and genesis of groundwater are helpful for groundwater mining in the study area. The genesis and evolution of groundwater hydrochemical characteristics are related to human activities, and affected by geological factors. As an index of groundwater properties, the groundwater composition content changes when interacting with different water-bearing media along its flow. Therefore to study the formation and evolution of groundwater hydrochemical composition, it is necessary to study the water-rock interaction between groundwater and water-bearing media.

Many scholars have proposed numerous methods for studying the genesis and evolution of groundwater chemistry, such as Piper diagram, Gibbs diagram, ion ratio coefficient, stable isotope [2], multivariate statistical analysis, chlor-alkali indices and hydrogeochemical simulation. The Piper diagram can show the hydrochemical types of a

V. Achal et al. (Eds.): EPPC 2024, ESE, pp. 210–222, 2025.
https://doi.org/10.1007/978-3-031-92123-0_22

large amount of samples by representing the relative concentrations of chemical components in groundwater [3]. The Gibbs diagram can identify the factors controlling the hydrochemical characteristics of groundwater [4], and the factors include atmospheric precipitation, water-rock interaction and evaporation. The ion ratio analysis can be used to analyze the sources of groundwater components and the stable isotopes method can trace groundwater movement [5]. Correlation analysis and cluster analysis of the multivariate statistical analysis can help explore the relations between chemical components of groundwater [6]. The chlor-alkali index method can detect ion exchange between groundwater and aqueous media [7].

2 Overview of the Study Area

The strata around and in the Dongting Lake Plain are relatively intact with tertiary and quaternary systems from Proterozoic, Paleozoic to Cenozoic. This study focuses on the quaternary strata as major outcrops. The Quaternary system is well-developed in the plain with complete layers fluvial-lacustrine clay, sand, and sand and gravel. From the Lower Pleistocene to the Holocene, each series has a considerable thickness and stability, and the thickness increases gradually from south to north. Most places also have Neogene system. The thickness of loose accumulation layer is generally more than 100–200 m, locally to 334 m.

Quaternary strata in the plain is the main occurrence of groundwater. According to difference of lithology and depositional environment. The Quaternary can be divided into four aquifers, including Holocene pore phreatic water aquifer, shallow late and middle pleistocene pore confined water aquifer, deep early pleistocene pore fracture and confined water aquifer, deep clastic rock pore fracture and confined water aquifer. The main recharge sources of groundwater are atmospheric precipitation, leakage recharge of surface water, and infiltration of irrigation water. The main ways of groundwater drainage are evaporation drainage, lateral recharge to surface water and artificial drainage. The quaternary system is widely developed in the study area. Sandy clay is mainly the water-bearing medium of pore water where groundwater depth is shallow, therefore evaporation is the main way of drainage. Different aquifers have indirect or direct hydraulic connection. In the natural state, the shallow pore phreatic water often has a higher water level than the underlying confined water, so as to replenish and discharge the latter. Dongting Lake Plain is rich in surface water systems. Songzi, Taiping and Ouchi rivers, which are tributaries the Yangtze River [8], flow in from the north. The south and west sides accommodate the Xiangjiang, Zijiang, Yuanjiang and Lishui rivers of Hunan Province. Many small rivers, such as Miluo River and Xinqiang river, flow in from the east. Consequently a unique centripetal water pattern is formed, Pore phreatic water flow field diagram is shown in Fig. 1.

Fig. 1. Pore phreatic water flow field diagram

3 Sample Collection and Analysis

This study focused on the spatial-temporal hydrochemical characteristics of groundwater in the loose-rock porous media in Dongting Lake plain. The targeted water-bearing rock groups were shallow Holocene pore phreatic water aquifer with a depth of 0–20 m and deep late and middle Pleistocene pore and fractured confined water aquifer with a depth of 20–120 m.

Groundwater and surface water samples were collected from the study area over a period of 6 years, totaling 693 samples, and conventional hydrochemical and isotopic testing were conducted and carried out the conventional hydrochemical and isotopic testing for these samples. The groundwater samples for conventional hydrochemical testing were mainly taken from pumping wells and large open wells. The sampling bottles were made of polyethylene plastic and were rinsed with deionized water for 3 times. The residual water in the well was drained first and the sampling bottle was fully rinsed with fresh well water for 3 times before collecting groundwater samples. Each sampling site collected one 500 mL bottle and two 50 mL bottles of water sample for regular hydrochemistry test and one 100 mL bottle for hydrogen and oxygen stable isotope test. DZB-718 portable multi-parameter water analysis meter was used to measure pH, dissolved oxygen, conductivity and water temperature of the collected samples at sampling sites. Within 24 h, the 500 mL bottles of water sample were used for basicity analysis (titration), and the two 50 mL bottles of water sample were pumped with 0.45 μm microporous membrane before testing, then one of them used for conventional cation analysis was acidified with pure nitric acid, and was sent to the State Key Laboratory for testing and analysis [9].

4 Results and Discussion

4.1 Hydrochemical Characteristics of Groundwater

The statistical characteristics can represent the value range and dispersion degree of the data. Based on all the data sampled in the study area for 6 years, the statistical values of each index are calculated as shown in Table 1.

Table 1. Mathematical statistics values of groundwater parameters

Aquifer	Shallow					Deep				
Parameter	Max	Min	Mean	SD	CV	Max	Min	Mean	SD	CV
TDS	1192.6	10.00	592.19	259.63	0.44	1369.47	7.39	540.58	253.98	0.47
PH	8.5	5.6	7.17	0.56	0.08	9.17	5.60	7.41	0.58	0.08
K^+	25.72	0.08	3.32	3.80	1.14	26.35	0.21	2.66	3.02	1.14
Na+	347.33	7.47	91.00	89.50	0.98	301.00	3.58	78.58	76.74	0.98
Ca2+	232.00	2.85	69.19	53.80	0.78	222.46	3.32	65.53	42.92	0.65
Mg2+	75.07	1.39	23.66	16.02	0.68	75.53	0.00	21.86	13.87	0.63
HCO_3^-	1031.00	6.84	337.83	217.24	0.64	1010.78	27.36	338.93	192.97	0.57
SO_4^{2-}	705.00	1.78	161.76	186.82	1.15	636.00	0.40	118.96	154.57	1.30
CI^-	139.96	0.71	16.00	21.79	1.36	128.76	0.35	11.74	16.05	1.37
$NO_3{}^-$	14.80	0.00	0.77	1.78	2.32	28.97	0.00	1.10	3.39	3.10
$NO_2{}^-$	4.3	0.00	0.12	0.47	3.83	6.00	0.00	0.08	0.48	5.96
$NH_4{}^-$	45.00	0.00	1.26	6.11	4.83	11.00	0.00	0.41	1.49	3.61
F^-	0.80	0.00	0.13	0.14	1.10	0.80	0.00	0.19	0.16	0.86
Fe	8.2220	0.0000	0.4780	1.1176	2.3382	33.6380	0.0000	1.1928	3.6957	3.0983
Mn	8.7913	0.0006	0.7894	1.4178	1.7962	9.3511	0.0004	0.4996	1.0413	2.0842

All values are in 'mg/L' except pH 1.1176

The TDS of shallow groundwater in the study area ranged from 10 mg/L to 1192.69 mg/L, with an average value of 592.19 mg/L. The TDS of deep groundwater ranged from 7.39 mg/L to 1369.47 mg/L, averaging 540.58 mg/L. The TDS values were generally low, and the groundwater was fresh water (TDS < 1 g/L) in most zones with brackish water (1 g/L < TDS < 3 g/L) sporadically distributed and no salt water (3 g/L < TDS < 10 g/L). The TDS zoning characteristics are shown in Figs. 2 and 3. The TDS values in the east and west of study area were below 550 mg/L and lower than the average, while the TDS values of Lixian County, Jinshi City, Anxiang County, Nanxian County and Hanshou City were higher than the average value [10].

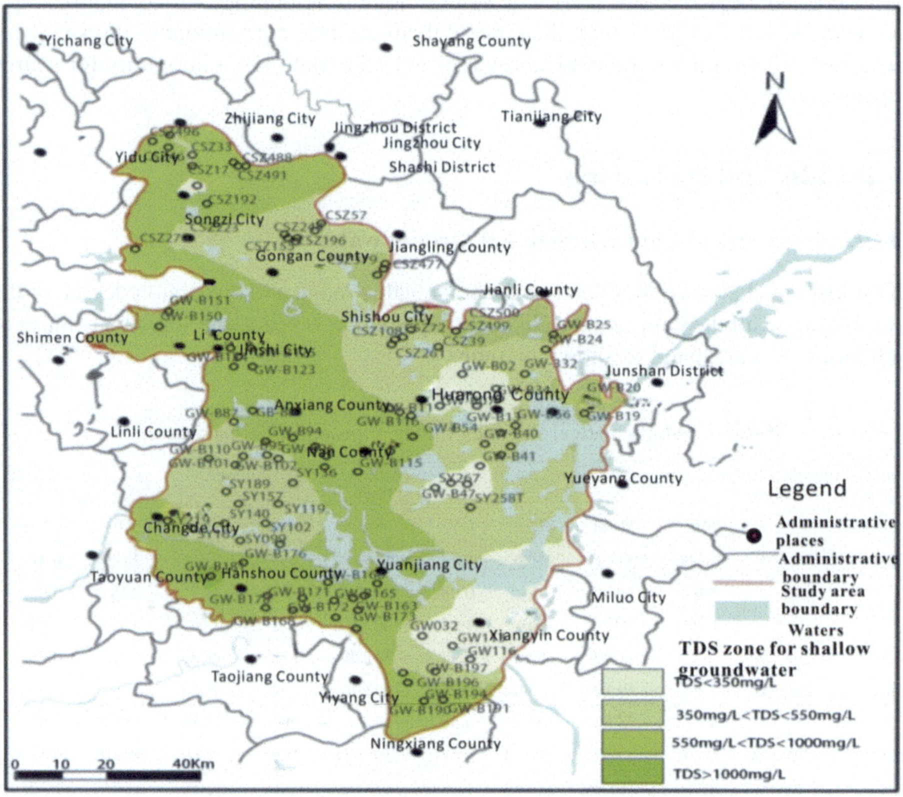

Fig. 2. The TDS zoning map for Shallow groundwater

4.2 Hydrochemical Type of Groundwater [11]

This study used Schukalev classification method to show the hydrochemical types of groundwater. The hydrochemical types of shallow and deep groundwater in the study area were relatively consistent, mainly HCO_3, $HCO_3{\cdot}SO_4$ and SO_4 type. Na^+ and Ca^{2+} were main cations in $HCO_3{\cdot}SO_4$ and SO_4 type water, and Mg^{2+}, Ca^{2+} and Na^+ in HCO_3 type water.

Based on the characteristics of groundwater flow field, this study took the hilly region at the boundary of the study area as the recharge zone, the sedimentary region of eastern Dongting Lake as the discharge zone, and the remaining regions as the runoff

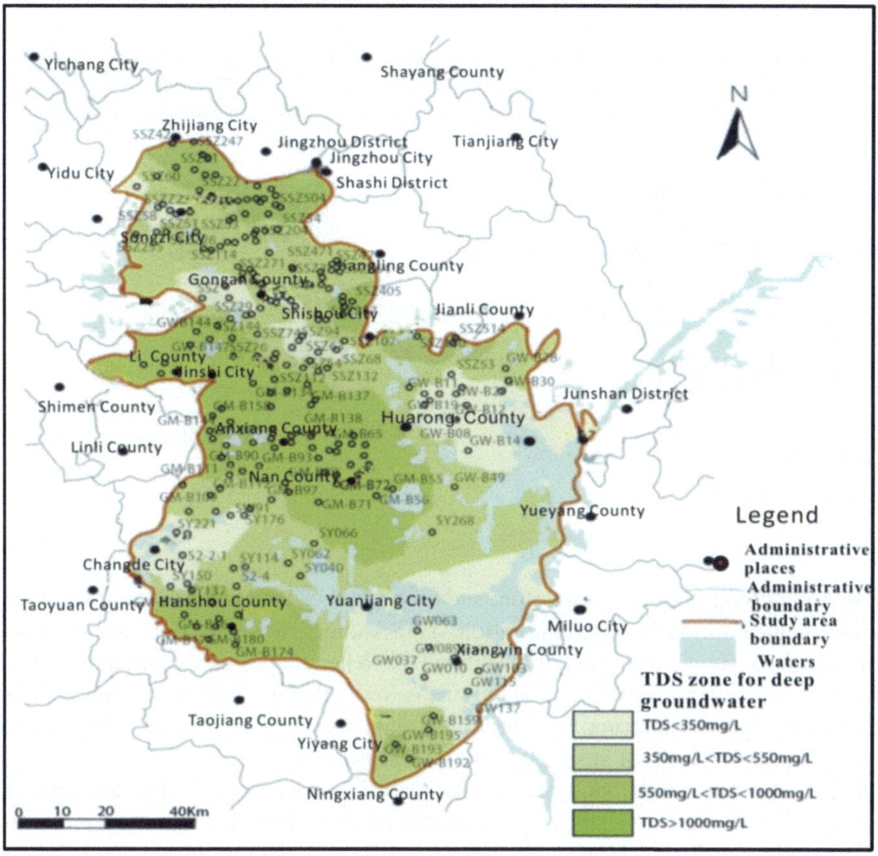

Fig. 3. The TDS zoning map for deep groundwater

zone. Groundwater chemical characteristics in different zones were shown by Piper diagram, in Figs. 4 and 5. The figure shows that the groundwater in the recharge zone was mostly $HCO_3{\cdot}SO_4$ and SO_4 type water with SO_4^{2-} and Na^+ as the main components. The study area was one of the main acid rain polluted areas in China, and the content of SO_4^{2-} and NO^{3-} in atmospheric precipitation was high. As the main recharge source of groundwater, atmospheric precipitation was rich in sulfate ions and thus becoming one of the sources of SO_4^{2-} in shallow groundwater. With the movement of groundwater in the study area, the contents of SO_4^{2-} and Na^+ in groundwater gradually decreased, and those of HCO_3^- and Mg^{2+} are growing as the major components. The sampling points in the discharge zone had mostly HCO_3-Ca type water.

4.3 Analysis of Groundwater Chemical Genesis

4.3.1 Analysis of Hydrochemical Formation Processes of Groundwater

Gibbs diagram can show the changing trend and hydrochemical formation mechanism of main ions in groundwater. The controlling factors are divided into three categories:

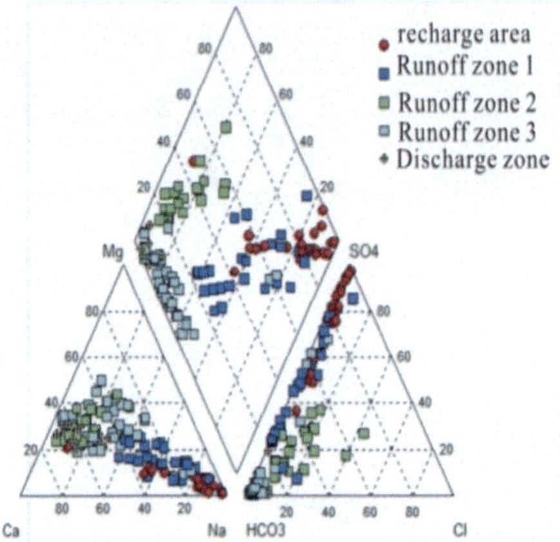

Fig. 4. The Piper diagram for shallow groundwater

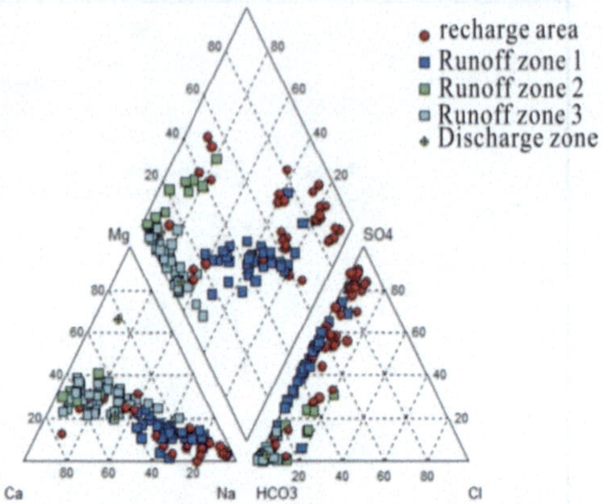

Fig. 5. The Piper diagram for deep groundwater

atmospheric precipitation, evaporation, and rock weathering. When the controlling factors are mainly atmospheric precipitation, TDS value is low, and the ratio of cations or anions mass concentration is higher than 0.5. When TDS value is medium and the mass concentration ratio of anions or cations is lower than 0.5, the formation is mainly controlled by rock weathering. Evaporation dominates when TDS value is high and the mass concentration ratio of anions or cations is close to 1.

Gibbs diagrams of the study area are shown in Figs. 6 and 7. The average TDS value was 592.19 mg/L in shallow groundwater and 540.5788 mg/L in deep groundwater. The study area had a subtropical monsoon climate with concentrated and abundant rainfall and well-conditioned groundwater runoff. There was no groundwater generated from evaporation, so the ratio of $Cl^-/(Cl^-+HCO_3^-)$ was low. The value range of $Cl^-/(Cl^-+HCO_3^-)$ was between 0.002 to 0.52 in shallow groundwater and 0.002 to 0.45 in deep groundwater, respectively averaging 0.065 and 0.053. The ratios of $Cl^-/(Cl^-+HCO_3^-)$ in most sampling points were lower than 0.5 and the value range of $Na^+/(Na^++Ca^{2+})$ in groundwater was wide. Based on groundwater types, the ratio of $Na^+/(Na^++Ca^{2+})$ in HCO_3 type water was lower than 0.5, and that of $HCO_3{\cdot}SO_4$ and SO_4 type water mostly higher than 0.5 while the latter was the highest.

Based on Figs. 6 and 7, groundwater composition in the study area was mainly affected by water-rock interaction, and less affected by evaporation and atmospheric precipitation.

4.3.2 Correlation-Based Analysis of Groundwater Hydrochemical Components

In the various interactions occurring in groundwater, ions mutually restrict each other and also exhibit certain correlations. By analyzing the correlation coefficients between ions, it is possible to determine whether different chemical components have a consistent source.

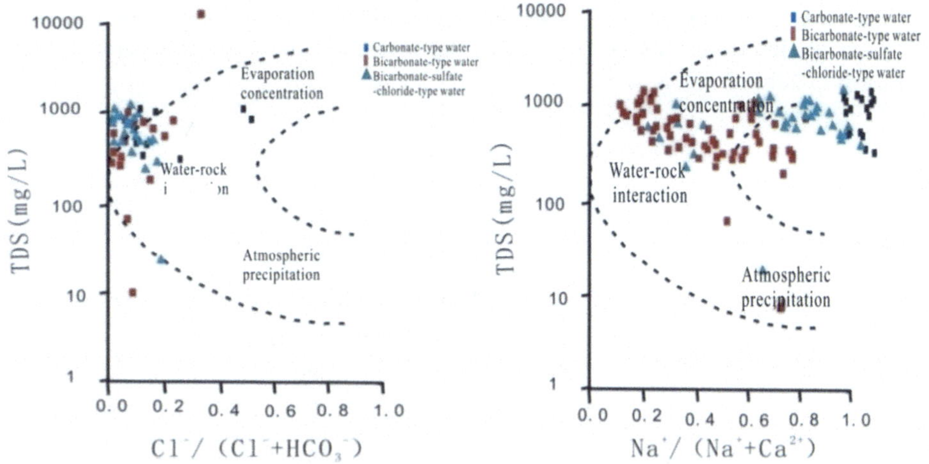

Fig. 6. The Gibbs diagram for shallow groundwater

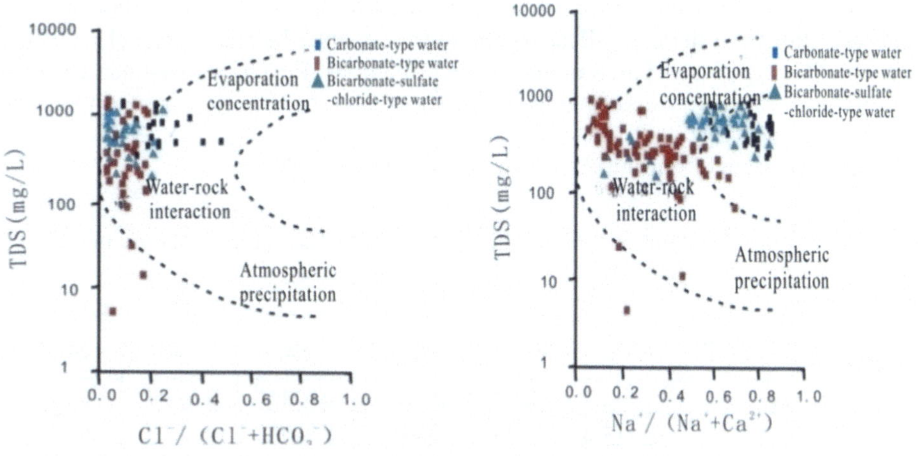

Fig. 7. The Gibbs diagram for deep groundwater

The Pearson correlation coefficient can represent the relationships between different ions. The calculation formula is as follows:

$$\gamma_{ij} = \frac{\sum_{i=1}^{n}(x_i - \overline{x})\sum_{j=1}^{n}(y_i - \overline{y})}{\sqrt{\sum_{i=1}^{n}(x_i - \overline{x})^2\sum_{j=1}^{n}(y_i - \overline{y})^2}} \quad (1)$$

Where (1), r_{ij} for the correlation coefficient between variable i and variable j; x_i for the value of the ith variable; y_j for the value of the jth variable; for the average value of the ith variable; for the average value of the jth variable; i, j for the selected variables (1, 2, 3, ... p); n for the total number of samples.

The Person correlation coefficient can represent the correlation between ions with "r" value ranging from -1 to $+1$ to measure the degree of correlation between variables, as shown in Table 2.

Table 2. Scale of correlation coefficient

r	Scale of correlation
$0.8 \leq |\gamma| \leq 1$	Very high correlation
$0.5 \leq |\gamma| < 0.8$	Moderate correlation
$0.3 \leq |\gamma| < 0.5$	Low correlation
$0 \leq |\gamma| < 0.3$	Very low correlation

SPSS software was used to analyse the correlation of mass concentration of each component in the groundwater samples. If the overall ion correlation coefficient of the research area was used, Since it was difficult to analyze the correlation between ions in different groundwater types by calculating the overall ion correlation coefficient of the study area, this study divided collected groundwater samples into HCO_3 type, $HCO_3 \cdot SO_4$ type and SO_4 type water based on the content of anions. Results showed that Na^+ was concentrated on shallow and deep SO_4 type water, and the SO_4 type groundwater was SO_4-Na type water .

The correlation of ions in SO_4-Na water is shown in Table 3.

Table 3 The relational numbers table for main ions of SO_4-Na type groundwater in the study area

deep layer / shallow layer	HCO_3^-	SO_4^{2-}	Cl^-	Na^+	Ca^{2+}	Mg^{2+}
HCO_3^-		0.629**	0.456*	0.624**	0.672**	0.310
SO_4^{2-}	-0.114		0.317	0.986**	0.719**	0.131
Cl^-	-0.074	0.176		0.267	0.531*	0.692**
Na^+	0.041	0.954	-0.042		0.640**	0.037
Ca^{2+}	0.359	0.311	0.732	0.143		0.528*
Mg^{2+}	0.309	0.209	0.601**	0.093	0.665**	

"*" and "" represent significant levels of 0.05 and 0.01 respectively**

According to Table 1, SO_4^{2-} had a high correlation with Na^+ in shallow and deep SO_4-Na type water.

The correlation of ions in HCO_3-Na·Ca·Mg water is shown in Table 4.

As can be seen from Table 3, in shallow HCO_3^- Na·Ca·Mg type groundwater, HCO_3^- had a high correlation with Ca^{2+}, Mg^{2+}, Na^+ and SO_4^{2-}, and a moderate negative correlation with Cl^-. Na^+ was highly correlated with Mg^{2+} and Ca^{2+} in deep HCO_3^- Na·Ca·Mg water.

Table 4 The correlation coefficient for main ions of HCO_3-Na·Ca·Mg type groundwater in the study area

shallow layer \ deep layer	HCO_3^-	SO_4^{2-}	Cl^-	Na^+	Ca^{2+}	Mg^{2+}
HCO_3^-		0.638**	0.187	0.950**	0.956**	0.962**
SO_4^{2-}	0.826**		0.503**	0.773**	0.599**	0.696**
Cl^-	-0.651*	-0.325		0.299	0.337	0.299
Na^+	0.951**	0.934**	-0.511		0.885**	0.909**
Ca^{2+}	0.962**	0.841**	0.654*	0.923**		0.930**
Mg^{2+}	0.940**	0.786**	0.502	0.881**	0.845**	

"*" and "**" represent significant levels of 0.05 and 0.01 respectively

The correlation of ions in HCO_3·SO_4-Na·Ca groundwater is shown in Table 5.

Table 5 The correlation coefficient for main ions of HCO_3·SO_4-Na·Ca type groundwater in the study area

shallow layer \ deep layer	HCO_3^-	SO_4^{2-}	Cl^-	Na^+	Ca^{2+}	Mg^{2+}
HCO_3^-		-0.321	-0.275	0.056	0.653**	0.758**
SO_4^{2-}	-0.142		0.513**	0.833**	0.363	-0.028
Cl^-	-0.308	0.817**		0.367	0.054	0.007
Na^+	0.037	0.865**	0.804**		0.472*	0.041
Ca^{2+}	0.521	0.685*	0. 465	0.568		0.639**
Mg^{2+}	0.825**	0.334	0.009	0.363	0.755**	

"*" and "**" represent significant levels of 0.05 and 0.01 respectively

"*" and "**" represent significant levels of 0.05 and 0.01 respectively

It can be seen from Table 5 that in shallow HCO_3·SO_4-Na·Ca type groundwater, correlations between HCO_3^- and Mg^{2+}. SO_4^{2-} and Na^+, SO_4^{2-} and Cl^-, and Cl^- and Na^+ were high in shallow HCO_3·SO_4-Na·Ca type water. The correlation between SO_4^{2-} and Na^+ was high in the deep layer.

4.3.3 Analysis of Component Sources Based on Ion Ratio Coefficients

The sources of various ions in groundwater are diverse [12]. The ion ratio coefficient method is used to determine the origin of groundwater and the sources or formation processes of its chemical components by calculating the milliequivalent ratio coefficients of different ions. The $\gamma Na^+/\gamma Cl^-$ coefficient is a parameter that characterizes the degree of Na^+ enrichment in groundwater and serves as an indicator of the intensity of salinity leaching and accumulation in groundwater. The average γ $Na^+/\gamma Cl^-$ coefficient for standard seawater is 0.85. Low salinity water has a higher $\gamma Na^+/\gamma Cl^-$

coefficient, $\gamma Na^+/\gamma Cl^-$ >0.85, while high salinity water has a lower $\gamma Na^+/\gamma Cl^-$ coefficient, $\gamma Na^+/\gamma Cl^-$ <0.85.In natural groundwater, Cl^- often originates from rock salt. The $\gamma Na^+/\gamma Cl^-$ values of groundwater in the study area are much greater than 1, indicating that, in addition to rock salt, there are other significant sources of Na^+. often originates from rock salt. The $\gamma Na^+/\gamma Cl^-$ values of groundwater in the study area are much greater than 1, indicating that, in addition to rock salt, there are other significant sources of Na^+.

Ca^{2+}, Mg^{2+}, HCO_3^- and SO_4^{2-} in groundwater often originate from carbonate and sulfate minerals such as calcite, dolomite, and gypsum present in the aquifer medium. The ratio of $\gamma(Ca^{2+} + Mg^{2+})/\gamma(HCO_3^- + SO_4^{2-})$ can reflect the sources of these ions in groundwater. The ratio for bicarbonate-type water approaches the 1:1 equivalence line, indicating that Ca^{2+}, Mg^{2+}, HCO_3^- and SO_4^{2-} in bicarbonate-type water mainly originate from the dissolution of calcite, dolomite, and gypsum.Most of the water sample points for sulfate-type water and bicarbonate-sulfate-type water are below the 1:1 equivalence line, with $\gamma(HCO_3^- + SO_4^{2-}) > \gamma(Ca^{2+} + Mg^{2+})$. This suggests that HCO_3^- and SO_4^{2-} in sulfate-type water and bicarbonate-sulfate-type water may originate from other carbonate and sulfate minerals such as mirabilite.

The $\gamma Ca^{2+}/\gamma SO_4^{2-}$ ratio can reflect the ion sources in groundwater. If Ca^{2+} and SO_4^{2-} mainly originate from the dissolution of gypsum ($CaSO_4$, $CaSO_4{\cdot}2H_2O$), their ratio would be close to 1, gypsum is not the sole source of Ca^{2+} and SO_4^{2-} in the groundwater.

The ratio of $\gamma(Na^+\text{-}Cl^-)/\gamma(HCO_3^- + SO_4^{2-}\text{-}Ca^{2+}\text{-}Mg^{2+})$ can reflect whether the composition of groundwater includes the dissolution of mirabilite (Na SO_4). The term $\gamma(Na^+\text{-}Cl^-)$ can indicate the sources of Na^+ excluding halite, where $\gamma(Na^+\text{-}Cl^-)$ >0 signifies an increase in Na^+ relative to halite dissolution. Similarly, $\gamma(HCO_3^- + SO_4^{2-}\text{-}Ca^{2+}\text{-}Mg^{2+})$ can represent the sources of SO_4^{2-} excluding gypsum. When the ratio of $\gamma(Na^+\text{-}Cl^-)$ to $\gamma(HCO_3^- + SO_4^{2-}\text{-}Ca^{2+}\text{-}Mg^{2+})$ approaches 1:1, it indicates the presence of mirabilite dissolution.

From the correlation graph of $(Na^+\text{-}Cl^-)$ and $\gamma(HCO_3^- + SO_4^{2-}\text{-}Ca^{2+}\text{-}Mg^{2+})$, it can be seen that a large number of samples in the study area have a ratio of $\gamma(Na^+\text{-}Cl^-)$ to $\gamma(HCO_3^- + SO_4^{2-}\text{-}Ca^{2+}\text{-}Mg^{2+})$ close to 1:1. This indicates the widespread presence of mirabilite dissolution in the study area, which is consistent with the mineral detection results in the strata.

5 Conclusions

This study further investigates the distribution characteristics, hydrochemical characteristics, and formation mechanisms of groundwater in the Dongting Plain through various numerical statistical methods such as Piper diagrams, Gibbs diagrams, and ion ratio coefficients [13].

1. The Piper diagram clarifies that the groundwater types of shallow phreatic water and shallow confined water in the Dongting Plain are mainly bicarbonate type water, bicarbonate-sulfate type water, and sulfate type water. Spatially, from the hilly areas at the basin edge to the plain areas in the central region, there is a transition from SO_4 type and $HCO_3{\cdot}SO_4$ type water to HCO_3 type water.

2. The Gibbs diagram indicates that the study area is in the rock weathering zone, far from the evaporation-concentration zone and the atmospheric precipitation zone. This suggests that the hydrochemical ions in this area are minimally affected by evaporation-concentration and atmospheric precipitation, with water-rock interaction being the controlling factor for hydrochemical ions. By calculating the chlor-alkali index, it was analyzed that there is widespread cation exchange adsorption in the study area.
3. By calculating the Pearson correlation coefficients, the relationships between various components in different types of groundwater in the study area were determined. In groundwater where Na^+ is the main cation, Na^+ and SO_4^{2-} exhibit a strong positive correlation. In groundwater where HCO^{3-} is the main anion, HCO_3^- shows a strong positive correlation with Mg^{2+} and Ca^{2+}. In $HCO_3^- \cdot SO_4^{2-}$ type water, SO_4^{2-} has a strong positive correlation with Na^+, Mg^{2+}, Ca^{2+}.
4. Based on the lithology of the strata and the ion ratio coefficient method, it was determined that the sources of HCO_3^- and SO_4^{2-} in SO_4type water and $HCO_3 \cdot SO_4$ type water are other carbonates and sulfates such as mirabilite. The Ca^{2+} and Mg^{2+} in the groundwater mainly originate from the dissolution of carbonates and sulfates. Na^+ is derived not only from halite but also from the dissolution of mirabilite and cation exchange in the soil. Additionally, based on the analysis of rainfall components, atmospheric precipitation is one of the sources of sulfate in shallow groundwater.

Acknowledgments. 2023 Hubei Provincial Construction Science and Technology Plan project and Building Energy Conservation Demonstration Project [2023] 1656 59.

References

1. Yi, B., Liu, J., Lu, X., He, W., et al.: Hydro-chemical and isotopic evidence for groundwater conversion of surface water in alpine arid areas. A case study of the Datong River Basin. **44**, 752–760 (2023)
2. Hu, Y., Zhou, R., Song, W., et al.: Hydrochemical evolution and genesis analysis of shallow groundwater east of Yongding River in Fengtai District, Beijing. Environ. Sci. **45**, 2651–2664 (2024)
3. Said, I., Salman, S.A., et al.: Salinization of groundwater during 20 years of agricultural irrigation, Luxor. Egypt. Environ. Geochem. Health. **11**, 3821–3835 (2021)
4. Yuan, M.-H., Chiueh, P.-T., Lo, S.-L.: Measuring urban food-energy-water nexus sustainability: finding solutions for cities. Sci. Total Environ. **752**, 141954 (2021)
5. Kaur, L., Rishi, M.S., et al.: Hydrogeochemical characterization of groundwater in alluvial plains of River Yamuna in Northern India: an insight of controlling processes. J. King Saud Univ. Sci. **31**(4), 1245–1253 (2019)
6. Andreoli, R., Van Haaren, B.: Ecosystemic services assessment: application to forests for the preservation of water resources in Tropical Islands. QGIS Appl. Terri. Plan. **3**, 169–237 (2018)
7. Wei, R., Tang, S., Ouyang, Q., et al.: Study on the groundwater system of the Dongting Lake Plain, central-south China: a tectonic perspective. **30**, 707–721 (2022)
8. Huang, W.W., Jiang, C.L., Chen, X., et al.: Chemical characteristics and genesis of deep groundwater in the Xinji mining area. Earth Environ. **48**, 432–442 (2020)

9. Gao, Z., Wan, Z., He, K., et al.: Hydro-chemical characteristics and controlling factors of karst groundwater in middle and upper reaches of Dawen River basin. **41**, 264–272 (2022)
10. Chen, J., Jiang, S., Yang, X., et al.: Study on hydro-chemical characteristics and controlling factors of the upper reach of Degu River basin. **68**, 1853–1862 (2022)
11. Wang, G., Zhang, Y., Gou, Q., et al.: Hydro-chemical characteristics of surface water and groundwater in oasis edge in the middle reaches of the Heihe River basin. **42**, 1818–1828 (2022)
12. Avgoustaki, D.D., Xydis, G.: Plant factories in the water-food-energy nexus era: a systematic bibliographical review. Food Secur. **12**, 253–268 (2020)
13. Hu, Z., Yong, P., Richen, X., et al.: Study on the proportion and flow path of Yangtze River water diversion into Taihu Lake. Water Supple. **22**(2), 1820–1834 (2022)

A New Model of Spray Deposition Distribution of Canopy in the Greenhouse under Air-Assisted Application

Chunhua Gao, Zhen Li, Yueyang Li(✉), and Xiaofei Cheng

College of Mechanical and Electronic Engineering, Shandong Agriculture and Engineering University, Jinan, China
37969913@qq.com

Abstract. This research addresses the complex interactions between plant canopies and air velocity in enclosed greenhouse environments during wind-driven pesticide application. Traditional methods struggle to analyze the patterns of aerosol deposition within the canopy. This thesis develops a model for the distribution of aerosol deposition across crop canopies based on the SIMPLE algorithm and Lagrange discrete phase models. It simulates the trajectory changes of droplets under five airflow angles: 5 m/s, 10 m/s, 15 m/s, 20 m/s, 25 m/s and five spraying speeds: 0°, 4°, 8°, 12°, 16°. Subsequently, the paper constructs a regression model linking droplet deposition rates to the airflow velocity at the intake. Analysis of droplet trajectories reveals that droplet size significantly influences the distribution of aerosol deposition; smaller droplets travel further and gradually decrease in number due to evaporation. Optimal airflow velocities yield higher deposition rates per unit area. The initial speed of airflow at the intake directly impacts deposition rates: there is a positive correlation between airflow velocity at the intake and droplet deposition rates; increasing the initial airflow speed enhances the deposition rate.

Keywords: Spray · Deposition distribution · Air-assisted

1 Introduction

Pests and diseases are among the key factors limiting vegetable yield and quality, particularly in the warm and humid conditions of greenhouses [1]. To enhance the uniformity of droplet deposition distribution and address the challenge of reaching inner canopy areas, wind-driven spraying technology is employed. Llop et al. [2] conducted experiments to study the droplet deposition distribution of four different sprayers on traditional greenhouse tomatoes. Duga et al. [3] utilized Computational Fluid Dynamics (CFD) to simulate the airflow fields and droplet distributions of three designed wind-driven sprayers. Qiu et al. [4] designed a wind curtain spraying device suitable for greenhouse environments to reduce the amount of pesticides used and address uneven spraying issues. Delele et al. [5] developed a CFD model to predict the three-dimensional air velocity distribution of sprayers. Endalew et al. [6] simulated and validated the process

V. Achal et al. (Eds.): EPPC 2024, ESE, pp. 223–230, 2025.
https://doi.org/10.1007/978-3-031-92123-0_23

of sprayer airflow through a pear tree canopy using a CFD model. However, further research is required into the trajectories of droplets within the canopy and the patterns of their deposition. This thesis employs CFD technology and the SIMPLE algorithm to explore the patterns of droplet deposition and movement within the crop canopy under the microclimatic conditions of a greenhouse.

2 Establishment of the Physical Model

A geometric model of a simulated greenhouse area is established in the ICEM environment, Suppose the greenhouse is sealed, the temperature of the greenhouse is 25 °C, and the humidity is 60%. As illustrated in Fig. 1, the simulation area is a rectangular space measuring 6 m × 2 m × 3 m. The selected plant (Flowering tomatoes) is positioned 1.5 m from the sprayer's intake, with a height of 1.2 m. The physical size of the sprayer is negligible relative to the entire simulation area and does not affect the distribution of airflow velocity or droplet deposition. The sprayer nozzle is simplified to a circular intake.

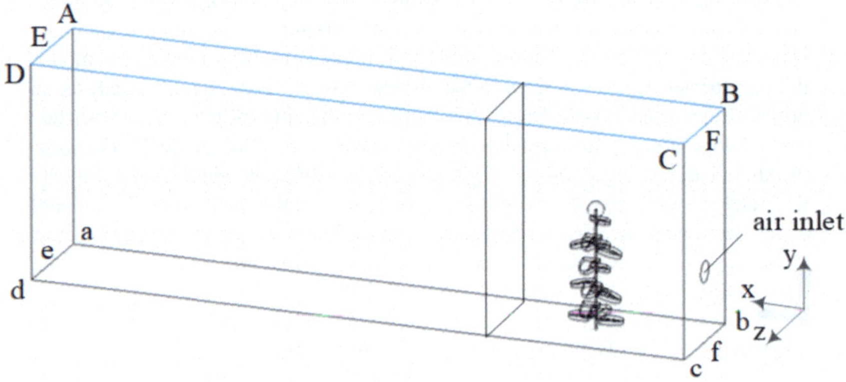

Fig. 1. Schematic diagram of the geometric model of the simulation area

3 Computational Fluid Dynamics

It is a problem of gas-liquid two-phase flow that the droplets are blown to crops and deposited under the airflow stress during the spraying process of greenhouse sprayer, the volume ratio of droplets in the flow field is less than 10%, satisfy the basic assumptions of Lagrange discrete phase model. The air flow was set as a continuous phase, the droplets as a discrete phase, the boundary condition of the discrete phase was the wall, the flow type inlet was "velocity-inlet," and the temperature during the simulation was 25 °C. The simulation calculation employs a standard $k - \varepsilon$ turbulence model. The movement of droplets is simulated using the Lagrangian discrete phase model to approximate the actual motion of droplets. The SIMPLE algorithm iteratively gradually approximates the real pressure and velocity fields, and solves the pressure-velocity coupling problem in Navier-Stokes equation. The SIMPLE algorithm is used to calculate the velocity distribution in the velocity field in airfield of spraying. The number of iterations of the calculation time is set to 300. The discrete phase material is water with a density of 998.2 kg/m^3, a viscosity of 0.001 Pa s, a surface tension of 0.072 N/m. It is calculated that the droplets at the inlet and the air flow enter the air flow field at the same speed. The model is:

$$\frac{\partial \alpha_q}{\partial t} + \overline{u} \cdot p \nabla \alpha_q = \frac{S \alpha_q}{\rho_q} \tag{1}$$

$$\rho = \alpha_q \cdot p \rho_q + \left(1 - \alpha_q\right) \rho_q \tag{2}$$

$$\frac{\partial}{\partial t}(\rho \cdot p\overline{u}) + \nabla(\rho \cdot p\overline{u} \cdot p\overline{u}) = -\nabla p + \nabla\left[\mu\left(\nabla\overline{u} + \nabla\overline{u}^T\right)\right] + \rho \cdot p\overline{g} + \overline{F} \tag{3}$$

Among them: α_q is the volume ratio of the qth phase flow; S_q is a generalized source term; ρ(kg/m^3) denotes the density; ρ_q(kg/m^3) is the density of the qth phase; μ(Pa·s) represents the turbulent dynamics viscosity; $\overline{u}$(m/s) is the velocity vector; t(s) is the time; $\overline{g}$(m·s^{-2}) stands for the gravitational acceleration;$\nabla$is the gradient operator; p(Pa) expresses the pressure; $\overline{F}$(N) is the volume force term.

4 Simulation Parameter Settings

Discrete phase particles are introduced upon convergence of the continuous phase iterative calculations. The velocities of the discrete phase particles are set at 5 m/s, 10 m/s, 15 m/s, 20 m/s, and 25 m/s. The initial airflow velocity and the velocities of the discrete particles are aligned. Spray angles are set at 0°, 4°, 8°, 12°, and 16°. The calculations account for interactions between the continuous and discrete phases, resulting in fifty iterations for the continuous phase for every single discrete phase calculation. Upon convergence of the discrete phase, the next step of the continuous phase is computed. The maximum iteration count for the discrete phase is set at 105. The material of the discrete phase droplets is water, as specified in the material database. The type of material is classified as 'droplet,' with water as the evaporating component. The droplet size distribution follows the Rosin-Rammler distribution. The mass flow rate of droplets at the inlet is 0.0033 kg/s. The inlet type for spraying is surface-based. The median volume diameter of the droplets is 60 μm, with the largest droplet size being 1×10^{-4} m and the smallest being 2×10^{-5} m. To more accurately simulate droplet motion, a random walk model is employed for the discrete phase.

5 Analysis of Simulation Results

5.1 Trajectories of Droplets at Different Airflow Speeds

Figure 2 illustrates the trajectories of droplets when the spray angle is set at 4°, under airflow speeds of 5 m/s, 10 m/s, 15 m/s, 20 m/s, and 25 m/s. Based on droplet size, droplet diameters ranging from 0 μm to 100 μm are categorized into 20 levels, with colors transitioning from blue to red to represent corresponding droplet sizes. Droplet sizes from 0 μm to 30 μm are depicted in blue. Sizes from 70 μm to 100 μm are shown in red, while those from 30 μm to 70 μm are marked in green. This research categorizes droplets above 70 μm as large droplets and those below 30 μm as small droplets. Other droplet sizes are considered medium. Figure 2 demonstrates that droplet size significantly affects the distribution of deposition. Larger droplets, influenced by their own weight, generally settle below the airflow stream. Small droplets, affected by turbulent movement, disperse widely. Medium droplets' deposition distribution aligns with the airflow pattern.

Figure 2 demonstrates that droplet diameter affects their behavior in the airflow field generated by the sprayer, influenced by gravity and drag. At the same airflow speed, larger droplets travel shorter distances from the nozzle; smaller droplets travel further. Airflow speed significantly impacts the distribution of droplet deposition. As airflow speed increases, droplet momentum increases. Large droplets are carried further away from the sprayer by the airflow. During their motion, the droplets slow down due to resistance and gravity until they deposit or remain suspended in the air. With increasing airflow speeds, the intensity of eddies, formed by air waves rebounding off greenhouse walls, also increases. At an airflow speed of 15 m/s, droplets distribute throughout the greenhouse space due to these eddies. As airflow speed increases, the number of small droplets decreases. This reduction is due to complete evaporation of small droplets into the air, influenced by temperature, humidity, and airflow speed during their movement.

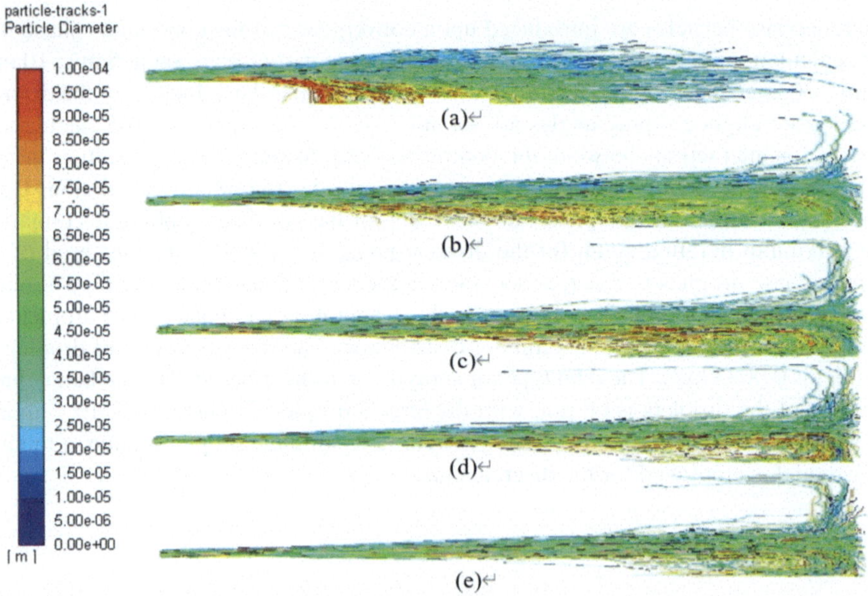

Fig. 2. Spray angle at 4°, trajectories of droplets at different airflow speeds. (**a**) v = 5 m/s (**b**) v = 10 m/s (**c**) v = 15 m/s (**d**) v = 20 m/s (**e**) v = 25 m/s

5.2 Trajectories of Droplets at Different Spray Angles

Figure 3 shows the trajectories of droplets at an airflow speed of 15 m/s for spray angles of 0°, 4°, 8°, 12°, and 16°. The diagram illustrates that the central axis of the droplet stream shifts upward as the spray angle increases. At a spray angle of 12°, droplets engage in rotational movements, carried by the swirling airflow. As the spray angle increases, larger droplets deposit further toward the far end of the greenhouse. At spray angles of 0° and 4°, droplets are widely distributed vertically. At a spray angle of 16°, droplets are more concentrated in the vertical distribution; larger droplets predominantly settle at the far end of the greenhouse. When the spray angle is not 0°, large droplets experience an upward movement trend due to the drag force in the positive direction of the y-axis and form eddies in the downwind direction of the airflow.

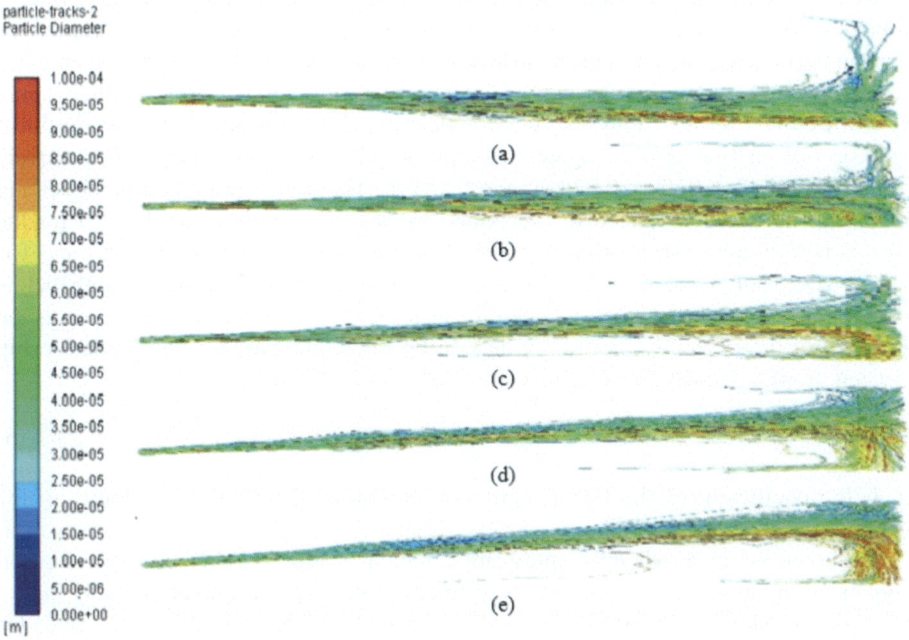

Fig. 3. Trajectories of droplets at varying spray angles. (**a**) 0° (**b**) 4° (**c**) 8° (**d**) 12° (**e**) 16°

5.3 Impact of Airflow Speed on Droplet Deposition Rate

In the CFD model, steady-state iterative calculations are conducted. Once residuals converge, discrete phase particles reach an equilibrium state where no further change in the mass of components occurs. The deposition quantity of droplets is defined as the flow rate per unit target leaf area per unit time. The deposition rate in the crop canopy is defined as the ratio of the average deposition within the canopy per unit time to the average volume of droplets at the inlet. The formula is as follows:

$$R = \frac{\sum_{i=1}^{3} m_{ci}(1 - \gamma_i)/A_c}{m_i/A_i} \tag{4}$$

R is the droplet deposition rate; m_{ci} (kg/s) is the mass flow rate through the crop canopy; m_i(kg/s) is the mass flow rate of droplets sprayed from the nozzle; γ_i is the crop canopy porosity; A_c (mm^2) is the canopy area; A_i (mm^2) is the inlet area.

Droplets colliding with any surface, whether the ground or the canopy cross-section, are assumed to be fully retained (restitution coefficient = 0%). This means that all droplets landing on surfaces remain there. Consequently, all droplets passing through the canopy surface are considered to be deposited on the leaf surfaces. Figure 4 presents the variation curves of droplet deposition rates within the canopy under different spray angles. The graph demonstrates that, at various spray angles, the deposition rate of droplets increases with the airflow speed. A linear regression analysis was conducted between the deposition rates and the airflow velocity at the inlet. The regression model is shown in Eq. (5), and the coefficients of the model are presented in Table 1. Under spray angles of 0°, 4°, 8°, 12°, and 16°, the coefficients of determination R2 for the linear regression of airflow velocity against canopy droplet deposition rate are respectively 0.98988, 0.98775, 0.97991, 0.97815, and 0.95196. The regression analysis model and its coefficients were subjected to t-tests and F-tests. The significance level α for the F-test is 0.05, with corresponding F-values for spray angles of 0°, 4°, 8°, 12°, and 16° being 2.81325E-4, 3.7456E-4, 7.88343E-4, 8.94983E-4, 2.93E-3, all significantly below the significance level α. The established linear regression model effectively reflects the impact of airflow velocity on droplet deposition rates, indicating a linear relationship between droplet deposition rate and airflow velocity at the inlet.

$$y = a + bv_i \tag{5}$$

a,b is the coefficient of the linear regression equation. v_i(m/s) is the velocity of the airflow.

As shown in Fig. 4, when the spray angle is set at 8° and 12°, the deposition rates of droplets are relatively consistent across different airflow speeds and are notably higher than those at other spray angles. At an airflow speed of 5 m/s, the deposition rates for each spray angle fluctuate around 0.0028. With an increase in airflow speed to 25 m/s, the maximum deposition rate reaches 0.0609 at a spray angle of 12°. This indicates that changing the airflow speed can achieve higher deposition rates per unit inlet area.

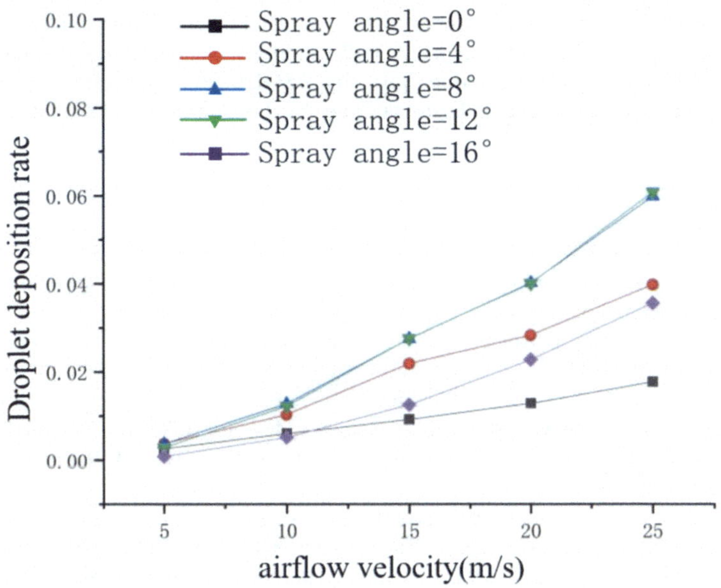

Fig. 4. Droplet deposition rates under different operational conditions

Table 1. Coefficients of linear regression between speed and deposition rate

Spray angle (°)	Intercept (*b*)	Intercept standard error	Gradient (*a*)	Gradient standard error	R^2
0	−0.00144	6.22145E-4	7.43E-4	3.75168E-5	0.98988
4	−0.00623	0.00166	0.0018	1.0023E-4	0.98775
8	−0.01312	0.00331	0.00279	1.9953E-4	0.97991
12	−0.01445	0.00356	0.00288	2.14477E-4	0.97815
16	−0.01078	0.00323	0.00175	1.9493E-4	0.95196

6 Conclusions

This thesis has developed a model for the distribution of droplet deposition in crop canopies based on the SIMPLE algorithm and Lagrange models. It simulated the impact of airflow angles and velocities on droplet trajectories. Further, a linear regression model was established to relate droplet deposition rates to the airflow velocity at the intake. The analysis leads to the following conclusions: Droplet size significantly affects the distribution of deposition. Smaller droplets travel greater distances and their numbers decrease due to evaporation. Optimal airflow speeds result in higher deposition rates per unit area. The airflow velocity at the wind-driven sprayer's intake is positively correlated with droplet deposition rates. Increasing the initial airflow speed enhances the deposition rate of droplets.

Acknowledgments. This work was supported by Horizontal research project of Shandong Agriculture and Engineering University (sgyhx2023-12), High-level Talent Research Funding Project of Shandong Agricultural and Engineering University (BSQJ202302), Agricultural machinery R & D, manufacturing, promotion and application of integrated pilot project (NJYTHSD-202307-1).

References

1. Zhang, X.: Vegetable planting technology and pest control strategy in greenhouses. Hebei Agric. Mach. **08**, 63–65 (2024)
2. Llop, J.: Influence of air-assistance on spray application for tomato plants in greenhouses. Crop Prot. **78**, 293–301 (2015)
3. Duga, A.T.: Numerical analysis of the effects of wind and sprayer type on spray distribution in different orchard training systems. Bound.-Layer Meteorol. **157**, 517–535 (2015)
4. Qiu, W.: Design and test of greenhouse air curtain application device. J. Nanjing Agric. Univ. **46**(02), 405–415 (2023)
5. Delele, M.A.: Modelling and validation of the air flow generated by a cross flow air sprayer as affected by travel speed and fan speed. Biosyst. Eng. **92**(2), 165–174 (2005)
6. Endalew, A.M.: A new integrated CFD modelling approach towards air-assisted orchard spraying. Part I. Model development and effect of wind speed and direction on sprayer airflow. Comput. Electron. Agric. **71**(2), 128–136 (2010)

Study on the Physicochemical Properties of Biochar from Different Raw Materials and Its Adsorption of Cadmium in Water

Guizhai Zhang(✉), Dongman Zhou, Xinli Li, Jing Li, and Lihua Jiang

Shandong Agriculture and Engineering University, Jinan 251100, Shandong, China
zgzok2005@163.com

Abstract. In this paper, biochar was prepared from plateau livestock and poultry manure and corn straw at different temperatures, and its adsorption of cadmium in water was studied. The physicochemical properties, optimal adsorption pH value, adsorption kinetics test and adsorption isotherm test of biochar prepared at different temperatures were studied. The research results indicate that as the preparation temperature increases, the yield of biochar decreases, the ash content increases, and the pH value increases. When the pH is 8, the adsorption effect of biochar on cadmium is better. 5 h is the best adsorption time, and kinetic adsorption belongs to a quasi-secondary model, which is dominated by chemisorption. The optimal preparation temperature is 700 °C.

Keywords: Plateau livestock and poultry manure · corn straw · biochar · cadmium

1 Introduction

With the acceleration development of China's industry, a large number of wastewater was discharged into the rivers, seas and soil, which was caused heavy metal pollution [1]. Cadmium-polluted wastewater enters the water body and endangers human health through the food chain [2]. Therefore, it is necessary to remove cadmium from water.

In recent years, China's animal husbandry and planting industry have developed and grown, and the output of livestock manure and corn straw in the plateau has increased, while aggravating the pollution of soil, air and water resources. Although China already has manure, straw feed, energy and fertilizer treatment methods [3], it is still necessary to find a new treatment method to solve the pollution problem and increase its added value. The preparation of biochar is an important way to recycle biomass waste. Compared with the traditional methods of manure and straw treatment, the use of agricultural and forestry waste to prepare biochar to adsorb pollutants in water bodies has the advantages of green environmental protection, renewable, and circular development of resources.

In this paper, according to the ratio of 1:1, livestock manure and corn straw in plateau were weighed and burned at different temperatures to produce biochar. The effects of yield, pH, ash content and initial pH of solution on cadmium adsorption by biochar were studied. The adsorption kinetics experiment and adsorption isotherm experiment were carried out to find the best adsorption conditions of biochar for cadmium in water.

V. Achal et al. (Eds.): EPPC 2024, ESE, pp. 231–236, 2025.
https://doi.org/10.1007/978-3-031-92123-0_24

2 Experiments

2.1 Preparation of Biochar

In this research, livestock and poultry manure and corn straw were used as raw materials at a ratio of 1:1 at 300 °C, 400 °C, 500 °C, 600 °C, 700 °C, respectively, and anaerobic pyrolysis for 3 h to produce biochar [4].

2.2 Determination of Physicochemical Properties and Yield of Biochar

In this study, the yield, ash content, pH and other physicochemical properties and yield of different temperatures biochar were determined. The method [4] was used to determine the yield.

2.3 Adsorption Kinetics Experiment

The biochar of 0.1 g and 25 mL of 75 mg/L cadmium solution were oscillated. The thermostat oscillator temperature is 25.0 °C; the rotational speed is 150 r/min. Oscillate for 30, 60, 120, 180, 240, 300, 360, 480, 600 min, respectively. The filtrate was to be subsequent used.

2.4 Isothermal Adsorption Experiment

0.1 g biochar was adsorbed 5 mg/L, 25 mg/L, 40 mg/L, 50 mg/L, 75 mg/L cadmium solution, respectively, under the adsorption conditions of 25.0 °C and 150 r/min, and the oscillatory adsorption was carried out for 5 h. The filtrate was to be subsequent used.

2.5 Determination of Cadmium Content

The cadmium concentration was determined by TAS-990 atomic absorption spectrophotometer.

3 Results and Discussion

3.1 Physicochemical Properties of Biochar from Different Raw Materials

Biochar yield, ash content, pH and other physicochemical properties of different raw materials are shown in Table 1.

As can be indicated from the above table, when the temperature of biochar preparation increases, the yield of biochar will decrease, from 300 °C to 700 °C, and the yield will decrease from 46.93% to 23.64%. Because temperature plays a very important role in the pyrolysis process, it has a great influence on the yield, composition and characteristics of the obtained products [5]. With the increase of temperature, the dissolution degree of livestock and poultry feces in the plateau increased, and the liquefaction/gasification reaction was promoted, and the liquefaction/gasification products were increased. The hydrogen bond between cellulose and hemicellulose in corn straw breaks, forming a

Table 1. Physical and chemical properties of biochar from different raw materials.

Item	HC-300	HC-400	HC-500	HC-600	HC-700
Yield (%)	46.93	36.46	31.52	29.81	23.64
Ash (%)	48.09	61.94	63.33	78.89	88.42
pH	8.18	8.72	9.02	9.06	9.23

large number of volatile components. The combined effect of the two is to reduce the yield.

The ash content of biochar increased from 48.09% to 88.42% with the increase of preparation temperature. It may be caused by minerals contained in livestock and poultry manure on the plateau. It is also caused by the decomposition of cellulose and hemicellulose in straw. In terms of elemental composition, the proportion of carbon in biochar increases with the increase of pyrolysis temperature.

The pH of biochar showed an increasing trend, from 8.18 to 9.23. This is because pyrolysis can release alkali metal salts from livestock and poultry droppings on the plateau, and the acidic functional groups on the surface of biochar decrease while the alkaline functional groups increase. At the same time, high temperature promotes the formation of inorganic components (such as hydroxides, carbonates, etc.). Therefore, the pH will go up.

3.2 Effect of Initial pH of Solution on Cadmium Adsorption by Biochar

The initial pH value of the solution will affect the adsorption of cadmium by biochar in water. The results of the initial pH effects are shown in Table 2.

Table 2. The adsorption capacity of biochar for cadmium at different pH values.

pH	Concentration	Adsorbing capacity
4	8.053	8.544
6	6.545	8.921
8	1.58	10.163
10	3.686	9.636

According to the data in the table and the trend of change, it showed that with the increase of pH, the cadmium adsorption capacity first increases and then decreases. When pH is 8, the maximum cadmium adsorption capacity is 10.163. When pH is 10, the adsorption capacity is reduced because of the formation of a complex or cadmium hydroxide precipitation, which impedes adsorption. Therefore, the pH of the solution was adjusted to 8 in the subsequent test.

3.3 Adsorption Kinetics Study

The adsorption kinetics results are shown in Table 3.

Table 3. Adsorption kinetics data.

Time (min)	30	60	120	180	240	300	360	480	600
C_t (mg/L)	3.31	3.37	3.57	2.16	2.84	1.33	1.54	1.67	1.90
q_t (mg/g)	9.73	9.72	9.67	10.02	9.85	10.23	10.17	10.14	10.08

Table 3 showed that in the first 300 minutes the concentration of cadmium in water increases slightly at first and then decreases, and the overall trend of adsorption capacity increases. It may be because at the early stage of the reaction, the adsorption sites on the biochar surface are sufficient, and the competition between cadmium ions is small [6]. At 300 min, the maximum biochar adsorption capacity of cadmium was 10.23. The best adsorption time was 300 min.

The fitting adsorption kinetics results are shown in Table 4.

Table 4. Adsorption kinetic model parameters.

Actual adsorbing capacity (mg/g)	Quasi-first-order kinetic model			Quasi-second-order kinetic model		
	q_e (mg/g)	k_1	R^2	q_e (mg/g)	k_2	R^2
10.23	2.102	0.196	0.3435	10.173	1.934	0.9997

It can be seen from the table that the R^2 value of quasi-first-order kinetic fitting is 0.3435. The R^2 value of the quasi-second-order kinetic fitting is 0.9997, and the adsorption capacity is 10.173, which is close to the real measurement 10.23. Therefore, the adsorption is more in line with the quasi-second-order kinetic equation and is mainly chemisorbed.

3.4 Isothermal Adsorption Study

Adsorption isotherm test data are shown in Fig. 1.

According to the data and charts in the table, as the initial concentration of cadmium solution increases, the adsorption capacity of cadmium by the different temperatures biochar gradually increases, and the degree of biochar fitting at each temperature is higher. In order to further analyze the adsorption capacity of cadmium by different preparation temperatures biochar, the isothermal adsorption model was fitted.

By fitting the model with the adsorption isotherm, corresponding parameters are shown in Table 5.

For biochar prepared at 300 °C, 400 °C, 600 °C and 700 °C, the correlation coefficients (R^2) of their fitting with Freundlich model were 0.9880, 0.9529, 0.9564 and

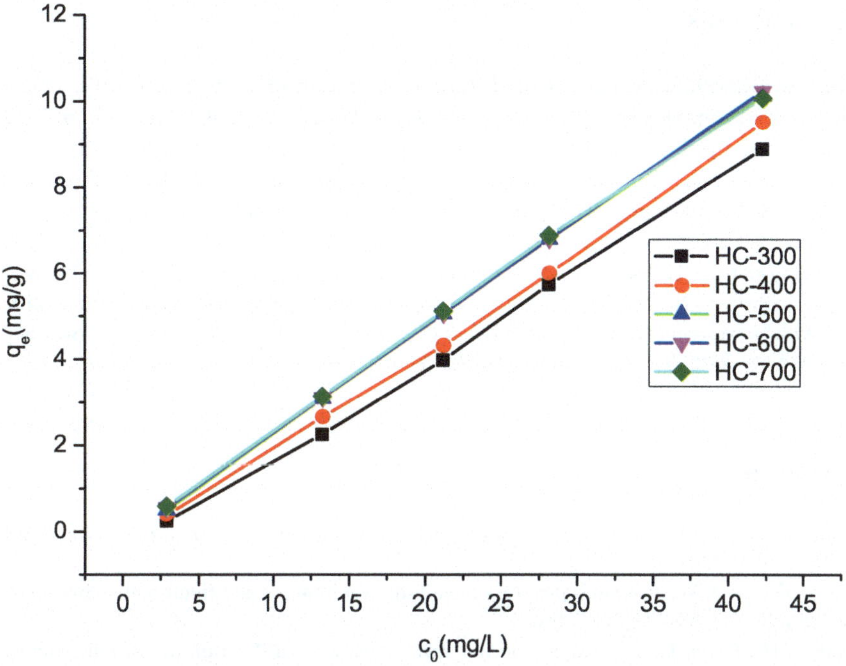

Fig. 1. The adsorption capacity of cadmium by biochar from different raw materials varies with the initial concentration of cadmium in the solution.

Table 5. Adsorption kinetic model parameters.

Biochar	Langmuir			Freundlich		
	K_L	q_m	R^2	K_F	1/n	R^2
HC-300	0.1583	0.6680	0.8638	0.0404	2.8501	0.9880
HC-400	0.2122	1.233	0.817	0.2357	2.3837	0.9529
HC-500	0.566	1.403	0.116	3.279	2.6207	0.8604
HC-600	0.865	0.742	0.5252	5.0759	4.1368	0.9564
HC-700	0.385	5.456	0.0956	6.496	1.4755	0.9729

0.9729, respectively. The adsorption of biochar prepared at five temperatures was dominated by multi-molecular layers. In the Freundlich model, the adsorption capacity of biochar increases with the increase of K_F. The adsorption decreases as 1/n increases. According to the results, the adsorption capacity has a corresponding increase as the temperature rises. Under the 5 temperatures, best adsorption effect is the biochar of 700 °C.

4 Conclusions

The physicochemical properties of different temperatures biochar are different. With the increase of temperature, the yield decreased, the ash content increased, and the pH increased. The optimum adsorption pH is 8. According to the adsorption kinetics test, this experiment can fit the quasi-second-order kinetic equation well, and the chemical adsorption is the main one. The biochar of 700 °C has a good adsorption effect.

Acknowledgments. The work was supported by Shandong-Chongqing Science and Technology Cooperation Project (2022LYXZ-029); the central government guides local funds for scientific and technological development (YDZX2023013, YDZX2022152); Key R&D Program (Soft Science Project) of Shandong Province, China (2023RKY06020).

References

1. Zhang, J.T., Liu, J.L.: Harm of heavy metal pollution in water and its treatment. Shandong Ind. Technol. **8**, 35 (2019)
2. Dai, S.M., Lv, X.W.: Research progress of cadmium polluted water treatment technology. Saf. Environ. Eng. **3**, 63–65+71 (2006)
3. Yang, Y.H., Chen, H.L.: Analysis on the way of resource utilization of animal manure and straw. Anim. Husb. Vete. Sci. Inform. **10**, 31–33 (2023)
4. Zhang, X.M., Wang, X.Y., Li, S.J.: Study on chemical components of different organs of Larch and their biochar physicochemical properties. For. Ecol. Scie. **4**, 391–398 (2023)
5. Wilson, A.J., Wilson, H.W.: Colloldal phenomena and the adsorption formula. J. Am. Chem. Soc. **6**, 886–895 (2002)
6. Park, J.H., Wang, J.J., Kim, S.H., Kang, S.W., Jeong, C.Y., Jeon, J.R.: Cadmium adsorption characteristics of biochars derived using various pine tree residues and pyrolysis temperatures. J. Colloid Interface Sci. **553**, 298–307 (2019)

Research Progress of Heterogeneous Photocatalysts in Wastewater Treatment

Chengjun He(✉)

Jinqiu International Hangzhou School, Hangzhou 310000, Zhejiang, China
362866542@qq.com

Abstract. With the rapid development of industrialization and urbanization, water pollution has become a major global challenge. Organic compounds emitted by various industries have become difficult-to-treat pollutants in water bodies. Recent research has shown that semiconductors based on metal oxides (such as TiO_2, SnO_2, CeO_2, ZrO_2, WO_3, and ZnO) exhibit excellent photo-catalytic effects for decomposing organic pollutants in wastewater. However, their photocatalytic performance is limited by their own high energy gaps (requiring ultraviolet light activation) and the rapid recombination of photo-generated electron-hole pairs. This article explores strategies to enhance the application performance of these metal oxides in the field of photo-catalysis. Researchers have successfully improved their photo-catalytic activity through doping, preparing nano-composite materials, and constructing core-shell structures of metal oxides. The main purpose of this article is to propose a series of efficient, cost-effective, and industrially viable integrated photo-catalytic solutions.

Keywords: Photocatalysts · Wastewater · Titanium dioxide · Core-shell composite

1 Introduction

With economic development and population growth, protecting water resources has become an urgent task. Approximately two million tons of waste pollutants are indiscriminately discharged into public water sources daily, severely impacting human health and leading to numerous fatalities [1]. Notably, these pollutants include residual toxic dyes, monocyclic aromatic hydrocarbons, and antibiotics found in industrial waste, such as Rhodamine B (RhB), 4-nitrophenol (4-NP), and naproxen (NPX), which are resistant to thermal and photolytic degradation. It is recommended that the concentration of 4-NP in water should be maintained below 10 ppb [2]. Annually, over 175,000 tons of dyes and 54,000 tons of antibiotics are recklessly discharged into global water resources [3]. The growing concern over water pollution has become a formidable challenge, leading to the proposal and implementation of various wastewater purification methods including ion exchange, adsorption, membrane separation, and electrochemical purification [4]. Despite significant efforts to address these issues, challenges persist, such as limited adsorption capacity, easy deactivation, secondary pollutant formation, and high energy consumption [5].

V. Achal et al. (Eds.): EPPC 2024, ESE, pp. 237–242, 2025.
https://doi.org/10.1007/978-3-031-92123-0_25

The photocatalytic degradation technology induced by light has recently attracted great attention [3]. Solar energy, being the most environmentally friendly, cost-effective, and sustainable source of energy, aligns well with this technology. Advances in nanoscience and nanotechnology have enabled the rational design and production of innovative semiconductor photocatalysts that degrade organic pollutants using reactive free radicals [6]. The photocatalytic process also offers other advantages, including cost-effectiveness, environmental friendliness, and low energy consumption. However, despite its enormous potential, the efficiency of photocatalysts in converting solar energy is far lower than expected. This is mainly because finding a single photocatalyst that can absorb the entire solar spectrum while maintaining low photocarrier recombination rate and high light stability is very challenging.

Composite co-catalysts, which exhibit synergistic effects, are considered capable of meeting some of these demanding requirements. Various materials are suitable for photocatalytic processes, but specific metal oxides such as titanium dioxide (TiO_2), cerium oxide (CeO_2), zinc oxide (ZnO), zirconium oxide (ZrO_2), manganese oxide (MnO_2), and tin oxide (SnO_2) are regarded as excellent photocatalysts due to their high stability across a wide pH range, high efficiency, low cost, easy availability, low toxicity, environmental friendliness, and high oxidative photogenerated holes [7]. However, these metal oxides have drawbacks in photocatalytic degradation, such as incomplete mineralization and wide band gaps, which significantly limit their photocatalytic activity. Furthermore, rapid transmission of electrons and holes is a key factor in dye degradation, but their rapid recombination greatly reduces their photocatalytic potential. To improve their efficiency, researchers have explored innovative approaches such as rare earth-doped metal oxides, metal oxide nanocomposites, and metal oxide core-shell nanostructures [8]. This paper extends these developments by categorizing the materials used into three types: simple metal oxides, doped, and composite photocatalysts.

2 Photocatalysts in Water Pollutant Treatment

2.1 Doped Catalyst

Generally, the efficiency of photocatalysts can be enhanced by doping them with rare earth elements or metals [9]. Such as Gd_2O_3-TiO_2, Ag-TiO_2, $LaVO_4$-TiO_2 and others. After doping, the dopant energy levels reside within the valence band, causing the conduction band bottom to shift towards the Fermi level, resulting in a narrower band gap width. This reduction in the energy required to excite valence band electrons to the conduction band expands the spectral response range. Consequently, it decreases the recombination of photon-electron/hole pairs, thereby enhancing the photocatalytic performance of mesoporous materials. For instance, under light irradiation which wavelength at 325 nm, when the molar fraction of Ag doping was 5%, the band gap width of mesoporous TiO_2 corresponded to 2.82 eV, at this point, the decomposition rate of RhB reached its maximum, which was about 98% (Table 1). Moreover, as the molar fraction of Ag doping increased, the decomposition rate of RhB by mesoporous TiO_2 showed an initial increase followed by a decrease trend. This was because an excessive amount of elemental Ag deposits on the surface of mesoporous TiO_2 formed defect recombination centers that accelerated the recombination of photogenerated electron-hole pairs, thereby leading to a decrease in photocatalytic performance [9]. At an RhB concentration of 2 mg/L, using pure TiO_2 as a catalyst achieved a decolorization efficiency of 87.5%. When doped with Gd_2O_3, the decolorization efficiency for RhB improved to 93.75%. The introduction of Gd increased the number of hydroxyl groups on the catalyst surface, where hydroxyl radicals could suppress the recombination of photogenerated electron-hole pairs, thus enhancing the photocatalytic activity of the composite catalyst [10]. Additionally, $LaVO_4$ significantly enhanced the photocatalytic effect of TiO_2, with the catalytic performance influenced by pH and the amount of $LaVO_4$ doping. The $LaVO_4/TiO_2$ composite photocatalyst exhibited reduced recombination of photogenerated electrons and holes, resulting in more photogenerated electron-hole pairs available for photocatalysis. However, excessive doping of lanthanum vanadate reduced the number of hydroxyl groups adsorbed on the TiO_2 surface, increasing the probability of recombination of photogenerated electron-hole pairs and thereby reducing the photocatalytic efficiency [11].

Table 1. Degradation rate of water pollutant by different doped catalysts.

Catalysts	Condition	Pollutant	Concentration (mg/L)	Degradation rate (%)	References
Ag-TiO_2	/	RhB	/	98.76	[9]
Gd_2O_3-TiO_2	Visible light	RhB	2	93.75	[10]
$LaVO_4$-TiO_2	Xenon lamp light	Methyl orange	20	/	[11]

2.2 Composite Materials

Doping method can improve the photocatalytic activity of TiO_2, and the modified TiO_2 is non-toxic and has good photostability, and can effectively remove pollutants in wastewater. However, TiO_2 is a powder, which is easily dispersed in water and difficult to recycle. Therefore, selecting a suitable carrier for TiO_2 to prepare composite materials can solve this problem. For example, the N-TiO_2/PP melt-blown composite material can be obtained by loading N-TiO_2 on the surface of PP melt-blown nonwoven material. Under 30 min dark adsorption and 90 min light conditions, the degradation efficiency of methylene blue (MB) exceeded 98%. N-TiO_2 in the composite can produce superoxide free radicals and hydroxyl free radicals under light conditions, and the two can degrade MB together. In addition, the N-TiO_2/PP melt-blown composite material can still maintain its intact structure after photocatalytic degradation, and the photocatalytic degradation efficiency of MB decreased to 58% after four photocatalytic degradation times [12]. Composite metal oxide catalyst ($TiO_2/ZnO/SnO_2$) is an excellent composite photocatalytic material. Compared with traditional photocatalytic and ozone catalytic treatment of polluted water, TiO_2/ZnO/SnO2 and photocatalytic ozone oxidation method can significantly improve the treatment ability of COD pollutants in water. When the pollutant concentration is about 110 mg/L, the COD removal rate exceeded 80%, and compared with traditional TiO_2 single catalyst to treat wastewater, $TiO_2/ZnO/SnO_2$ has strong stability. When $TiO_2/ZnO/SnO_2$ is reused to treat COD pollutants in water. The removal rate of COD has not changed much [13]. In addition to TiO_2 metal oxide composites, $ZnWO_4/WO_3$ composites also have a strong ability to treat sewage. Due to the energy band structure of $ZnWO_4$ and WO_3, the separation efficiency of photogenerated carrier can be improved, so that the photocatalytic activity can be improved. The degradation rate of RhB reached more than 96%, and the degradation rate of ciprofloxacin and oily sewage was also more than 80%, and the sewage treatment efficiency was significantly improved compared with the traditional single $ZnWO_4$ and single WO_3 [14]. Furthermore, TiO_2 and biochar composites can also treat polluted water efficiently. Among them, the degradation rate of methyl orange of TiO_2/walnut shell biochar composite reached 96%, and the degradation rate of methyl orange of Ag/TiO_2/walnut shell biochar composite reached 97.48%. TiO_2/Reed straw biochar can also treat some pollutants in water well, because biochar can increase the specific surface area and the number of active sites of nano-TiO_2 materials, and biochar has a porous structure and adsorption capacity, which can enrich pollutants around TiO_2, thereby improving the degradation efficiency of pollutants. On the other hand, biochar has good electrical conductivity and electron transport ability, which can reduce the photogenerated electron-hole pair recombination and accelerate the degradation process of pollutants. In addition, biochar can also cooperate to reduce the band gap width of TiO_2, so that the band gap energy of the composite material is reduced, the light response range is increased, and finally effectively promote the degradation of pollutants.

2.3 Core-Shell Composite Photocatalyst

Core-shell nanoparticles are a composite structure formed by spherical particles (in the size of micron to nanometer) and coated with uniform nano films on their surface. The core and shell are connected by physical or chemical interaction. The core-shell structure is conducive to the organic combination of the properties of the nuclear material and the shell material, and the composite properties can be controlled by controlling its composition and thickness. Nano-ferrite materials can be separated and enriched in the external magnetic field, and a class of nano-ferrite composites with core-shell structure can be prepared by combining them with the high photocatalytic activity of nano-semiconductor materials. For example, the anatase $Fe_3O_4@TiO_2$ core-shell nanoparticles were prepared using Fe_3O_4 nanoparticles with a particle size of about 15 nm as the core. Using it as a photocatalyst to degrade 4-nitrophenol in wastewater, the degradation rate could reach 98.6%, which was significantly higher than that of pure TiO_2, and the catalytic efficiency has little change after repeated use [14]. At the same time, α-$Fe_2O_3@SiO_2@Cu_2O$ core-shell structure complex also had the ability to degrade pollutants. When high-pressure mercury lamp was used as the light source, the methylene blue degradation rate of α-$Fe_2O_3@SiO_2@Cu_2O$ exceeded 47% when the pollutant concentration was about 15 mg/L [15]. In addition, Fe_3O_4 @ SiO_2@ TiO_2/Ag is a composite catalyst with a bivalve structure, which has a good inhibition effect on some bacterial pollution. Compared with $Fe_3O_4@SiO_2@TiO_2$, the addition of Ag improved the photocatalytic antibacterial activity, and Ag can be used to capture photogenerated electrons. Thus, the composite of photogenerated electrons and holes in TiO_2 was inhibited, and more strong oxidizing ·OH radicals and $O^-{}_2$ radicals were formed, and the photocatalytic activity was improved, thus enhancing its antibacterial property [16].

3 Conclusion

Photocatalytic oxidation technology has the advantages of high degradation efficiency, low energy consumption, low cost and low secondary pollution. It is a new green environmental treatment technology and has been widely used in sewage treatment. TiO_2 is the most common photocatalyst. In order to obtain more efficient and ideal photocatalytic materials, TiO_2 is usually doped with metal to improve its photocatalytic performance. In order to solve the problem of TiO_2 recovery and reuse in water, suitable carriers are generally selected to prepare composite materials with TiO_2, such as N-TiO_2/PP composite melt-blown materials, TiO_2/ZnO/SnO_2, TiO_2/walnut shell biochar composites, etc. The degradation of methylene blue, COD, methyl orange and other pollutants in sewage has achieved good results. In addition, nano-ferrite materials combined with nano-semiconductor materials can prepare a kind of nano-ferrite composite materials with core-shell structure, which has a high degradation rate of 4-nitrophenol and RhB in sewage, and has the characteristics of photocatalytic oxidation of shell materials, but also can realize efficient recycling.

4 Outlook

Although photocatalysis technology has made some progress in wastewater treatment, there are still some problems and challenges. At present, most of the research on photocatalytic treatment of antibiotic wastewater is still in the laboratory stage. In the future, the research on the reaction mechanism of photocatalytic treatment of wastewater should be strengthened to lay a foundation for practical application. The preparation of biochar and photocatalytic composite materials is still dominated by powder materials, and there are few studies on the recycling of this material, which limits its popularization and application. In addition, the preparation process of core-shell nanocomposites is complicated, and agglomeration and reduction of specific surface area are easy to occur in the process, resulting in decreased photocatalytic activity. How to prepare core-shell composite materials with good dispersion at low cost and high efficiency is an urgent problem to be solved.

References

1. Xuyan, Y., Jifei, F., Peixian, L., Jingjun, L., Baohui, R., Shuiying, G., Rong, C., et al.: Fast and efficient removal of mercury ions using zirconium-based metal–organic framework filter membranes. Inorg. Chem. Commun. **131**, 108796 (2021)
2. Seema, D., Suman, C.S., Saurabh, S., Ankita, K.P., Sandeep, K., Ashok, K., Vinod, K., et al.: Magnetic Fe2O3/CNT nanocomposites: characterization and photocatalytic application towards the degradation of Rose Bengal dye. Ceram. Int. **49**(12), 20071–20079 (2023)

3. Xingsheng, H., Da-Peng, Y., Xiaoyan, Z., Minghuan, L., Zewen, K., Chenfei, L., Nengqin, J., Rafael, L., et al.: Waste eggshell membrane-templated CuO-ZnO nanocomposites with enhanced adsorption, catalysis and antibacterial properties for water purification. Chem. Eng. J. **369**, 621–633 (2019)
4. Junhua, L., Linsen, H., Xiaoqing, J., Liangxing, Z., Xiaosong, S., et al.: Preparation and characterization of ternary Cu/Cu2O/C composite: An extraordinary adsorbent for removing anionic organic dyes from water. Chem. Eng. J. **404**, 127091 (2021)
5. Feng Yuhao, S., Xuewei, C.Y., Yanyan, L., Zhao Xiaona, L., Chenggang, M.Y., Guixia, L., Mingliang, M., et al.: Research progress of graphene oxide-based magnetic composites in adsorption and photocatalytic degradation of pollutants: a review. Mater. Res. Bull. **162**, 112207 (2023)
6. Keerthana, S.P., Yuvakkumar, R., Kumar, P., Senthil, R.G., Dhayalan, V., et al.: Rare earth metal (Sm) doped zinc ferrite (ZnFe2O4) for improved photocatalytic elimination of toxic dye from aquatic system. Environ. Res. **197**, 111047 (2021)
7. Chandra, J.N., Gururani, P., Prasad, G.S.: Metal oxide nanoparticles and their nanocomposite-based materials as photocatalysts in the degradation of dyes. biointerface research in applied. Chemistry. **12**(5), 6557–6579 (2021)
8. Chunying, B., Sarentuy.: Preparation of modified mesoporous TiO2 Photocatalytic materials and study on their photocatalytic degradation of wastewater. Funct. Mater. **10**(54), 10231–10236 (2023)
9. Wei, Z., Peng, D., Zheran, Y., et al.: Preparation and visible light activity of Gd-doped TiO2 photocatalyst [J]. J. Changchun Normal Univ. **43**(2), 182–185 (2024)
10. Xie, B., Zunhong, H., Xiuguang, Y., Luping, Z.: Photocatalytic performance of LaVO4/TiO2 in the decomposition of methyl orange. J. Jinggangshan Univ. **43**(6), 27–32 (2022)
11. Rongxuan, C., Sun Hui, Y., Bin.: Preparation and photocatalytic properties of N-TiO2/polypropylenemelt-brown nonwovens. J. Text. Res. **45**(3), 138–147 (2024)
12. Xili, S., Huang Xizhen, D., Ping, D.Y., Zhanghai, L., Bo, L.: Preparation and photocatalytic ozonization of TiO2/ZnO/SnO2 for advanced treatment of tannery wastewater. Technol. Water Treat. **50**(04), 61–65 (2024)
13. Hong, X., Te, W., An, W., Lihong, Z.: Construction of ZnWO4/WO3 heterojunction and its application in industrial wastewater treatment. Chem. Eng. **37**(09), 5–10 (2023)
14. Morigen, B., Rigejile, M., Hongye, Z., Siqin, Z., Shan, Z.: Application status of metal oxid-based biochar composites in wastewater treatment. New Chem. Mater. **52**(05), 213–220 (2024)
15. Xiaofeng, G., Shiyu, L., Yang Lei, F., Yufan, L.X., Fenghan, Y.: Synthesis and photocatalytic properties of α-Fe2O3 @SiO2 @Cu2O core-shell structure complexes. Appl. Chem. Ind. (2024)
16. Huiyu, S., Miaomiao, L., Ruidan, L., Qiao, R.: Preparation of Fe3O4@ SiO2@ TiO2/Ag core/ shell-structured microspheres and their photocatalytic antibacterial property. J. Zhejiang Normal Univ. (2024)

Spatiotemporal Pattern and Drivers of Carbon Storage from 2000 to 2020 in the Poyang Lake Basin, China

Yitong Chen[1] and Hua Zhu[1,2](✉)

[1] Zhejiang University of Water Resources and Electric Power, Hangzhou, China
zhuhua18@mails.ucas.ac.cn

[2] Key Laboratory of Poyang Lake Wetland and Watershed Research, Ministry of Education, Jiangxi Normal University, Nanchang, China

Abstract. This study employed the InVEST model and various statistical methods to explore the spatiotemporal variations and the underlying mechanisms affecting carbon storage in the Poyang Lake Basin (PYLB) from 2000 to 2020. The results revealed that 93.29% of the PYLB exhibited a decline in carbon stock, with a distinct spatial pattern of "high in the south and low in the north, as well as high around the periphery and low in the central region." Climatic factors and human activities primarily negatively impact carbon storage, with the degree of influence being more pronounced in the northern regions than the southern ones. Meanwhile, soil moisture, elevation, and NDVI all contribute positively to carbon storage, with their influence gradually diminishing from the water system towards its surroundings. The overall impact of topography, soil, and NDVI on carbon storage was significant, with coefficients of 0.570, 0.153, and 0.269, respectively. In contrast, climate and human activities had adverse effects, with negative coefficients of −0.119 and −0.092, respectively.

Keywords: Carbon storage · InVEST model · Spatial heterogeneity · Drivers

1 Introduction

With the acceleration of urbanization and industrialization, emissions of greenhouse gases have surged significantly, thereby inflicting severe impacts on both the ecological environment and socio-economic development [1]. The carbon cycle is a crucial material exchange process on the Earth's surface. It plays an essential part in conserving biodiversity and enhancing sustainable progress, thereby contributing to the sustainability of terrestrial ecosystems [2]. Studies have shown that terrestrial carbon sequestration function and rate have significantly changed under the intertwined effects of climatic transformation and human endeavors [3]. Therefore, it is imperative to conduct research into carbon storage dynamics and underlying driving mechanisms.

V. Achal et al. (Eds.): EPPC 2024, ESE, pp. 243–252, 2025.
https://doi.org/10.1007/978-3-031-92123-0_26

Due to the advantages of requiring relatively fewer data and high estimation accuracy, the InVEST model has garnered extensive utilization in carbon storage estimation. Many studies have combined the InVEST model with other methodologies to simulate and forecast carbon storage across various regions and under diverse development scenarios [4–6]. Additionally, some research has employed various statistical methods to investigate the spatiotemporal evolution of regional carbon storage [7, 8]. Land use and land cover change (LUCC) exerts a substantial influence on the terrestrial carbon cycle, which further leads to variations in carbon storage. Numerous studies have examined the impact of LUCC on the decline in carbon sequestration. These studies have also employed methods such as structural equation model, geographical detector, and machine learning to delve into the multiple drivers of carbon storage variations from distinct perspectives [8, 9]. Nevertheless, research on the spatial heterogeneity of driving factors and their underlying mechanisms remains inadequate.

The Poyang Lake Basin (PYLB), a pivotal ecosystem in Southeastern China, significantly mitigates the greenhouse effect. However, with the intensification of human activities, the PYLB is facing immense ecological pressure. To tackle this concern, the goals of this research endeavor entail (1) Investigating spatiotemporal dynamics of carbon stock in the PYLB from 2000 to 2020, (2) Analyzing spatial heterogeneity of factors influencing carbon stock within the basin, and (3) Identifying the influencing mechanisms of carbon stock. The findings are expected to inform policies that aim to achieve dual carbon goals and enhance regional carbon sequestration capacity.

2 Materials

2.1 Study Area

The PYLB (113°35′–118°29′E, 24°29′–30°05′N), lies to the north of the Yangtze River and spans an extensive area of roughly 162,200 km^2. The PYLB experiences a mean annual temperature of 18 °C and a mean rainfall of 1600 mm. The east, west, and south mountainous regions give rise to five rivers that ultimately meet at Poyang Lake, with the entire basin enveloped by expansive forest.

2.2 Data Sources

The data utilized in the study consist of land use and drivers of carbon stock. All data are unified to the same resolution. The pertinent specifics are outlined in Table 1.

Table 1. Summary of data.

Data	Resolution	Sources
Land use	30 m	CLCD, http://www.ncdc.ac.cn
GDP and population (POP)	1 km	https://data.tpdc.ac.cn/
DEM	90 m	https://www.gscloud.cn
Slope	/	Calculated from DEM
NDVI	1 km	GEE cloud platform
Soil moisture (SM)		
Vapor pressure deficit (VPD)		
Relative humidity (RH)		
Solar radiation (SR)		
Temperature (TEM)		https://www.geodata.cn
Precipitation (PRE)		

3 Methods

3.1 The InVEST Model

The InVEST model segments terrestrial carbon storage into three components: aboveground, belowground, and soil organic carbon storage. The total carbon stock is determined by summing up the products of the area occupied by each land use and its respective carbon density [10]. The formula is outlined below:

$$C_{total} = \sum\nolimits_{k=1}^{n} A_k \times C_k \tag{1}$$

$$C = C_{above} + C_{below} + C_{soil} \tag{2}$$

where C represents the carbon storage per unit area of certain land use (Mg C); C_{above}, C_{below}, and C_{soil} are the aboveground, belowground, and soil organic carbon storage (Mg C), respectively. C_{total} denotes the regional carbon storage (Mg C); A_k (ha) and C_k (Mg C/ha) are the area of k-th land use and respective carbon density.

Annual dynamic adjustments [11] are applied to the estimated carbon densities for each land use.

3.2 Theil-Sen Median Slope and the Mann-Kendall Test

Theil-Sen Median Slope and the Mann-Kendall Test are commonly used to determine the significance of variable trends, without requiring the data samples to follow a specific distribution. The specific formulas are detailed in references [8, 11].

3.3 MGWR Model

The MGWR model enables the variation between variables across different geographical locations, assigning distinct spatial weights accordingly [12]. The calculation formula for implementing the MGWR model is as follows:

$$y_i = \sum\nolimits_{j=0}^{n} \beta_{bwj}(u_i, v_i)x_{ij} + \varepsilon_i \tag{3}$$

where y_i represents the carbon storage of pixel i; (ui, vi) represents the spatial coordinates of pixel i; x_{ij} denotes the value of the j-th driver for pixel i; β_{bwj} represents the local regression coefficient of the j-th driver for pixel i; ε_i is the residual term.

3.4 PLS-SEM Model

PLS-SEM model is a model for studying the causal relationships between multivariate variables, including measurement models and structural models. The measurement model delineates the linkages between observable variables and their underlying latent constructs, whereas the structural model explicates the causal interdependencies among these latent variables. The details can be found in reference [13].

4 Results

4.1 Spatial Distribution of Carbon Storage

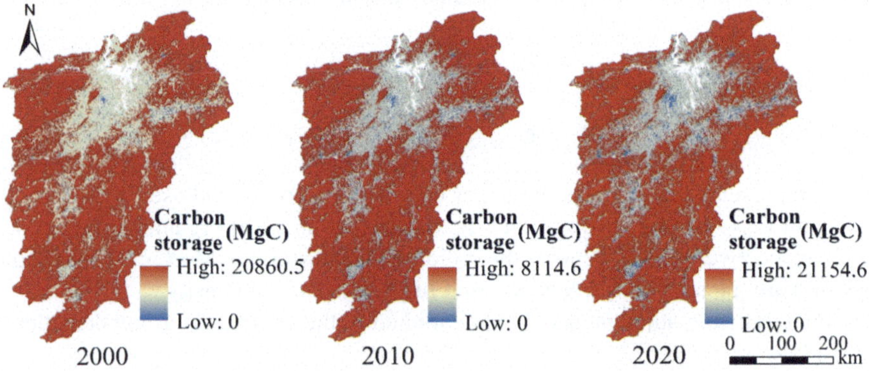

Fig. 1. Carbon Stocks in PYLB in 2000, 2010, and 2020.

Figure 1 illustrates the estimation of the carbon storage. The overall carbon storage in the basin displays a distinctive "high in the south and low in the north, high around and low in the middle" spatial distribution characteristic. More specifically, the southern part of the basin has higher carbon storage (> 10,000 MgC), distributed in patches, while near river systems, the carbon storage is lower (<6000 MgC), distributed linearly.

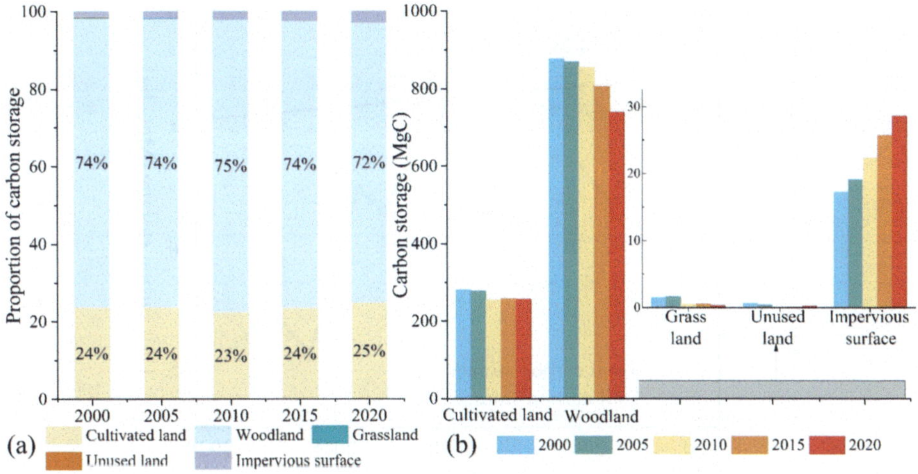

Fig. 2. Carbon storage and its proportion in different land uses.

Figure 2 demonstrates the marked disparities in carbon storage across various land uses within the PYLB. Woodland exhibits the highest carbon storage, constituting over 70% of the total; cultivated land follows at approximately 25%; whereas impervious surfaces, grasslands, and bare lands collectively store less than 5% of the carbon. During the study period, there was a decline in carbon stock for woodland, cultivated land, grassland, and bare land, while impervious surfaces demonstrated an increase.

4.2 Temporal Variations of Carbon Storage

Throughout the study period, the PYLB witnessed a marked decline in its carbon stock (Fig. 3), decreasing from 1176.85 Tg C in 2000 to 1026.26 Tg C in 2020, with a decline rate of 4.66 Tg C/a, and a cumulative reduction of 150.59 Tg C. The carbon storage per unit area exhibited a fluctuating downward trend, which may be related to changes in land use within the unit area.

4.3 Identification of Key Factors Affecting Carbon Storage

The partial correlation coefficients and Variance Inflation Factors (VIF) were calculated to assess the relationship between carbon stock and 11 contributing factors. A higher VIF value suggests stronger multicollinearity among variables. Therefore, by removing the factors with larger VIF values, seven main determinants affecting carbon storage in the PYLB were identified. As shown in Table 2, carbon storage in the PYLB positively correlated with SM, DEM, and NDVI, with the highest partial correlation coefficient being 0.28 for NDVI. On the other hand, it has a negative correlation with TEM, PRE, GDP, and POP, with the largest partial correlation coefficient being −0.10 for POP.

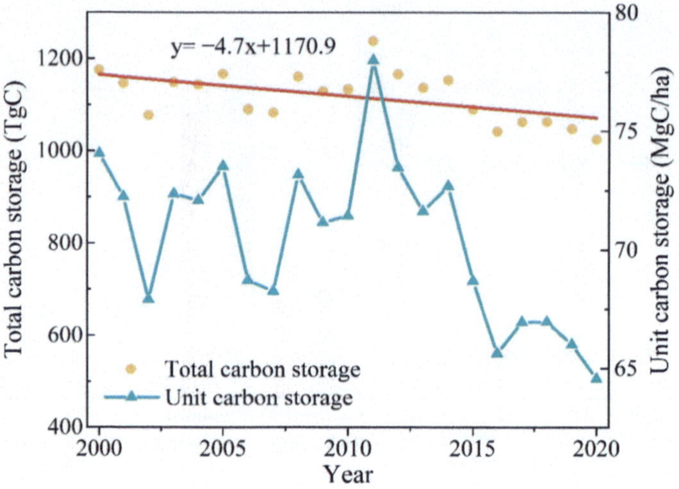

Fig. 3. Total and per-unit area carbon storage.

Table 2. Partial correlation and multicollinearity test of carbon stocks and its influencing factors.

Driving factors	Partial correlation coefficient (r)	VIF
TEM	−0.095501***	1.854
PRE	−0.071639***	1.151
SM	0.150835***	1.190
GDP	−0.004948***	1.119
POP	−0.100642***	1.207
DEM	0.271886***	1.826
NDVI	0.277564***	1.297

Note: *** denotes $p < 0.001$

4.4 Spatial Heterogeneity of the Dominant Factors of Carbon Storage

The MGWR model boasts excellent capability in analyzing the spatial heterogeneity of the primary controls of carbon storage. Figure 4 presents the standardized regression coefficients and local R^2 values for various influencing factors. The MGWR model's overall coefficient of determination, namely R^2, is 0.95, indicating that the regression outcomes are highly explanatory. Judging from the average coefficients of different influencing factors, the most significant factors affecting carbon storage are DEM and NDVI, with average coefficients of 0.58 and 0.46, respectively. PRE and SM follow closely, with average coefficients of −0.23 and 0.21, respectively. TEM, GDP, and POP exert only a minor influence on variations in carbon storage.

As shown in Fig. 4a, b, the standardized regression coefficients of temperature and precipitation range from −0.35 to 1.22 and − 0.85 to 0.33, respectively. The regression

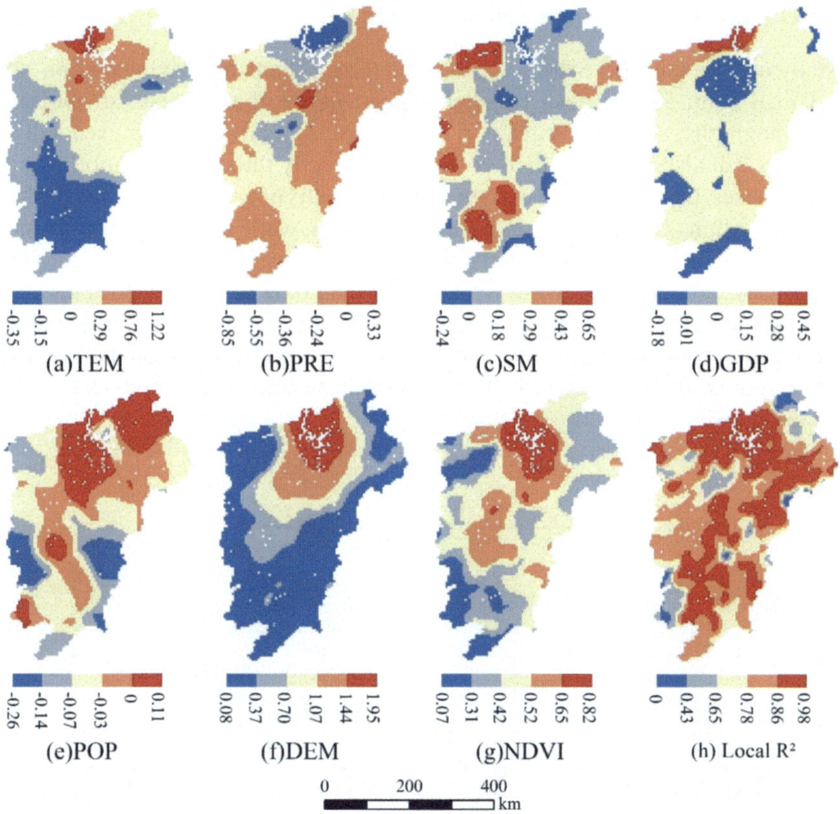

Fig. 4. Spatial distribution of regression coefficients in the MGWR model.

coefficient of temperature presents a "high in the north (+) and low in the south (−)" distribution pattern, indicating that temperature is positively correlated with carbon storage in the northern PYLB, while they are negatively correlated in the southern part. The distribution pattern of precipitation's regression coefficient typically exhibits a pronounced east-to-west gradient, characterized by high values in the eastern regions and low, often negative, values in the west. Figure 4c illustrates that the standardized regression coefficients for soil moisture vary between −0.24 and 0.65, with a predominantly positive correlation. The highest values are observed in the southern and northwestern regions. Figure 4d, e reveal that the standardized regression coefficients for GDP and population size range from −0.18 to 0.45 and −0.26 to 0.11, respectively. The distribution of the GDP regression coefficient is low in the center and high around the periphery. Conversely, the population size regression coefficient exhibits a negative correlation in most areas, with a few exceptions in the northern region where it demonstrates a positive correlation. As shown in Fig. 4f, g, the standardized regression coefficients of elevation and NDVI range between 0.08–1.95 and 0.07–0.82 respectively, both of which have a positive impact on carbon storage, presenting a distribution pattern that gradually decreases from the lake to its surroundings.

4.5 Mechanisms Influencing Carbon Storage

This study utilizes the PLS-SEM to investigate the mechanisms driving carbon storage in the PYLB (Fig. 5). We posit two hypotheses: firstly, that climate, socioeconomic factors, topography, soil, and vegetation directly impact carbon storage; secondly, that these factors can also influence carbon storage indirectly through vegetation. Based on these hypotheses, the study constructs the influence paths of carbon storage in the PYLB. The PLS algorithm is employed to estimate parameters, while the Bootstrapping method is utilized to validate the model results. The findings reveal that all influencing factors have relatively large loadings with T-values less than 0.01 and VIF values less than 3. These results suggest that the selected main influencing factors are highly credible and the hypotheses proposed are reliable for testing. Furthermore, the model's explanatory power coefficient (R^2) reaches 0.552, indicating that the constructed model has moderate explanatory power.

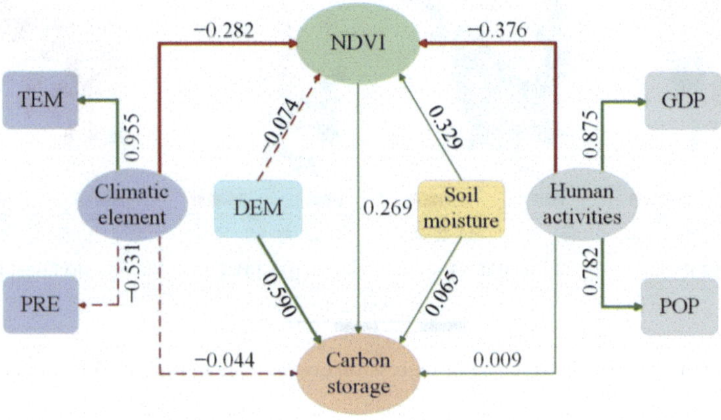

Fig. 5. Results of the PLS-SEM model.

According to Fig. 5, climate and human activities have an inhibitory effect on carbon storage, with comprehensive effects of −0.119 and −0.092 respectively, indicating that economic development, urban construction, or high temperatures may trigger a decrease in carbon stock. In contrast, topography, soil, and vegetation have positive promoting effects on carbon storage, with comprehensive effects of 0.570, 0.153, and 0.269 respectively. From the perspective of the interaction between direct and indirect effects of influencing factors, DEM emerges as the primary factor exerting the strongest influence on variations in carbon storage. Overall, DEM has a positive promoting effect on carbon storage, as higher altitudes are dominated by land use types with higher carbon density, such as forests, leading to higher carbon storage. However, the indirect effect of DEM on carbon storage is negative, indicating that at higher altitudes, the NDVI of vegetation decreases, resulting in reduced carbon storage. Nevertheless, from a comprehensive effect perspective, the positive influencing effect dominates.

5 Discussion

From 2000 to 2020, the spatial distribution of carbon storage within the PYLB region did not undergo notable alterations, which generally exhibited a pattern of "higher in the south than in the north, and higher around the perimeter than in the center." Most regions within the PYLB showed a significant decreasing trend in carbon storage, which is projected to continue. The trends in carbon storage changes across different land uses indicate that the predominant cause for the decline in carbon storage is the encroachment of urbanization on forests, grasslands, and agricultural land.

The factors affecting carbon storage in the PYLB also exhibit significant spatial heterogeneity. Vegetation is essential for maintaining carbon stocks, and a decline in vegetation coverage drastically reduces regional carbon storage. There is a negative correlation between carbon storage and temperature, indicating that High temperatures hinder the effective accumulation of carbon storage. Carbon storage is negatively correlated with GDP and POP, suggesting that with the economic development in recent years, urban expansion and transportation development have exerted substantial adverse effects on carbon storage capacities. Elevation is positively correlated with carbon storage, indicating that as altitude increases and the vegetation type is predominantly forest, carbon storage increases significantly compared to urban areas.

The comprehensive impact effects of climate and human activities on carbon storage are −0.119 and −0.092, respectively, indicating that climate and human activities have an inhibitory effect on carbon storage in the PYLB. The comprehensive impact effect coefficients of topography and soil on carbon storage are 0.570 and 0.153, respectively, suggesting that topography and soil have a promotional effect on carbon stock in the PYLB. The direct impact coefficient of vegetation on carbon storage is 0.269, indicating that vegetation growth or recovery can facilitate an increase in carbon storage.

This study did not consider the carbon stock in the water area of the PYLB. However, water bodies accumulate a small amount of carbon, which has a potential carbon neutrality value. Interactions likely exist between climate, human activities, soil, and topography, but this study did not delve into the possible impacts of the interactions among influencing factors, which warrant further exploration in the future.

6 Conclusion

1. Carbon storage across the PYLB exhibits a characteristic pattern: high in the south and low in the north, with high values around the perimeter and low values in the center. From 2000 to 2020, 93.31% of the PYLB experienced significant reductions in carbon storage.
2. Temperature was positively correlated with carbon storage in the north, whereas it was negatively correlated in the south. Precipitation is negatively correlated with carbon storage in most regions, with the intensity of correlation following a "high in the east and low in the west" pattern. Both GDP and POP adversely impact carbon storage, with the northern regions experiencing a more pronounced effect compared to the southern regions. SM, DEM, and NDVI all contribute positively to enhancing carbon storage, with the degree of impact presenting a gradually decreasing distribution pattern from lakes to their surrounding areas.

3. Terrain, soil, and vegetation factors exhibit a promotional effect on carbon storage in the PYLB, with respective comprehensive effect coefficients of 0.570, 0.153, and 0.269. Both climate and human activities exert inhibitory effects on the carbon storage in the PYLB, with notable comprehensive effect coefficients of −0.119 and −0.092 respectively.

Acknowledgments. This research was supported by the National College Students' Innovative Entrepreneurial Training Plan Program (grant no. 202411481021) and the Joint Funds of the Zhejiang Provincial Natural Science Foundation of China (grant no. LZJWY23E090004).

References

1. Ikram, M., Zhang, Q., Sroufe, R., Shah, S.Z.A.: Towards a sustainable environment: the nexus between ISO 14001, renewable energy consumption, access to electricity, agriculture and CO2 emissions in SAARC countries. Sustain. Product. Consumpt. **22**, 218–230 (2020)
2. Strassburg, B.B., Kelly, A., Balmford, A., Davies, R.G., Gibbs, H.K., Lovett, A., Miles, L., Orme, C.D.L., Price, J., Turner, R.K.: Global congruence of carbon storage and biodiversity in terrestrial ecosystems. Conserv. Lett. **3**(2), 98–105 (2010)
3. Fang, J., Yu, G., Liu, L., Hu, S., Chapin III, F.S.: Climate change, human impacts, and carbon sequestration in China. Proc. Natl. Acad. Sci. **115**(16), 4015–4020 (2018)
4. Nyamari, N., Cabral, P.: Impact of land cover changes on carbon stock trends in Kenya for spatial implementation of REDD+ policy. Appl. Geogr. **133**, 102479 (2021)
5. Li, J., Chen, X., Kurban, A., Van de Voorde, T., De Maeyer, P., Zhang, C.: Coupled SSPs-RCPs scenarios to project the future dynamic variations of water-soil-carbon-biodiversity services in Central Asia. Ecol. Indic. **129**, 107936 (2021)
6. Lin, W., Sun, Y., Nijhuis, S., Wang, Z.: Scenario-based flood risk assessment for urbanizing deltas using future land-use simulation (FLUS): Guangzhou metropolitan area as a case study. Sci. Total Environ. **739**, 139899 (2020)
7. Cao, J., Chen, Y., Hu, Y., Zhang, J., Chen, Y., Yang, B., Wu, S.: Spatiotemporal trends of Forest carbon stock and its response to environmental factors in the Yangtze River Basin during 2005–2020. Forests. **14**(9), 1793 (2023)
8. Li, T., Li, M.-Y., Tian, L.: Dynamics of carbon storage and its drivers in Guangdong Province from 1979 to 2012. Forests. **12**(11), 1482 (2021)
9. Setiawan, O., Rahayu, A.A.D., Samawandana, G., Tata, H.L., Dharmawan, I.W.S., Rachmat, H.H., Suharti, S., Windyoningrum, A., Khotimah, H.: Unraveling land use land cover change, their driving factors, and implication on carbon storage through an integrated modelling approach. Egypt. J. Remote Sens. Space Sci. **27**(4), 615–627 (2024)
10. Piyathilake, I.D.U.H., Udayakumara, E.P.N., Ranaweera, L.V., Gunatilake, S.K.: Modeling predictive assessment of carbon storage using InVEST model in Uva province, Sri Lanka. Model. Earth Syst. Environ. **8**, 2213–2223 (2022)
11. Li, H., Wu, Y., Chen, J., Zhao, F., Wang, F., Sun, Y., Zhang, G., Qiu, L.: Responses of soil organic carbon to climate change in the Qilian Mountains and its future projection. J. Hydrol. **596**, 126110 (2021)
12. Fotheringham, A.S., Yang, W., Kang, W.: Multiscale geographically weighted regression (MGWR). Ann. Am. Assoc. Geogr. **107**(6), 1247–1265 (2017)
13. Chin, W.W.: The partial least squares approach to structural equation modeling. In: Modern Methods for Business Research. Lawrence Erlbaum Associates (1998)

Bioenhanced Treatment of Simulated Phenol-Containing Wastewater by Phenol-Tolerant Aerobic Denitrifying Bacteria

Hongxu Liu(✉), Hongyan Wei, and Lizhu Chen

Changchun University of Architecture and Civil Engineering, Changchun 130607, China
187104041@qq.com

Abstract. In this study, an aerobic denitrifying bacterial strain with phenol as the only carbon source was used to make a composite bacterial agent and combined with the SBR process to strengthen the activated sludge treatment of simulated phenol-containing wastewater, and to analyze the effect of the composite bacterial agent on the efficacy of simultaneous nitrogen removal and phenol removal. The results showed that the phenol loading capacity of the SBR system was improved after biofortification, and the ammonia nitrogen, phenol and total nitrogen removal rates were increased by 5.17%, 6.16% and 9.49%, respectively. The bacterial agent used for biofortification belongs to the phylum Aspergillus, and the high-throughput sequencing results showed that its abundance tended to increase in the fortified reactor, indicating that the composite bacterial agent has a good adaptive ability in the system, and it may play a role in the process of nitrogen removal and phenol removal.

Keywords: Heterotrophic nitrification aerobic denitrification · SBR · Simultaneous nitrogen and phenol removal · High-throughput sequencing technology

1 Introduction

Influenced by human activities, the large accumulation of nitrogen-containing compounds such as NH_4^+, NO_3^-, NO_2^- in water during nitrogen cycling and the transformation process in nature not only cause the aggravation of climate change [1], eutrophication of the water body, but also impede the long-term development of agriculture and fishery and even tourism. The traditional wastewater nitrogen removal process consists of nitrification and denitrification to form a complete nitrogen removal process, and this process is still the mainstream process [2]. However, due to the different physiological mechanisms of nitrifying bacteria and denitrifying bacteria, such as carbon source, dissolved oxygen, resistance to shock loading capacity and other conditions of the variability of the factors [3], resulting in the actual operation of both time and space can not be unified, and thus nitrification under aerobic conditions and denitrification under anoxic conditions need to be carried out in different reaction units. In

V. Achal et al. (Eds.): EPPC 2024, ESE, pp. 253–259, 2025.
https://doi.org/10.1007/978-3-031-92123-0_27

recent years, Simultaneous Nitrification and Denitrification (SND) has received a lot of attention from many researchers, including Heterotrophic Nitrifier and Aerobic Denitrification (HN-Aerobic Denitrifier (HN-AD). HN-AD bacteria can convert NH_4^+, NO_3^-, and NO_2^- into nitrogenous gases under aerobic conditions [4] to remove total nitrogen. The advantages of aerobic denitrification as a promising denitrification process include fast nitrogen removal and high adaptability [5], short growth cycle [6], the ability to simplify the process and reduce energy consumption [7], and so on.

Phenol, as an organic substance that is toxic and inhibitory to most organisms, is widely present in the wastewater of coal chemical, coking, refining and pharmaceutical industries [8]. In the natural environment phenol not only causes contamination of soil and water, but also has a strong inhibitory effect on the activity of some microorganisms [9]. In previous studies of biological denitrification, organic matter has been used as both an electron donor and a substrate for the reaction [10]. Phenol as a carbon source also fulfills that condition. However, it is utilized as a growth carbon source by heterotrophic nitrifying aerobic denitrifying bacteria and participates in the metabolic activities of this system [11], which can reduce the amount of carbon source added to the conventional nitrification process, and also achieves the purpose of treating waste with waste.

In this study, we used heterotrophic nitrifying aerobic denitrifying strains that have been screened for their ability to utilize phenol as a carbon source, and made them into a composite bacterial agent to carry out biofortification technology on Sequencing Batch Reactor (SBR) to investigate the removal effect of ammonia nitrogen and phenol after biofortification. The effects of phenol and biofortification on the characteristics of microbial populations in the system were also investigated.

2 Materials and Methods

2.1 Source of Inoculated Sludge and Bacterial Agent

The inoculated sludge was taken from the biochemical tank of Jilin wastewater treatment plant, and the initial MLSS value was determined. The two strains of phenol-resistant aerobic denitrifying bacteria screened in the laboratory were Fusobacterium spp. (Registration No. HQ538658) and Pseudomonas spp. (Registration No.MF276642), and the screened strains of aerobic denitrifying bacteria were connected to the enrichment medium, and cultured for 24 h at 30 °C, 140 rpm, and then centrifuged to collect the bacteria, and then rinsed twice with inorganic salt medium. Centrifugation to collect the bacterial body, and finally resuspend the bacterial body with sterile water to the bacterial solution with OD600 of 0.5. The composite bacterial agent was made by mixing the two bacterial liquids 1:1.

2.2 Reactor Unit and Operating Conditions

The experiment used two parallel operation of the SBR as a reactor as shown in Fig. 1, respectively named SBR1 and SBR2. SBR reactor material is organic glass, the whole is cylindrical, the effective volume of the reactor is 2.0 L, the internal effective depth of the water is 20 cm, the diameter of the reactor is 10 cm. The outer layer of the water bath compartment circulating water to control the reaction temperature. The reactor is pumped by a peristaltic pump into the feed water, the stirring device is a mechanical stirrer, the reactor is connected to the internal aeration device, the intensity of the gas flow meter to regulate the whole process through the program controller control to achieve autonomous operation. The reaction process is divided into five stages: water inlet, reaction, sedimentation, water and idle.

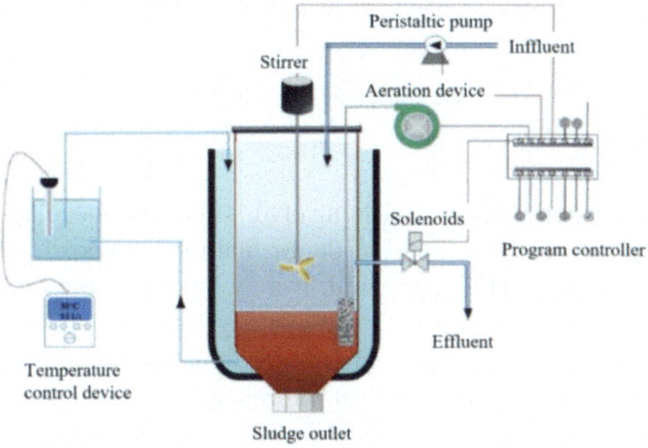

Fig. 1. Schematic diagram of SBR reactor

2.3 Measurement Methods

Phenol was determined by 4-aminoantipyrine direct photometry, NH_4^+-N was determined by Nano reagent photometry, NO_2^- -N was determined by N-(1-naphthyl)-ethylenediamine photometry, NO_3^- -N was determined by muscimol spectrophotometry, TN was determined by alkaline potassium persulphate digestion and ultraviolet spectrophotometry, COD was determined by rapid elimination spectrophotometry.

2.4 Analysis of Microbial Community Structure

In order to investigate the changes of microbial community structure in the SBR reactor, sludge samples were taken to determine the biological community composition by high-throughput sequencing analysis. The activated sludge samples under different working conditions of the SBR process were selected, and the specific naming and sampling locations are shown in Table 1:

3 Results and Discussion

3.1 Efficacy of the Bio-Enhanced SBR Process for Simultaneous Nitrogen and Phenol Removal

The reactor operation effect is shown in Fig. 2. From the data changes, the 31st-35th cycle after the dosing of bacterial agents, the effluent water quality fluctuated slightly, and the effluent nitrite nitrogen and total nitrogen appeared to increase slightly, which

Table 1. Activated sludge sample naming

Sample name	Reactor conditions
LSBR1B	Raw sludge from reactor 1
LSBR1A1	Activated sludge after addition of phenol to reactor 1
LSBR1A2	Activated sludge after bioaugmentation in reactor 1
LSBR2B	Raw sludge from reactor 2
LSBR2A1	Activated sludge after addition of phenol to reactor 2
LSBR2A2	Activated sludge in late reactor 2

phenomenon also appeared in Chen Haisheng et al.'s [12] enhanced aerobic SBR reactor using heterotrophic nitrification- aerobic denitrification composite strains Delftia sp. YH01 and Acidovorax sp. YH02. appeared. It may be due to the fact that the addition of the composite strains competed with the existing microorganisms in the system in terms of nutrition and living space, which changed the microbial community structure in the SBR, and therefore the efficiency of the reactor in removing the pollutants was temporarily affected.

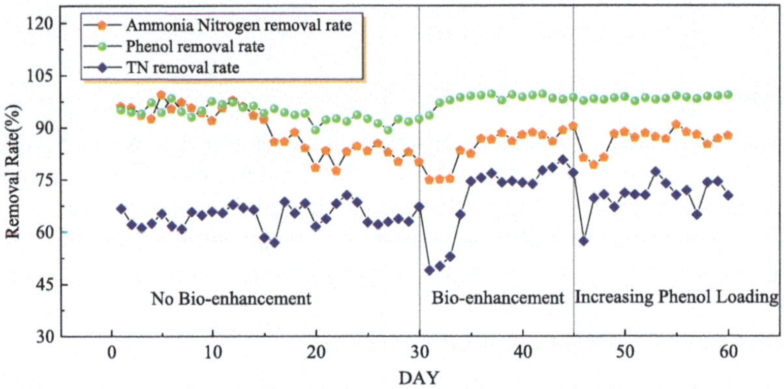

Fig. 2. Effect of simultaneous nitrogen and phenol removal before and after SBR1 biofortification

In the 35th–45th cycles after the acclimatization period, the average ammonia nitrogen removal before and after bioaugmentation of SBR1 increased by 5.17%, phenol removal by 6.16%, and total nitrogen removal by 9.49%, and very little nitrite accumulation occurred. The MLSS at the 45th cycle was 9743.78 mg/L, which was slightly higher than that of 9589.31 mg/L in the non-bioreinforced stage. The reinforced activated sludge cut down the toxicity and inhibition of phenol to microorganisms and increased the biomass of the system.

After increasing the influent phenol load to 500 mg/L at the beginning of the 45th cycle, the bioenhanced SBR1 system could still maintain a high simultaneous nitrogen and phenol removal capacity, with an average phenol removal rate of 97.64% and total nitrogen removal rate of 71.44%. Although slightly lower than the total nitrogen removal before the addition of phenol, the phenol removal was 5.21% higher and the total nitrogen removal was 5.39% higher than SBR2 when comparing to the unenhanced SBR2 system. It shows that the bio-enhanced system still has good removal ability for higher concentration of phenol and can maintain a certain effluent quality, probably because the synergistic effect between the composite strains makes them more adaptive to the environment and more tolerant to the environment, so the composite strains are more suitable for the biosynchronous nitrogen removal and phenol enhancement.

3.2 Changes in Biome Structure

The microbial community was analyzed using high-throughput sequencing, and the phylum-level bacterial structure is shown in Fig. 3. As shown in the figure, Proteobacteria, Actinobacteria, Bacteroidetes, Firmicutes, Nitrospirae Acidobacteria, Chloroflexi were more advanced in abundance within the reactor. And the ones with higher relative abundance at the phylum level were selected to make a community structure Heatmap diagram (Fig. 4).

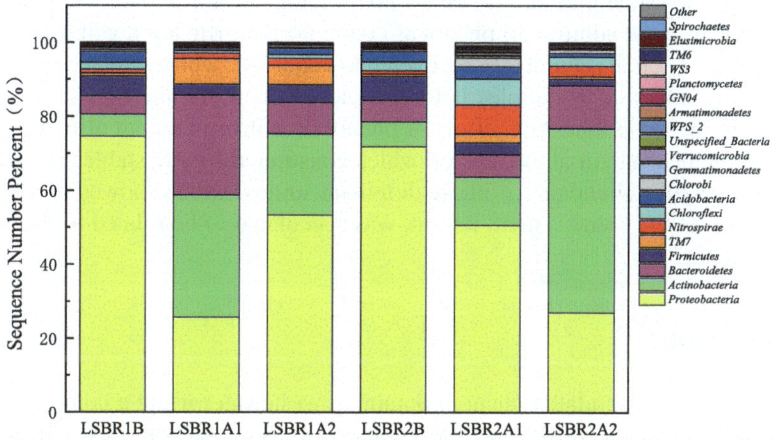

Fig. 3. Relative abundance of bacteria at the phylum level in different samples

Among them, Ascomycetes was always the main bacterial phylum inside the sludge, with the percentage of all of them above 20%, while most of the ammonia-oxidising and nitrifying bacteria related to denitrification belonged to Proteobacteria, which are widely distributed in aerobic denitrification systems [13]. After the addition of phenol, the percentage of Proteobacteria and Actinobacteria in the two reactors decreased in different magnitudes, and the structure of microbial community in the reactors changed, which was guessed to be the main reason for the degradation of NH_4^+-N removal rate in the reactors. After the addition of biofortified bacteria, the Proteobacteria percentage rebounded and re-emerged as the most abundant and dominant phylum within the reactor, allowing the fortified process to still have a very good nitrogen removal performance.

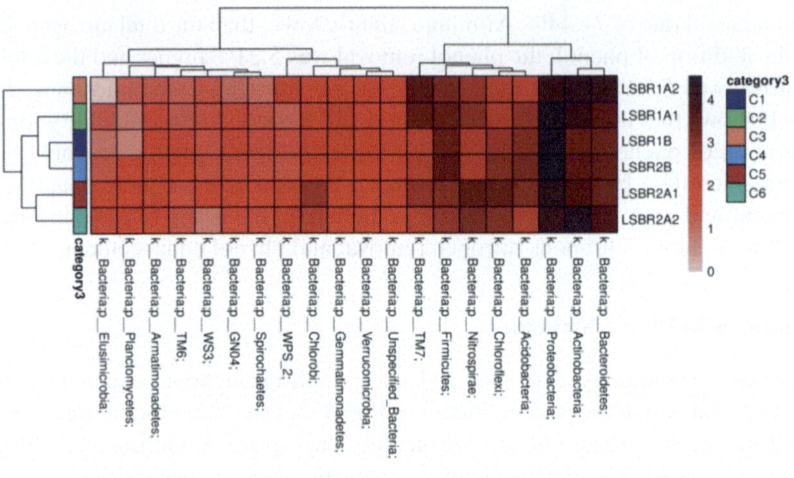

Fig. 4. Heatmap of abundance in the phylum Mycobacterium

Similarly, similar results were found in the enhanced reactor denitrification process using XH02 bacteria by Huang et al. [14] Nitrospirae, a typical nitrite bacterium, decreased in abundance after the addition of phenol affecting the denitrification within the system. After biofortification, its abundance recovered to about 2%, which was even higher than that of the original sludge samples before. It can be seen that the biofortification not only reduced the toxic inhibition effect of phenol on Nitrospirae, but also adjusted the denitrification mechanism of the reactor, which constituted a more stable denitrification system. The relative abundance of these dominant strains, which showed an increasing trend throughout the enhancement period, was speculated to be related to the addition of composite bacteria.

4 Conclusions

1. The treatment of simulated phenol-containing wastewater using a composite bacterial agent bio-enhanced SBR reactor improved the simultaneous denitrification and phenol removal efficiency of the reactor, and the removal rates of phenol, ammonia nitrogen and total nitrogen before and after bio-enhancement were increased by 6.16%, 5.17% and 9.49%, respectively.
2. The microbial community diversity in the reactor decreased after the phenol load was increased in the influent water, revealing the toxicity and inhibitory effect of phenol on microorganisms; the microbial community abundance in the reactor increased after bioaugmentation and was higher than that in the case of non-bioaugmentation, and all the synchronous denitrification and phenol removal efficiencies were also rebounded.
3. At the phylum classification level, the dominant phyla in the SBR reactor were Proteobacteria, Actinobacteria and Bacteroidetes. With the biological enhancement, the abundance of these phyla showed an upward trend, and the composite bacterial agent had a better ability to adapt to the environment in the reactor, which played an important role in the simultaneous denitrification and phenol removal.

Acknowledgments. Science and Technology Research Project of Jilin Provincial Department of Education, 2023 (NO.JJKH20231487KJ).

Natural Science Research Project of Changchun University of Architecture and Civil Engineering, 2025 (NO.CJKJ202511).

References

1. Ramírez-Melgarejo, M., Reyes-Figueroa, A.D., Gassó-Domingo, S., Güereca, L.P.: Analysis of empirical methods for the quantification of N2O emissions in wastewater treatment plants: comparison of emission results obtained from the IPCC Tier 1 methodology and the methodologies that integrate operational data[j]. Sci. Total Environ. **747**, 141288 (2020)
2. Zhang, M., Jun, G., Liu, Y.: Engineering feasibility, economic viability and environmental sustainability of energy recovery from nitrous oxide in biological wastewater treatment plant[J]. Bioresour. Technol. **282** (2019)
3. Wang, Q., He, J.: Complete nitrogen removal via simultaneous nitrification and denitrification by a novel phosphate accumulating Thauera sp. strain SND5[J]. Water Res. **185**, 116300–116300 (2020)
4. Xie, F., Thiri, M., Wang, H.: Simultaneous heterotrophic nitrification and aerobic denitrification by a novel isolated Pseudomonas mendocina X49[J]. Bioresour. Technol. **319**, 124198 (2021)
5. Yang, J.R., Wang, Y., Chen, H., et al.: Ammonium removal characteristics of an acid-resistant bacterium Acinetobacter sp. JR1 from pharmaceutical wastewater capable of heterotrophic nitrification-aerobic denitrification[J]. Bioresour. Technol. **274**, 56–64 (2019)
6. He, T.X., Li, Z.L., Sun, Q., et al.: Heterotrophic nitrification and aerobic denitrification by Pseudomonas tolaasii Y-11 without nitrite accumulation during nitrogen conversion[J]. Bioresour. Technol. **200**, 493–499 (2016)
7. Lauchnor, E.G., Semprini, L.: Inhibition of phenol on the rates of ammonia oxidation by nitrosomonas europaea grown under batch, continuous fed, and biofilm conditions[J]. Water Res. **47**(13), 4692–4700 (2013)
8. Chunyan, X., Han, H., Jia, S., Zhao, Q.: Influence of phenol on ammonia removal in an intermittent aeration bioreactor treating biologically pretreated coal gasification wastewater[J]. J. Environ. Sci. **43**(05), 99–105 (2016)
9. Rajta, A., Bhatia, R., Setia, H., Pathania, P.: Role of heterotrophic aerobic denitrifying bacteria in nitrate removal from wastewater. J. Appl. Microbiol. **128**(5), 1261-1278 (2020)
10. Ge, Q., Yue, X., Wang, G.: Simultaneous heterotrophic nitrification and aerobic denitrification at high initial phenol concentration by isolated bacterium Diaphorobacter sp. PD-7[J]. Chin. J. Chem. Eng. **23**(5), 835–841 (2015)
11. Goulding, C., Gillen, C.J., Bolton, E.: Biodegradation of substituted benzenes [J]. J. Appl. Bacteriol. **65**(1), 1–5 (1988)
12. Haisheng, C., Gang, C., et al.: Aerobic denitrification and microbial community shift in SBR bioaugmented with strains YH01 and YH02[J]. Environ. Sci. **39**(04), 1773–1781 (2018)
13. Du, C., Cui, C.W., Qiu, S., et al.: Nitrogen removal and microbial community shift in an aerobic denitrification reactor bioaugmented with a Pseudomonas strain for coal-based ethylene glycol industry wastewater treatment[J]. Environ. Sci. Pollut. Res. **24**(12), 11435–11445 (2017)
14. Huang, Z.Z., Cao, G., Li, Z.H., et al.: High-throughput sequencing analysis of community structure in reactor enhanced by heterotrophic nitrification-aerobic denitrification bacteria XH02[J]. China Environ. Sci. **37**(05), 1922–1929 (2017)

Study on Salinity Tolerance of *Cyclotella cryptica* Isolated from Sediment of Ballast Tank of Entering Ships

Jing Li[1,2], Ru Lan[3], and Xiangzhi Lin[1,2](✉)

[1] Ocean and Food School, Quanzhou Normal University, Quanzhou, China
{leekin,xzlin}@qztc.edu.cn
[2] Key Laboratory of Coastal Resource Biotechnology in Fujian Higher Institution, Quanzhou, China
[3] China Waterborne Transport Research Institute, Beijing, China
lanru@wti.ac.cn

Abstract. The large discharge of ballast sediment of entering ships is one of the important reasons for the frequent occurrence of algae red tide disasters in China. Salinity is an important environmental factor affecting algal growth. In this study, the salinity tolerance of *Cyclotella cryptica* which was isolated from the sediment of the entering ship was observed in the cell density, photosynthetic efficiency and symbiotic bacteria in salinity 0, 5, 10, 15, 20, 25 and 30 culturing medium. The results showed that *C. cryptica* could grow in the salinity range of 0–30, with high cell density under salinity 20 and 25, low cell density under salinity 5 and 10, and photosynthetic efficiency decreased with the increase of salinity. In this experiment, *C. cryptica* symbiotic bacteria were detected under different salinity conditions, and the relative abundance of *Marinobacter*, *Limnobacter*, *Celeribacter*, unclassified *Bacteroidetes*, and unclassified Rhodobacteriaceae were high. *Seobacterium* genus showed higher abundance at salinity 10, 15, 20, 25, 30 and lower abundance at salinity 0 and 5. The genus showed higher abundance at salinity 0 and lower abundance at other salinities. Tachyophyacter showed a higher abundance at salinity 5, 10 and a lower abundance under other salinity conditions. Unclassified Bacteroidetes showed higher abundance at salinity 0, 5, 10, 15, 20 and 25 and lower abundance at salinity 30. Unclassified Rhothrobacteriaceae showed higher abundance at salinity 30 and lower abundance at salinity 0, 5, 10, 15, 20, and 25. Through a variety of complex effects, symbiont bacteria show differences in cell density and photosynthetic rate under different salinity. In this study, by studying the growth changes of *C. cryptica* and its symbiotic bacteria under different salinity conditions, we can provide theoretical basis and data support for the prevention and control of *C. cryptica* in the sediment of ballast tanks of imported ships.

Keywords: *Cyclotella cryptica* · Symbiotic bacteria · Salinity tolerance · Sediment of ballast tank

V. Achal et al. (Eds.): EPPC 2024, ESE, pp. 260–270, 2025.
https://doi.org/10.1007/978-3-031-92123-0_28

1 Introduction

In recent years, the invasion by alien red tide algae species increased by more and more discharge of ballast water and sediment from entering ships [1–3]. With the quick development of China's economy, a high speed of coastal ports constructing came out since the 1980s, which led to an increasing discharge of sediment into China's coastal ports from entering ships' ballast tanks [4]. The sediment was composed by solid materials in ballast water precipitation, which entrained large biological groups, such as bacteria, viruses, algae, zooplankton and cyst, such as phytoplankton spores could reach 22,500 cells/cm^3 [5]. The sediment organisms could invade into the port waters, which may cause local biological extinction, or reduce biological genetic diversity, disturb the local ecosystem balance, damage the marine ecological environment [6].

In recent years, red tide occurred frequently in coastal China, and some red tide algae were adventive invasive species [7, 8]. It was found that algae in the sediment of the ballast tank were sometimes the same as the red tide ones along the coast of China. Therefore, the identification and physiology of algae species in the sediment of the ballast tank were getting more and more attention, and the related work is also increasing [9, 10]. It was estimated that at least 10,000 tons of sediment was discharged into the port every year, which gave a huge threat to the ecology, economy and human health of the discharge area [8].

Salinity was an important environmental factor that can affect the algae growth, either higher or lower salinity could inhibit the growth of algae [11]. Excessive salinity would kill algae, because salinity could affect their osmotic pressure, nutrient absorption, and photosynthesis processes. Suitable salinity was an important reason for the colonizing, spreading and finally blooming of the red tide [12, 13]. Through study on growth and symbiotic-bacteria of *Cyclotella cryptica* that lived in different salinity culturing medium, response to salinity tolerance was analyzed to provide theoretical basis and data support for the prevention and control of its invasion, and which could be helpful to marine ecological management in coastal waters.

2 Materials and Methods

2.1 Algal Isolation

The sediment sample was from a foreign cargo ship on the China-Australia route (sample no. 2022-DN-70, Jiangyin Customs). Natural seawater (salinity 35) was diluted to prepare f/2 medium with salinity 0, 8, 16, 35. Then sample 2022-DN-70 was cultured in these mediums with different salinity. *C. cryptica* only grew out in the salinity 16 medium, none of others. The *C. cryptica* was isolated by micropipette to do the salinity tolerance experiments.

2.2 Experimental Design

The *C. cryptica* was cultured in 7 different salinities (0, 5, 10, 15, 20, 25, 30), and each of them had 3 parallel sets. Experiments were performed in 250 mL erlenmeyer flasks with 100 mL medium. Algae were shaken three times a day during the experiment. The algae solution was collected at 10:00 am daily to detect cell density and maximum photosynthetic efficiency. After 14 days, the DNA of symbiotic-bacteria was extracted to analyze the species diversity.

2.3 Detection of Cell Density, the Maximum Photosynthetic Efficiency

After shaking, 3 mL of *C. cryptica* culturing solution was collected, fixed by 1.5% Lugol's solution, and then counted under microscope (Nikon ECLIPSE E200, Japan) to detect cell density. Another 3 mL of algae solution was collected and protected from light for 30 min, then the maximum photosynthetic efficiency Fv/Fm value was detected by algal chlorophyll fluorometer (WALZ Water-PAM, Germany).

2.4 Detection of Algal Symbiotic Bacteria

The 7 salinity cultured *C. cryptica* was filtered to make the algal cells stay on the filter membrane, then the algal symbiont bacteria was extracted with DNA extraction kit (FastDNA™ SPIN Kit for Soil, MP Biomedicals, USA) and sent to Biological Engineering (Shanghai) Co., Ltd. for testing to obtain information on the species' composition of algal symbiont bacteria.

2.5 Data Processing

The growth curve, maximum photosynthetic efficiency changes of *C. cryptica* cultured with different salinity were analyzed using Excel et al. The composition of algal symbiotic species cultured with different salinity was analyzed by Rank-abundance et al.

3 Results and Analysis

3.1 Effect of Salinity on the Cell Growth

Cell density was 2.8 × 103 cells/mL on day 0. On day 4, the cell density of all seven salinity cultures increased. On day 6, the cell density of the salinity 30 decreased, in opposite the other 6 salinity cultures increased. From day 6–12, the cell density of *C. cryptica* increased, meanwhile the cell density at salinity 0, 10, 15 decreased on day 8 and then continued to increase. On day 14, the cell density cultured at salinity 0, 5, 15 decreased when compared to day 12, though the other four increased. According to the change of cell density, the optimal salinity should be 20–25 (Fig. 1).

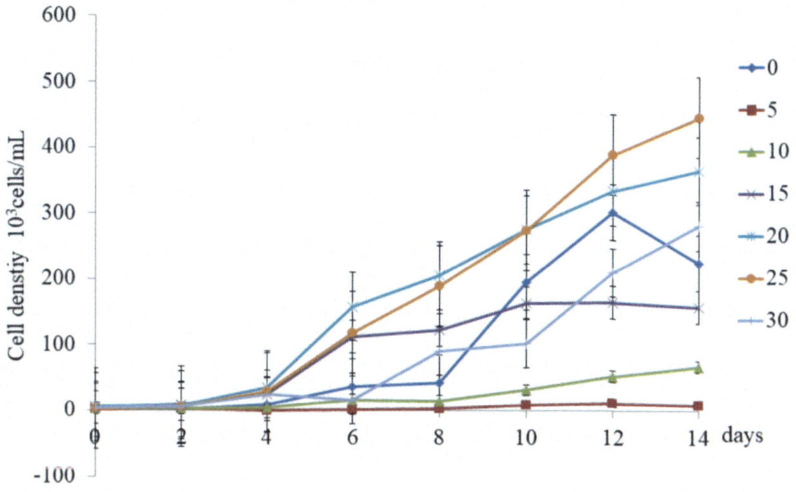

Fig. 1. Cell density of *C. cryptica* in 7 different salinity culturing mediums

3.2 Effect of Salinity on the Photosynthesis Efficiency of Cells

There were no significant differences in Fv/Fm value of *C. cryptica* under different salinity ($P < 10.01$). Fv/Fm values rarely changed during day 0–6. Though the Fv/Fm value of *C. cryptica* in salinity 0 increased on day 8, the others all decreased. Fv/Fm values in salinity 0, 5, 10, 15 increased while the rest decreased. Fv/Fm values decreased in all salinity cultures on day 12. And Fv/Fm values increased at 0, 5, 10 on day 14. In overall, the Fv/Fm values hinted a decreasing trend along with increasing salinity (Fig. 2).

3.3 Composition of Symbiotic Bacteria

According to Table 1, there were 11 main genera of *C. cryptica* symbiotic bacteria under 7 different salinity cultures. *Marinobacter* had the highest abundance 52.62 in salinity 10, abundance was lower at salinity 15, 20, 25, 30, and the lowest abundance was at salinity 0 and 5. *Limnobacter* had the highest abundance 50.84 in salinity 0, and much lower abundance in other 6 salinity (0.21–2.91). *Celeribacter* had a higher abundance in salinity 5 and 10, while lower in the other salinity cultures. *Bacteroides* had higher abundance in salinity 5, 15, 20, 25. Rhodobacteraceae had higher abundance in salinity 30, while lower abundance in other salinities. Low abundance (<10) of *Hydrogenophaga* and *Acinetobacter* were distributed in all 7 different salinity cultures.

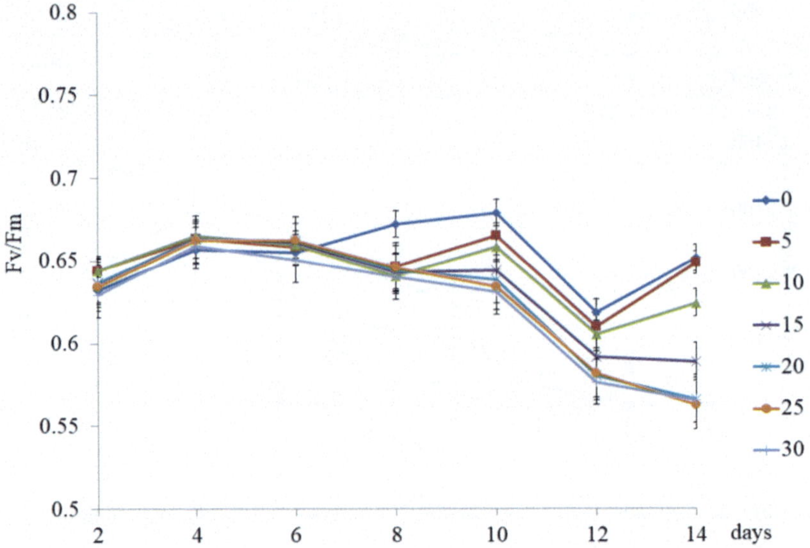

Fig. 2. Fv/Fm value of *C. cryptica* in 7 different salinity culturing mediums

Table 1. Abundance of *C. cryptica*'s dominant symbiotic bacteria in 7 different salinity culturing mediums

symbiotic bacteria	Salinity						
	0	5	10	15	20	25	30
Marinobacter	0.67	11.72	52.62	47.04	43.90	35.77	34.47
Limnobacter	50.84	2.91	2.05	0.32	0.82	0.23	0.21
Bacillariophyta	9.09	13.68	0.89	0.96	4.46	7.47	16.44
Celeribacter	0.63	14.43	10.48	1.21	0.91	0.94	0.82
Sandaracinus	0.04	0.31	0.27	2.93	4.45	5.18	4.08
Hydrogenophaga	1.83	8.62	4.57	0.66	0.34	0.15	0.11
Acinetobacter	1.57	1.02	1.93	1.82	1.88	1.75	1.88
Unclassified Bacteroidetes	19.43	32.03	19.93	33.23	32.07	35.30	9.18
Unclassified Rhodobacteraceae	0.36	3.39	4.92	2.83	1.63	4.66	23.89
Unclassified Xanthomonadaceae	3.46	1.66	1.07	0.36	0.14	0.06	0.03
Unclassified Weeksellaceae	0.58	1.20	0	1.27	1.18	1.21	0.29
Unclassified Comamonadaceae	4.67	0.57	0.20	0.05	0.06	0.04	0.04
Other	6.83	8.55	2.97	7.32	8.18	7.25	8.59

The composition differences of *C. cryptica* symbiotic bacteria varied greatly in 7 salinity cultures. We represented the differences by differential distances. The largest composition differences occurred between salinity 0 and the other 6 salinities with a distance ranged from 0.56 (salinity 5) to 0.74 (salinity 30). The second largest composition differences was salinity 30, which had a distance between 0.31 (salinity 25) and 0.74 (salinity 0) and have a positive correlation with salinity increasing. Symbiotic bacteria of salinity 5 and other 6 salinities had differential distances ranged from 0.36 (salinity 25) to 0.54 (salinity 30). The differential distances of salinity 10 and other 6 salinities were 0.25 (salinity 15) to 0.72 (salinity 0). The differential distances among salinity 15, 20, 25 were short which ranged from 0.07 to 0.14. Moreover, an increasing of differential distances were existed either when salinity increasing or decreasing with salinity 15, 20, 25 (Fig. 3).

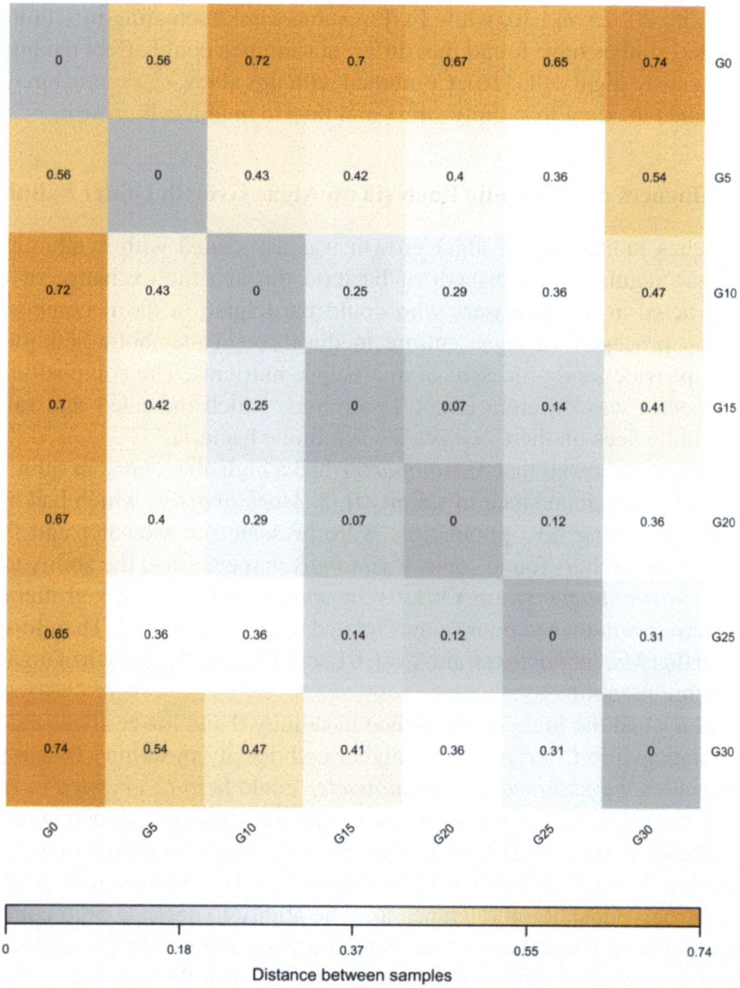

Fig. 3. Abundance distances of *C. cryptica*'s dominant symbiotic bacteria in 7 different salinity culturing mediums

4 Discussion

4.1 Different Salinity Tolerant Expressions After Algae Germinating

Salinity was an important factor in determining the distribution of marine or estuarine algae, and tolerance to salinity changes was a prerequisite for algal survival [13]. Studies have shown that some algae could endure to a wide range of salinity [14]. Our experimental results hinted that *C. cryptica* which germinated only in salinity 16 would adapt quickly then showed a wide range of salinity adaptability (0–30) after germinating.

Photosynthesis was a key reaction for algae survive, which would provide necessary material and energy [15]. In the experiment we found that the photosynthetic rate of *C. cryptica* decreased with increasing salinity from 0–30. Fv/Fm value showed decreasing trends in salinity 20, 25 and 30, while Fv/Fm value were fluctuating in salinity 0, 5, 10 and 15. Related studies have found that different salinities could affect the intracellular pigment contents of algal cells [16]. Combined with this study, *C. cryptica* may regulate pigment contents to adapt to salinity stress and thus maintain cell growth.

4.2 The Influences of Symbiotic Bacteria on Algae Growth Under Salinity Stress

Some researches indicated that algal growth was associated with symbiotic bacteria [17–21]. Algae regulated the growth of bacteria through the exchange of nutrients, and bacteria acted as decomposers who could participate in the decomposition and transformation processes in algae culture media through metabolic activities, which served as the provider and processor of microalgae nutrients. The composition of algal symbiotic bacteria was different under 7 salinities, which indicated that salinity had complicated influences on the *C. cryptica*'s symbiotic bacteria.

Our experiment showed that *Marinobacter* had a high abundance in salinity 10, 15, 20, 25, 30, and a low abundance in salinity 0, 5. *Marinobacter*, which had significant influence on controlling host population, were presented in seawater and freshwater [22–25]. Some researchers found some *Marinobacter* species had the ability to dissolve algae, such as *Marinobacter* strain YWL 01, by which algal cells cracked after infecting, and this bacteria strain owned many genes related to algae lysis [26]. Therefore, it could be speculated that *Marinobacter* strain YWL 01 could inhibit the growth of algae through algae-dissolving function.

Limnobacter had the highest abundance in salinity 0 and lower abundance in other 6 salinities, meanwhile *C. cryptica* had higher cell density in salinity 0 than salinity 5 or 10, from which we assumed that *Limnobacter* could help *C. cryptica* to reduce the low salinity stress. Unclassified *Bacteroidetes* had a relatively high potential ability of polysaccharide degradation [27], with which it could supply nutrients to the photosynthesis reaction of *C. cryptica*. Most of Rhodobacteraceae could take sulfide as the nutrient source, had strong phosphorus tolerance and the ability to degrade high concentration of organic matter. So Rhodobacteraceae would greatly reduce the phosphorus content in water and improve the water environment [28, 29]. The Rhodobacteraceae bacteria had high abundance in salinity 30, from which we could estimate that the function of reducing phosphorus content help *C. cryptica* to overcome the high salinity stress and increase the cell density.

Bacteria could inhibit algae growth through nutrient competition, production of chemical toxins, release of algolysin and other ways. It may also promote algae growth through secreting beneficially metabolic substances or extracellular products. *C. cryptica* symbiotic bacteria affected algal growth through complicated methods, resulting in fluctuating differences in cell density and photosynthetic rate at different salinity. The abundance of *C. cryptica* symbiotic bacteria varied under different salinity conditions, indicating that there was a certain connection between symbiotic bacteria and the environmental salinity. The effect of different symbiotic bacteria on the growth of *C. cryptica* and the variability of symbiotic bacteria community under different salinity conditions should be further investigated.

5 Conclusion

In this paper, we studied the effects of different salinity culture mediums on the cell growth, photosynthetic rate, composition of symbiont bacteria of *C. cryptica*. The results hinted that *C. cryptica* was adaptable to salinity range 0–30, and its optimal growth salinity was 20–30. There was a phenomenon of decreased photosynthetic rate along with increasing salinity. Furthermore, we believed that the divergent abundance of symbiotic bacteria had some effect on the growth and photosynthetic rate of *C. cryptica*. The specific influence mechanism of different symbiotic bacteria under different salinity stresses needed to be further studied.

This study provided some theoretical basis for the prevention and control of the invasion of alien algae in the sediment of the ballast tank of entering ships. In addition, the exploration of the relationship between *C. cryptica* and its' symbiotic bacteria could help us to understand the characteristics and distribution of water microorganisms, which was of great scientific significance to understand and maintain the balance of ecosystem and the sustainable utilization of biological resources of the port waters.

This study can analyze the possibility of the invasion of *C. cryptica* in the ballast sediment of entering ships, and provided theoretical basis and data support for the prevention and control of the invasion. The results could also provide reference for marine ecological management in coastal waters. In addition, the exploration of the relationship between *C. cryptica* and symbiotic bacteria was of great significance to understand the species structure distribution and characteristics of microorganisms in water, and maintain the balance of ecosystem and ensure the sustainable utilization of biological resources.

The research in this paper focused on the influence of salinity on the growth of *C. cryptica*, and preliminarily explored the influence of the symbiotic bacteria on its growth rate and photosynthetic rate, which can provide some help for exploring the mechanism of red tide outbreak of *C. cryptica*. However, there were still some deficiencies this study:

1. According to *C. cryptica* didn't germinate in 35 salinity, we set the salinity gradient 0–30, while the results showed the algae could grow in all 7 salinity mediums. So in further, the range of salinity should be expanded to explore the maximum range of salinity tolerant ability.

2. Algae photosynthesis was a complex mechanism. In this paper, due to the limitation of measurement conditions, the Fv/Fm value (an indicator of chlorophyll fluorescence parameters, also known as maximum photochemical efficiency) was selected to measure the experiment to discuss the effect of salinity on the photosynthesis rate of small algae. If possible, future experiments can increase the photosynthesis related enzyme activity, intracellular pigment content and other indicators to further explore the effect of salinity on the photosynthesis rate of *C. cryptica.*

Acknowledgments. The research was funded by Quanzhou high-level talents working project (2020D23049), Quanzhou Normal University high-level talents working project (2019H19036), Quanzhou CSATC project (2024S350521028). China Waterborne Transport Research Institute Basic Scientific Research Business Fee Project—Chief Researcher Project (wti62425): Research on the Ecological Safety Impact Mechanism and Innovative Control Technology of Ship Ballast Water.

References

1. Ardura, A., Borrell, Y.J., Fernández, S., González Arenales, M., Martínez, J.L., Garcia-Vazquez, E.: Nuisance algae in ballast water facing international conventions. Insights from DNA metabarcoding in ships arriving in Bay of Biscay. Water. **12**(8), 2168 (2020). https://doi.org/10.3390/w12082168
2. Amidi, R., Fatemi, S.M.R., Ghodousi, J., Javid, A.H.: Phytoplankton assemblages in the ballast water of the navigated ships from the Indian Ocean to Shahid Rajaee Port, Hormozgan, Persian Gulf, Iran. Iran. J. Fisher. Sci. **20**(5), 1483–1509 (2021). https://doi.org/10.22092/ijfs.2021.125168
3. Kumar, J.P.P.J., Ragumaran, S., Nandagopal, G., Ravichandran, V., Mallavarapu, R.M., Missimer, T.M.: Green method of stemming the tide of invasive marine and freshwater organisms by natural filtration of shipping ballast water. Environ. Sci. Pollut. Res. **28**, 5116–5125 (2021). https://doi.org/10.1007/s11356-020-10839-4
4. Wu, H., Chen, C., Wang, Q., Lin, J., Xue, J.: The biological content of ballast water in China: a review. Aquac. Fisher. **2**(6), 241–246 (2017). https://doi.org/10.1016/j.aaf.2017.03.002
5. Cuthbert, R.N., Pattison, Z., Taylor, N.G., Verbrugge, L., Diagne, C., Ahmed, D.A., Leroy, B., Angulo, E., Briski, E., Capinha, C., Catford, J.A., Dalu, T., Essl, F., Gozlan, R.E., Haubrock, P.J., Kourantidou, M., Kramer, A.M., Renault, D., Wasserman, R.J., Courchamp, F.: Global economic costs of aquatic invasive alien species. Sci. Total Environ. **775**, 145238 (2021). https://doi.org/10.1016/j.scitotenv.2021.145238
6. Zang, X., Yu, Z., Song, X., Cao, X., Jiang, K.: Insights into the differential removal of various red tide organisms using modified clay: influence of biocellular properties and mechanical interactions. Harmful Algae. **138**, 102695 (2024). https://doi.org/10.1016/j.hal.2024.102695
7. Wang, Q., Lin, L., Chen, X., Wu, W., Wu, H.: Transportation of bloom forming species in ballast water by commercial vessels at Yangshan deep water port. Ocean Coastal Manage. **219**, 106045 (2022). https://doi.org/10.1016/j.ocecoaman.2022.106045
8. Hou, W., Chen, X., Ba, M., Yu, J., Chen, T., Zhu, Y., Bai, J.: Characteristics of harmful algal species in the coastal waters of China from 1990 to 2017. Toxins. **14**(3), 160 (2022). https://doi.org/10.3390/toxins14030160
9. Zheng, X., Lin, L., Xue, M., Wang, Q., Wang, J., Wu, H.: Study on the diversity and germination of dinoflagellate cysts in the sediments of foreign ships in Shanghai port. Mar. Pollut. Bull. **205**, 116566 (2024). https://doi.org/10.1016/j.marpolbul.2024.116566

10. Chen, Y., Xue, J., Feng, W., Du, J., Wu, H.: Bloom forming species transported by ballast water under the management of D-1 and D-2 standards—implications for current ballast water regulations. Mar. Pollut. Bull. **194**, 115391 (2023). https://doi.org/10.1016/j.marpolbul.2023.115391
11. Olli, K., Tamminen, T., Ptacnik, R.: Predictable shifts in diversity and ecosystem function in phytoplankton communities along coastal salinity continua. Limnol. Oceanogr. Lett. **8**(1), 173–180 (2023). https://doi.org/10.1002/lol2.10242
12. Xue, Y., Kun, L.: Competition between *Ulva prolifera* and three red tide microalgae under different light conditions. China Environ. Sci. **39**(8), 3404–3411 (2019). https://doi.org/10.19674/j.cnki.issn1000-6923.2019.0405
13. Tominack, S.A., Coffey, K.Z., Yoskowitz, D., Sutton, G., Wetz, M.S.: An assessment of trends in the frequency and duration of Karenia brevis red tide blooms on the South Texas coast (western Gulf of Mexico). PLoS One. **15**(9), e0239309 (2020). https://doi.org/10.1371/journal.pone.0239309
14. Aidar, E., Gianesella-Galvão, S.M.F., Sigaud, T.C.S., Asano, C.S., Liang, T.H., Rezende, K.R.V., Oishi, M.K., Aranha, F.J., Milani, G.M., Sandes, M.A.L.: Effects of light quality on growth, biochemical composition and photosynthetic production in *Cyclotella caspia* Grunow and *Tetraselmis gracilis* (Kylin) Butcher. J. Exp. Mar. Biol. Ecol. **180**(2), 175–187 (1994). https://doi.org/10.1016/0022-0981(94)90065-5
15. Ping, L., Wei, W., Junhui, Y., Jianwen, L., Limei, Z., Rui, M., Pingzhou, D., Lixin, J.: Spatial and temporal distribution characteristics of phycobiliprotein in Dianchi Lake and analysis of its influencing factors. J. Environ. Eng. Technol. **13**(6), 2174–2183 (2023). https://doi.org/10.12153/j.issn.1674-991X.20230146
16. Shen, A.L., Li, D.J.: Effects of different nutrients levels on the growth of *Procentrum donghaiense* and *Karenia mikimotoi*. Marine Fish. **38**(4), 416–422 (2016). https://doi.org/10.13233/j.cnki.mar.fish.2016.04.009
17. Duan, L.L., Chen, W.L., Zhang, J.J., Ma, H.T., Ji, C.L., Cui, H.L., Wang, J.P., Li, R.Z.: Symbiotic bacteria facilitate algal growth and oil biosynthesis in *Scenedesmus obliquus*. J. Appl. Ecol. **31**(2), 625–633 (2020). https://doi.org/10.13287/j.1001-9332.202002.034
18. Peng, H., de Bashan, L.E., Higgins, B.T.: Comparison of algae growth and symbiotic mechanisms in the presence of plant growth promoting bacteria and non-plant growth promoting bacteria. Algal Res. **53**, 102156 (2021). https://doi.org/10.1016/j.algal.2020.102156
19. Nowruzi, B., Shishir, M.A., Porzani, S.J., Ferdous, U.T.: Exploring the interactions between algae and bacteria. Mini Rev. Med. Chem. **22**(20), 2596–2607 (2022). https://doi.org/10.2174/1389557522666220504141047
20. Palacios, O.A., López, B.R., de-Bashan, L.E.: Microalga growth-promoting bacteria (MGPB): a formal term proposed for beneficial bacteria involved in microalgal–bacterial interactions. Algal Res. **61**, 102585 (2022). https://doi.org/10.1016/j.algal.2021.102585
21. Sun, L., Lei, Y., Li, H.: Exploring the fundamental factors behind algal-bacterial symbiosis and their impact on ecological interactions. Front. Environ. Sci. **10**, 893412 (2022). https://doi.org/10.3389/fenvs.2022.893412
22. Su, J.Q., Yang, X.R., Zheng, T.L., Hong, H.S.: An efficient method to obtain axenic cultures of *Alexandrium tamarense*: a PSP-producing dinoflagellate. J. Microbiol. Methods. **69**(3), 425–430 (2007). https://doi.org/10.1016/j.mimet.2006.07.005
23. Su, J.Q., Yang, X.R., Zheng, T.L., Tian, Y., Jiao, N.Z., Cai, L.Z., Hong, H.S.: Isolation and characterization of a marine algicidal bacterium against the toxic dinoflagellate *Alexandrium tamarense*. Harmful Algae. **6**(6), 799–810 (2007). https://doi.org/10.1016/j.hal.2007.04.004
24. Gärdes, A., Iversen, M.H., Grossart, H.P., Passow, U., Ullrich, M.S.: Diatom-associated bacteria are required for aggregation of *Thalassiosira weissflogii*. ISME J. **5**(3), 436–445 (2011). https://doi.org/10.1038/ismej.2010.145

25. Sonnenschein, E.C., Gärdes, A., Seebah, S., Torres-Monroy, I., Grossart, H.P., Ullrich, M.S.: Development of a genetic system for *Marinobacter adhaerens* HP15 involved in marine aggregate formation by interacting with diatom cells. J. Microbiol. Methods. **872**, 176–183 (2011). https://doi.org/10.1016/j.mimet.2011.08.008
26. Wu, H., Li, D., Yan, X., Xu, R., Liu, F., Zhou, C., Liu, L.: Algicidal activity of *Marinobacter* sp. YWL01 and algicidal mechanism analysis based on its genome sequence. J. Ningbo Univ. (NSEE). **30**(1), 23–28 (2017). https://doi.org/10.3969/j.issn.1001-5132.2017.01.005
27. Jian, S., Wu, Y., Maripatay, T.D., Oren, A., Xu, X.: *Celeribacter ethanolicus* sp. nov., isolated from seawater of the South China Sea. Microbiol. China. **43**(5), 907–916 (2016). https://doi.org/10.13344/j.microbiol.china.160012
28. Zhang, L.S., Yu, D.C., Liu, H.L.: Purification effect of slow rate land treatment system on organic pollutants in municipal wastewater from West Shenyang. Chin. J. Appl. Ecol. **3**(4), 355–363 (1992). https://doi.org/10.1007/BF02005919
29. Liu, J., Xiong, Q.S., Li, Z., Dan, G.X.Q., Zhou, Z.Y., Wang, L.L.: Analysis of microbial composition and diversity in two efficient dissolved alga solutions. J. Chongqing Normal Univ. (Nat. Sci.). **35**(6), 97–104 (2018). https://doi.org/10.11721/cqnuj20180611

The Role of Sulfur-Driven Autotrophic Denitrification in Sustainable Wastewater Treatment

Jiabao Wendy Qi[1], Yi-Lu Sun[2], and Wei-Qin Zhuang[1](✉)

[1] Department of Civil and Environmental Engineering, University of Auckland, Auckland 1142, New Zealand
jqi305@aucklanduni.ac.nz

[2] Key Laboratory of Environmental Biotechnology, Research Center for Eco-Environmental Sciences, Chinese Academy of Sciences, Beijing 100085, China

Abstract. Sulfur-driven autotrophic denitrification (SdAD) utilizes reduced sulfur compounds as electron donors for nitrate reduction. Compared to conventional heterotrophic denitrification, SdAD can treat wastewater with low organic carbon to nitrogen ratios, such as dairy factory cleaning wastewater and post-anoxic municipal wastewater. Using sulfur compounds as electron donors avoids the need for external carbon sources for denitrification and reduces the overall carbon footprint of the denitrification process. In addition, previous studies have reported that SdAD produced less N_2O emissions than the conventional heterotrophic denitrification processes. N_2O is a potent greenhouse gas (GHG) with a global warming potential 295 times greater than carbon dioxide. The low GHG-emissions highlights that SdAD can be a more sustainable and environmentally friendly approach for the wastewater treatment industry. However, one of the major challenges of SdAD is the generation of a large quantity of protons, which can consume alkalinity, reduce pH, and inhibit denitrification. To address this challenge, this study evaluated insoluble pH buffering materials—$FeCO_3$ and sustainable $CaCO_3$ materials (e.g., mussel shells)—for their capabilities to provide alkalinity in the SdAD process. Batch cultures with elemental sulfur powder as the sole electron donor were set up. The control experiments contained baseline alkalinity using $NaHCO_3$. $FeCO_3$ (1.74 g/L) and ground green-lipped muscle shell (GLS) (2.13 g/L) were added to different batch cultures, respectively. The nitrate concentrations, pH, sulfate, and phosphorus concentrations were monitored during a time-course experiment. The results showed that New Zealand green-lip mussel shells, among other tested materials, can act as an efficient pH buffer. This study also investigated the prevalence of sulfur-utilizing denitrifiers in New Zealand Wastewater Treatment Plants (WWTPs). Return activated sludge samples were collected from three WWTPs in Auckland and used as seed sludge for batch experiments. Sodium nitrate and elemental sulfur powder were used as the sole electron acceptor and donor. The results showed that sulfur-utilizing denitrifies were ubiquitous in these WWTPs.

Keywords: Autotrophic Denitrification · Elemental Sulfur · Sustainable Wastewater Treatment · Net-zero-emissions · Low Carbon Footprint

V. Achal et al. (Eds.): EPPC 2024, ESE, pp. 271–280, 2025.
https://doi.org/10.1007/978-3-031-92123-0_29

1 Introduction

Wastewater treatment is a critical aspect of environmental engineering aimed at mitigating the impact of human activities on water bodies. Removing nitrogen compounds, such as nitrate, is essential in wastewater treatment to prevent eutrophication and potential harm to aquatic ecosystems. Nitrate and its metabolic by-product, nitrite, can be harmful to human health [1, 2]. Therefore, the World Health Organization (WHO) suggests the maximum acceptable level for nitrate in public drinking water shall be less than 11.3 mg/L NO_3–N [3, 4].

Biological denitrification can be carried out by both heterotrophic and autotrophic bacteria. Conventional heterotrophic denitrification has been the prevailing method for nitrogen removal, utilizing organic carbon sources as electron donors. Due to organic carbon being oxidized by heterotrophic denitrifying bacteria, carbon dioxide is released from denitrification processes [5–7]. Notably, in the post-anoxic denitrification processes, external organic carbon sources (e.g., methanol or acetate) may be added to boost denitrification efficiency, resulting in both higher CO_2 emissions and sludge production [1]. Furthermore, this process may face challenges when treating wastewater with low C:N ratios, such as wastewater from dairy processing plants or post-anoxic municipal wastewater.

Autotrophic denitrification is a metabolic process in which inorganic compounds such as elemental sulfur (S^0), reduced iron compounds (Fe^{2+}), or hydrogen gas (H_2) are utilized as electron donors with an inorganic carbon source [8–10]. Advantages of autotrophic denitrification include lower excess sludge production and less carry-over of organic substrate to the effluent than heterotrophic denitrification [11]. Notably, it can effectively treat wastewater with limited organic carbon content, reducing the need for external carbon sources and, consequently, the overall carbon footprint of the denitrification process. Among the different autotrophic denitrification processes, sulfur-driven autotrophic denitrification (SdAD) has emerged as an attractive alternative for heterotrophic denitrification due to its low cost and ease of use of elemental sulfur (S^0) as a packing medium and biofilm carrier in packed bed bioreactors [12–14].

As shown in the SdAD stoichiometry (1) below, the proton generation to nitrate reduction is almost 1:1 ratio (NO_3–N:H^+). Removing 1 mg N–NO_3^-/L produces 7.83 mg SO_4^{2-} /L and uses 3.36 mg/L of $CaCO_3$ as alkalinity. These protons can consume alkalinity and reduce pH levels, leading to the inhibition of denitrification efficiency. It has been observed in our study and reported previously that SdAD inhabitation occurred when the pH was below 6 [15]. To overcome this challenge, innovative approaches such as the use of low-cost, low-carbon footprint pH buffering materials need to be explored.

$$S + 0.876\ NO_3^- + 0.023\ CO_2 + 0.343\ H_2O + 0.08\ NH_4^+ + 0.379\ HCO_3^- \\ = 0.08\ C_5H_7O_2N + 0.44\ N_2 + SO_4^{2+} + 0.824\ H^+ \tag{1}$$

The specific aim of this report was to assess the performance and efficiency of insoluble chemicals as pH buffers. The New Zealand green-lipped mussel shell emerged as a promising low-cost and sustainable insoluble alkalinity option among some other materials. In addition, this study focused on using insoluble S_0 to determine if denitrifying sulfur-oxidizing denitrifying bacteria existed in WWTPs and assess the impact of operating factors on the bioreactor performance.

2 Materials and Methods

2.1 Synthetic Wastewater

Both trace element and mineral salts medium stock solutions were prepared using milli-Q water and chemicals of ACS grade and above [16]. The ingredients of the two stock solutions are shown in Tables 1 and 2. Synthetic wastewater (1 L) was prepared using 1 mL trace element stock solution, 10 mL mineral salt stock solution, and 989 mL milli-Q water.

Table 1. Mineral salts stock solution (100×)

Chemical formula	Concentration (g/L)
NaCl	100
$MgCl_2{\cdot}6H_2O$	50
KH_2PO_4	20
NH_4Cl	30
KCl	30
$CaCl_2{\cdot}2H_2O$	1.5
Na_2SO_4	5

Table 2. Trace element stock solution (100×)

Chemical formula	Concentration (g/L)
HCl (33% solution)	7.5 mL
$FeCl_2{\cdot}4H_2O$	1.50
$CoCl_2{\cdot}6H_2O$	0.19
$MnCl_2{\cdot}4H_2O$	0.10
$ZnCl_2$	0.07
H_3BO_3	0.006
$Na_2MoO_4{\cdot}2H_2O$	0.036
$NiCl_2{\cdot}6H_2O$	0.024
$CuCl_2{\cdot}2H_2O$	0.002

2.2 Batch Tests

To investigate the existence of sulfur-oxidizing denitrifying bacteria, return activated sludge (RAS) samples were taken from three major WWTPs in Auckland, New Zealand, namely the Mangere (MG) WWTP, the Rosedale (RD) WWTP, and the Army Bay (AB) WWTP. These three plants play a pivotal role, collectively managing over 90% of Auckland's daily wastewater [17] making them highly suitable candidates for investigating the potential ubiquity of sulfur-oxidizing denitrifying bacteria in New Zealand's WWTPs. The collected activated sludge samples were stored in 20 L carboys in a 4 °C refrigerator to preserve microbial activity.

A 200 mL of RAS was directly transferred into a 500 mL reagent bottle containing 200 mL of the culture medium (2× strength). For each WWTP RAS, three different pH buffers were used, soluble $NaHCO_3$ (as a control), mussels shell, and a sulfur composite granular packing media material (S^0-Composite). The total of nine reactors is summarized in Table 3.

Table 3. Detailed information of the batch tests conducted in this study

No.	WWTP sludge	pH buffer	Note
RD 1	Rosedale	$NaHCO_3$	Control reactor with a soluble alkalinity
RD 2	Rosedale	Mussels shell	Sustainable alkalinity
RD 3	Rosedale	Sulfur composite granular packing media material	(S^0-composite)
MG 1	Mangere	$NaHCO_3$	Control reactor with a soluble alkalinity
MG 2	Mangere	Mussels shell	Sustainable alkalinity
MG 3	Mangere	Sulfur composite granular packing media material	(S^0-composite)
AB 1	Army Bay	$NaHCO_3$	Control reactor with a soluble alkalinity
AB 2	Army Bay	Mussels shell	Sustainable alkalinity
AB 3	Army Bay	Sulfur composite granular packing media material	(S^0-composite)

Then the nine bottles were transferred onto an orbital shaker incubator set at 30 °C and incubated in the dark during the experimental period. The bottles were replenished with S^0 and nitrate when they had been depleted. Additional buffering chemicals were replenished to increase pH if it fell below 6.

2.3 Analytical Chemistry

Prior to taking the daily samples, pH measurements for each bottle were conducted using an Orion 3 stars pH meter to monitor any changes in pH over time.

Then, sample aliquots (1 mL) were taken periodically to analyse nitrate (NO_3^-), nitrite (NO_2^-), phosphate (PO_4^{3-}), and sulfate (SO_4^{2-}) concentrations using an Ion

Chromatography (IC) Instrument (Thermo Scientific ICS-2100). All samples were filtered using 0.45 μm syringe filters and stored at 4 °C if not measured immediately. At selected time intervals, a portion of 200 mL of spend medium was withdrawn from each reactor and was replenished with 200 mL of the fresh culture medium. The samples were also measured using HACH kits for nitrate (HACH method number 836) and nitrite (HACH method number 840) concentrations as an alternative to the IC methods.

3 Results and Discussion

3.1 Ubiquity of Sulfur-Driven Autotrophic Denitrifying Bacteria

Table 4 shows that sulfur-driven autotrophic denitrification happened in all three WWTPs' RAS samples without needing acclimatization. The results demonstrated that sulfur-oxidizing denitrifying bacteria are ubiquitous in all three major WWTPs in Auckland. However, the concentrations of sulfur-oxidizing denitrifying bacteria might be different in these WWTPs. Nevertheless, the results indicate that new SdAD treatment units installed in the future can be self-seeded using RAS from the local WWTP in New Zealand.

Table 4. Ubiquity of SdAD bacteria in three WWTPs' RAS samples.

	Average NO_3-N removal rates (mg/L·d) within the first 24-h incubation	pH on Day 1 ($NaHCO_3$)
MG	143	6.30
RD	1073	6.63
AB	419	6.49

3.2 Sustainable Insoluble Buffer

Batch culture experiments were conducted to search for effective and sustainable insoluble pH buffer chemicals. RAS samples from the three WWTPs were used to inoculate the bioreactors with different buffering chemicals. The experiment lasted for 6 days, with measurements taken on Day 1, Day 2, Day 5, and Day 6.

For the RAS from Mangere WWTP, the control experiment using the soluble alkalinity ($NaHCO_3$) was seen a reduction of nitrate nitrogen concentrations decreased from 424 mg/L NO_3–N to 182 mg/L NO_3–N in 6 days. No nitrite was detected using IC (0.5 mg/L NO_2^- detecting limit) during this period. The pH values of the water samples showed a gradual decline trend from 6.30 on Day 1 to 5.58 on Day 6. The results are shown in Fig. 1 below.

In the test using New Zealand Green-lipped mussels shell (GLS) as the pH buffer, the nitrate concentration is reduced from 440 mg/L NO_3–N to 24 mg/L NO_3–N in 6 days. Again, no nitrite was detected during this period. The pH values of the water samples

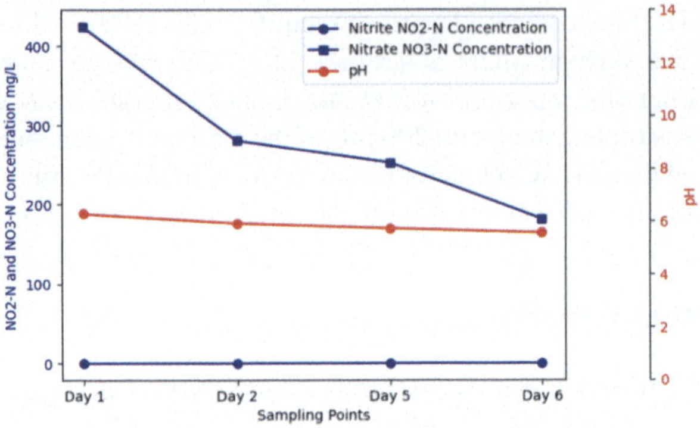

Fig. 1. Denitrification test results using Mangere WWTP sludge and $NaHCO_3$ as pH buffer.

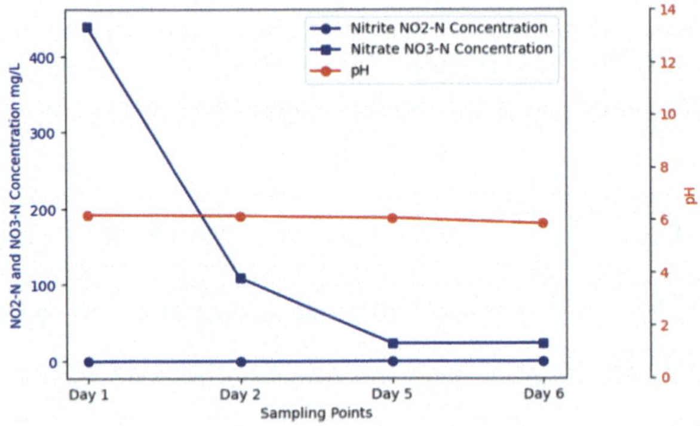

Fig. 2. Denitrification test results using Mangere WWTP sludge and GLM as a pH buffer.

showed a gradual decline trend from 6.22 on Day 1 to 5.88 on Day 6. Results can be seen in Fig. 2 below.

Based on the stoichiometric calculations Eq. (1), the pH would have been dropped to 3.22 ($NaHCO_3$ reactor) and 3.37 (GLS reactor) without these buffering chemicals. Notably, the GLS exhibited a nearly identical pH buffering capacity and performance when compared to the soluble alkalinity $NaHCO_3$. Furthermore, intriguingly, the GLS seemed to induce a slight enhancement in the denitrification process. Notably, both tests demonstrated a decreasing pH trend, indicating the acidification of the systems during the denitrification process. But their buffering capacity could keep the pH above 5.5 even with over 400 mg/L NO_3-N, an extremely high concentration used to get results quickly.

3.3 Green-Lipped Mussel Shell as a Sustainable Buffer

Because GLS showed relatively better buffering capacity in the previous experiments, we conducted experiments using sludge samples from three different WWTPs to have a holistic view. The experiments stopped when the nitrate removal rate reached 95% or more. As shown in Table 5 and Fig. 3, the GLS could buffer a stable pH which supported a consistently good nitrate removal performance across all three WWTPs sludge. These consistent patterns of nitrate reduction across different WWTPs underscore the robustness of sulfur-driven autotrophic denitrification processes when employing the mussel shell as a pH buffer.

Table 5. End-of-cycle pH values in the bioreactors using GLS as an insoluble and sustainable pH buffer.

WWTP	pH
MG	6.22
RD	6.25
AB	6.49

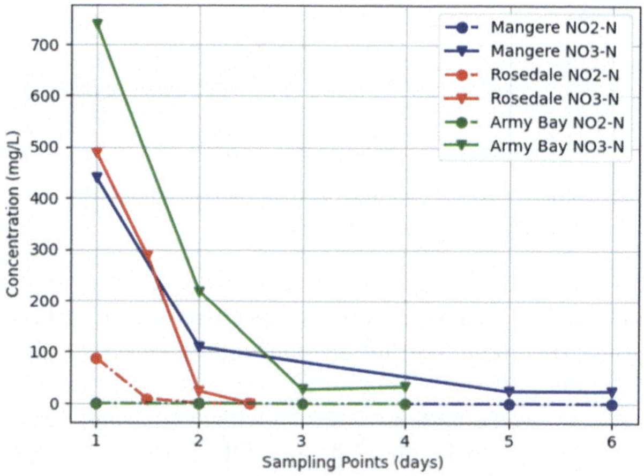

Fig. 3. Denitrification performance with different WWTP sludges and GLS as the sustainable pH buffer. Extremely high NO_3–N concentrations were used in these batch tests to obtain results quickly.

Several studies [18–20] have investigated the viability of using oyster shells as an alkalinity source for autotrophic denitrification. Adding crushed oyster shells also improved the microbial attachment as it increased the surface areas [20]. Green-lipped mussels are indigenous to New Zealand. Farming green-lipped mussels is accountable for the annual production of 90,000 tones of shell waste, and approximately half of these shells are disposed of at landfills [21]. Starting to look at using those shells as a buffer substrate in our SdAD processes could be a solution to reduce solid water and denitrification carbon footprint.

3.4 Sulfur Composite Granular Packing Media

Another study has been done to assess the sulfur composite granular packing media denitrification performance. Sodium bicarbonate and shell were also used to compare with the composite media for supporting the nitrate removal efficiency. The nitrate concentrations were measured periodically and compared for each buffer, as shown in and Fig. 4.

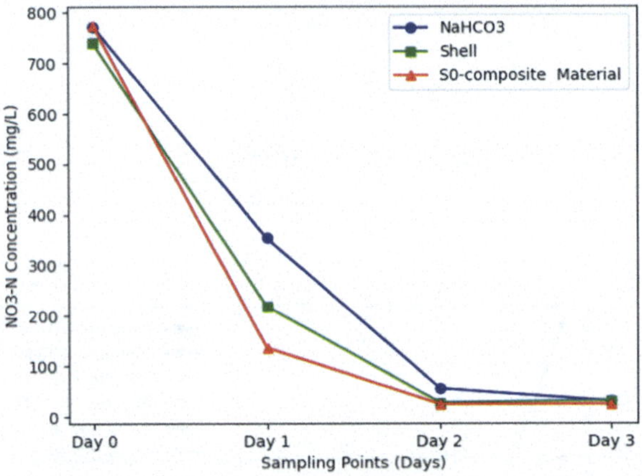

Fig. 4. Nitrate concentrations using Army Bay WWTP sludge with buffer of $NaHCO_3$, shell and the sulfur composite granular packing media material.

Overall, the data indicates that the sulfur composite granular packing media material consistently demonstrated the lowest nitrate concentrations across all sampling points, highlighting its efficient performance in nitrate removal during the denitrification process. The shell buffer showed competitive results with decreasing nitrate concentrations, suggesting its potential as a pH buffering material for denitrification. The $NaHCO_3$ buffer, although effective, exhibited comparatively higher nitrate concentrations at all sampling points, implying slightly less optimal nitrate removal efficiency compared to the other buffers.

4 Conclusions

The study investigated the role of sulfur-driven autotrophic denitrification (SdAD) in sustainable wastewater treatment, focusing on the utilization of different pH buffering materials to address the challenges associated with the process. Through a series of batch experiments using sludge from various WWTPs, the effectiveness of different pH buffers, including $NaHCO_3$, New Zealand green-lipped mussel shells, and a sulfur composite granular packing media material, were evaluated. The results underscore the buffering capacity of the GLS, which consistently exhibited its capability and even appeared to modestly enhance denitrification efficiency. This result suggested mussel shells as a sustainable alternative to conventional chemical alkalinities for SdAD systems.

The study reveals that sulfur-driven autotrophic denitrifiers are widely present in various WWTPs, as evidenced by their successful utilization of S^0 as electron donors in denitrification. This finding has significant implications for the potential implementation of SdAD as an effective and sustainable approach in wastewater treatment processes.

Acknowledgments. J.W.Q. thanks the University of Auckland for the Ph.D. scholarship to undertake the project. W.-Q.Z. thanks the funding from the NIWA-led MBIE contract C01X2203 Fund and the Warwick and Judy Smith Engineering Endowment Fund ENG-2022-01 W-Zhuang for their generous support.

References

1. Singh, S., Anil, A.G., Kumar, V., Kapoor, D., Subramanian, S., Singh, J., Ramamurthy, P.C.: Nitrates in the environment: a critical review of their distribution, sensing techniques, ecological effects and remediation. Chemosphere. **287**, 131996 (2022)
2. Srivastav, A.L., Patel, N., Rani, L., Kumar, P., Dutt, I., Maddodi, B.S., Chaudhary, V.K.: Sustainable options for fertilizer management in agriculture to prevent water contamination : a review. Environ. Dev. Sustain. **26**(4), 8303–8327 (2024)
3. Edition, F.: Guidelines for drinking-water quality. WHO Chron. **38**(4), 104–108 (2011)
4. World Health Organization, & World Health Organization Staff: Guidelines for Drinking-Water Quality, vol. Vol. 1. World Health Organization (2004)
5. Wang, T., Ji, C., Zhou, W., Chen, H., Chen, Y., Liu, Q., et al.: Dry season residual straw reduces nitrous oxide emissions during rice season in upland-paddy rotation systems by inhibiting soil denitrification. Biol. Fertil. Soils, 1–14 (2024)
6. Anand, A., Raghuvanshi, S., Gupta, S.: Sustainable approach for simultaneously reducing CO2 and NO emissions from synthetic industrial flue gases using bacterial consortium and domestic wastewater in a suspended glass bioreactor. Biomass Convers. Biorefinery. **14**(11), 11911–11926 (2024)
7. Patel, R.J., Nerurkar, A.S.: Thauera sp. for efficient nitrate removal in continuous denitrifying moving bed biofilm reactor. Bioprocess Biosyst. Eng. **47**(3), 429–442 (2024)
8. Xiao, J., Yue, Q., Gao, B., Sun, Y., Kong, J., Gao, Y., et al.: Performance of activated carbon/nanoscale zero-valent iron for removal of trihalomethanes (THMs) at infinitesimal concentration in drinking water. Chem. Eng. J. **253**, 63–72 (2014)
9. Yuan, Q., Gao, J., Liu, P., Huang, Z., Li, L.: Autotrophic denitrification based on sulfur-iron minerals: advanced wastewater treatment technology with simultaneous nitrogen and phosphorus removal. Environ. Sci. Pollut. Res. **31**(5), 6766–6781 (2024)
10. Ponce-Jahen, S.J., Cercado, B., Estrada-Arriaga, E.B., Rangel-Mendez, J.R., Cervantes, F.J.: Anammox with alternative electron acceptors: perspectives for nitrogen removal from wastewaters. Biodegradation. **35**(1), 47–70 (2024)
11. Tong, S., Rodriguez-Gonzalez, L.C., Feng, C., Ergas, S.J.: Comparison of particulate pyrite autotrophic denitrification (PPAD) and sulfur oxidizing denitrification (SOD) for treatment of nitrified wastewater. Water Sci. Technol. **75**(1), 239–246 (2017)
12. Sulfur-oxidizing autotrophic denitrification. Water Environ. Res. 79(13), 2519–2526
13. Di Capua, F., Papirio, S., Lens, P.N., Esposito, G.: Chemolithotrophic denitrification in biofilm reactors. Chem. Eng. J. **280**, 643–657 (2015)
14. Li, R., Morrison, L., Collins, G., Li, A., Zhan, X.: Simultaneous nitrate and phosphate removal from wastewater lacking organic matter through microbial oxidation of pyrrhotite coupled to nitrate reduction. Water Res. **96**, 32–41 (2016)

15. Di Capua, F., Pirozzi, F., Lens, P.N., Esposito, G.: Electron donors for autotrophic denitrification. Chem. Eng. J. **362**, 922–937 (2019)
16. Zhuang, W.-Q., Yi, S., Feng, X., Zinder, S.H., Tang, Y.J., Alvarez-Cohen, L.: Selective utilization of exogenous amino acids by Dehalococcoides ethenogenes strain 195 and its effects on growth and dechlorination activity. Appl. Environ. Microbiol. **77**(21), 7797–7803 (2011)
17. Watercare Ltd.: Wastewater collection and treatment Retrieved from https://www.watercare.co.nz/Water-and-wastewater/Wastewater-collection-and-treatment#:~:text=More%20than%2090%20per%20cent,and%20our%20coasts%20and%20harbours (2015)
18. Liang, J., Chen, N., Tong, S., Liu, Y., Feng, C.: Sulfur autotrophic denitrification (SAD) driven by homogeneous composite particles containing CaCO3-type kitchen waste for groundwater remediation. Chemosphere. **212**, 954–963 (2018)
19. Dong, S., Xu, T.: Analysis of the impact of urbanization on precipitation characteristics and trends in Shanghai. In: Civil Engineering and Urban Research, vol. 2, pp. 562–573. CRC Press (2023)
20. Sengupta, S., Ergas, S.J., Lopez-Luna, E.: Investigation of solid-phase buffers for sulfur-oxidizing autotrophic denitrification. Water Environ. Res. **79**(13), 2519–2526 (2007)
21. O'Sullivan, A.: New Zealand: Waste mussel shells to filter rivers and streams. Retrieved from https://waste-management-world.com/artikel/new-zealand-waste-mussel-shells-to-filter-rivers-and-streams/ (Jul 26, 2021)

Research and Application of Substation Wastewater Treatment System Based on Anaerobic Technology

Yanling Guan[1(✉)], Lijun Fu[1], Liang Zhang[1], Yanyan Guan[2], and Bing Wei[1]

[1] Electric Power Research Institute of State Grid Heilongjiang Electric Power Co., Ltd., Harbin 150031, China
Guanyanling1220@163.com

[2] State Grid Heilongjiang Electric Power Co., Ltd., Economic and Technological Research Institute, Harbin, China

Abstract. Transformer substations, as critical infrastructure in the power system, generate wastewater during operation, posing significant environmental and maintenance challenges. Traditional treatment methods, such as septic tanks and integrated powered sewage treatment devices, often fall short of meeting increasingly stringent environmental standards. With the rapid development of rural wastewater treatment technologies, there is a pressing need to develop efficient and economical wastewater treatment technologies suitable for transformer substations. This paper examines the limitations of existing treatment technologies based on the characteristics of substation wastewater. It explores innovative solutions through the study of anaerobic combined treatment processes and powered oxidation treatment processes. The aim is to provide theoretical support and practical guidance for the efficient treatment of wastewater in transformer substations.

Keywords: substation · wastewater · anaerobic Technology · process

1 Introduction

1.1 Research Background

As a key facility of the power system, the substation will produce a certain amount of domestic and industrial wastewater during its operation. If wastewater are not properly treated and directly discharged, they will cause serious pollution to the environment. Traditional wastewater treatment methods, such as septic tanks and power-based sewage treatment equipment, are widely used, but they expose many problems in actual operation and are difficult to meet the increasingly strict environmental protection standards. With the continuous improvement environmental protection regulations and the enhancement of people's environmental protection awareness, the upgrading of substation wastewater treatment technology is urgent. At the same time, the rapid development rural sewage treatment technology has provided new ideas and solutions for substation wastewater treatment. Therefore, the research on efficient and economical wastewater treatment technology suitable for substations is great significance for improving environmental quality and ensuring the green operation of power systems.

V. Achal et al. (Eds.): EPPC 2024, ESE, pp. 281–291, 2025.
https://doi.org/10.1007/978-3-031-92123-0_30

1.2 Research Purpose and Significance

By systematically analyzing the characteristics of substation wastewater discharge and the limitations of existing treatment technologies, this study proposes and verifies an efficient wastewater treatment plan suitable for substations. Specifically, the purpose of this research is to explore how to improve the efficiency of wastewater treatment and the quality of treated water to meet environmental protection requirements through anaerobic combined treatment processes and dynamic oxidation treatment processes. This study not only helps solve practical problems in substation wastewater treatment and enhances environmental performance of substations but also provides reference and guidance for wastewater treatment in other similar facilities. Additionally, this research can promote the modernization and green transformation of station operation and management, contributing to the construction of a resource-saving and environmentally friendly society.

1.3 Research Methods and Content

By conducting a literature review and field research, we will gain a comprehensive understanding of the current status and existing problems of wastewater treatment inations both domestically and internationally. Then, based on the emission characteristics of substation wastewater, we will select appropriate wastewater treatment technologies for experimental research. The specific research includes: performance evaluation of septic tanks and integrated wastewater treatment units, design and optimization of anaerobic combined treatment processes and dynamic oxidation treatment processes. The research include laboratory simulation experiments, on-site trials, and mathematical model analysis. By comparing the treatment effects, operating costs, and maintenance requirements of different treatment processes, we assess their applicability and economic viability. Based on the experimental results, we will propose an optimized wastewater treatment technology plan and verify its practical application. Through the above, we aim to provide substations with efficient, economical, and feasible wastewater treatment technology plans, promoting the development and application of substation wastewater treatment technology.

2 Current Situation of Wastewater Treatment in Substations

2.1 Research Status at Home and Abroad

Research on substation wastewater treatment technology has made certain progress both domestically and internationally. Abroad, research on substation wastewater treatment technology began early as the late twentieth century, mainly focusing on physical and chemical treatment methods for wastewater, such as magnetic flocculation and electrocatalytic. These technologies have shown excellent performance in terms of treatment efficiency and cost-effectiveness [1]. Domestically, the research and application of substation wastewater treatment began to gain attention in the early twenty-first century, with main research directions including biological treatment technology, physical and chemical treatment technology, and integrated treatment [2]. In recent years, with the increase in environmental protection requirements and technological advancements, scholars both domestically and internationally have been exploring high-efficiency, low-cost treatment technologies suitable for substations, in hopes of addressing the deficiencies of existing treatment technologies and meeting the increasingly strict environmental protection standards [3].

2.2 Current Processing Technologies and Their Application Status

At present, the treatment of wastewater from substations mainly adopts septic tanks and integrated wastewater treatment equipment with aeration. Septic tanks due to their simple structure and low cost, are widely used for the preliminary treatment of domestic wastewater from substations. However, their treatment effect is limited and it difficult to meet modern environmental protection requirements. The integrated wastewater treatment equipment with aeration treats wastewater through the biological contact oxidation process, featuring high treatment efficiency and good eff water quality [4]. In addition, in recent years, some new treatment technologies, such as the combination of magnetic flocculation and electrocatalytic oxidation have also been gradually applied to the treatment of wastewater from substations, to improve the treatment effect and economic efficiency [5]. However, these technologies still face certain and challenges in practical application, and further research and optimization are needed.

2.2.1 Septage Treatment Process

The septic tank is a common traditional treatment process in substation wastewater treatment, and its main principle is to degrade the organic matter in sewage through anaerobic fermentation. The structure of the septic tank is relatively simple, usually consisting of several compartments. Sewage goes through processes such as sediment and fermentation, removing some pollutants. However, the treatment effect of the septic tank is poor, with low removal rates, especially for suspended solids, nitrogen, phosphorus pollutants. Additionally, septic tanks require regular cleaning and maintenance, leading to high operating costs, and the treated sewage is difficult to meet discharge standards. The improvement of environmental protection standards, the limitations of septic tanks have gradually become apparent, and there is an urgent need for improvement and replacement [6].

2.2.2 Integrated Sewage Treatment Equipment

The dynamic integrated sewage treatment equipment uses the biological contact oxidation method, which involves suspending or filling the packing material to utilize microorganisms to the organic matter in the sewage. This process has high treatment efficiency and good effluent quality, effectively removing organic pollutants such as COD and BOD5. After through the sand settling and regulating pool, the sewage enters the contact oxidation tank for biochemical treatment, and finally, it is settled in the secondary sedimentation tank. Treated sewage can meet the environmental protection discharge standards [7]. However, the dynamic integrated treatment equipment also has some problems in practical application, such as high equipment failure and high maintenance costs. These issues need to be addressed through technical optimization and management improvement.

3 Problems with Substation Wastewater Treatment Technology

3.1 Problems with Septic Tank Treatment Technology

The septic tank treatment technology is a traditional method in substation wastewater treatment, widely used in the primary wastewater treatment of substations., with the improvement of environmental protection requirements, the treatment effect and operation efficiency of septic tanks have gradually exposed many problems. The main problems include poor treatment effect large maintenance workload, and inconvenient operation. Although septic tanks have the advantages of simple structure and low cost, they are difficult to meet the high standards of modern treatment in actual application. Moreover, the treatment process of septic tanks relies on natural sedimentation and anaerobic fermentation, which has low treatment efficiency and is to environmental pollution.

3.1.1 Poor Processing Performance

The main principle of a septic tank is to provide preliminary treatment of wastewater through sedimentation and anaerobic fermentation. However, due to simple design and lack of effective treatment methods, its removal efficiency is limited. Septic tanks perform poorly in removing pollutants such as suspended solids, nitrogen, andorus, often failing to meet environmental discharge standards. According to research, the removal rate of traditional septic tanks is generally only 50% to 60%, and the ability to remove organic matter is even lower, at only about 20%. This makes it difficult for the treated wastewater to meet discharge requirements, especially the face of increasingly stringent environmental standards, septic tank treatment technology appears to be inadequate.

3.1.2 Large Maintenance Workload

Although the septic tank has a simple structure, it requires frequent maintenance during operation. Due to the sludge deposition and fermentation process within theptic tank, a large amount of anaerobic sludge is produced, which needs to be regularly cleaned and treated. This not only increases the operating costs but brings operational inconvenience. Moreover, during long-term use, septic tanks are prone to problems such as blockages and leaks, requiring regular inspections and repairs, thereby the workload and complexity of operation and maintenance. Therefore, the application of septic tank treatment technology in substation wastewater treatment faces significant challenges in maintenance and management.

3.2 Problems of Dynamic Integrated Treatment Equipment

The dynamic integrated sewage treatment equipment has made significant technological advancements, but there are still some issues in practical applications. These problems are mainly reflected in operating conditions of the sludge system and frequent failures of the power equipment. Although the equipment is designed to meet the requirements of efficient treatment, it often leads to treatment results in actual operation due to various reasons. The existence of these problems not only affects the efficiency of the treatment equipment but also increases the costs of operation and maintenance.

3.2.1 The Sludge System Is Operating Poorly

The dynamic integrated treatment unit relies on the activated sludge system to degrade the organic matter in the wastewater. However, due to the small arduous discharge of substation wastewater, the operation of the activated sludge system is poor. The sludge loading is insufficient, the microbial activity decreases, and theudge system is prone to failure. Moreover, the concentration of pollutants in the wastewater fluctuates greatly, which further affects the treatment effect of the activated sludge. To these factors, it is difficult for the sludge system to maintain a stable operating condition, and the effluent water quality is difficult to meet the standards [8].

3.2.2 Frequent Failures of Power Equipment

The normal operation of the dynamic integrated treatment facility relies on various dynamic equipment, such as sewage lift pumps and aeration blowers. However, equipment are prone to failure in actual operation. Equipment failure leads to sewage that cannot be treated normally and may even overflow directly, causing environmental pollution. In addition, long-term operation of dynamic equipment is prone to aging and rusting, requiring frequent maintenance and replacement, which not only increases operating costs but also affects the continuity and of the treatment facility.

4 Research on Advanced Wastewater Treatment Technology

With the improvement of environmental protection standards and the advancement of technology, the technology for wastewater treatment in substations is also continuously developing and optimizing. Present, the dynamic oxidation treatment process and the anaerobic combination treatment process are considered to be more advanced wastewater treatment technologies with broad application prospects. These technologies performently in improving treatment efficiency, reducing operating costs and maintenance difficulties, and help to overcome the limitations of existing treatment technologies [9].

4.1 Oxidative Treatment Process

The dynamic oxidation treatment process mainly degrades organic matter and other pollutants in wastewater through oxidation-reduction reactions. This process uses the biological contact oxidation, utilizing microorganisms to biodegrade pollutants under aerobic conditions. The dynamic oxidation treatment process has the advantages of high treatment efficiency and good effluent water quality. is suitable for treating organic pollutants in substation wastewater. However, this process has high equipment requirements and needs to maintain the normal operation of the aeration system ensure the activity of microorganisms and treatment effectiveness.

4.1.1 Miniaturized Contact Oxidation Tank

The miniaturized contact oxidation tank is a treatment facility designed based on the small sewage discharge of the substation. By reducing the volume of contact oxidation tank, the device is more adapted to the intermittent discharge characteristics of the substation sewage. The miniaturized contact oxidation tank is filled with efficient biological, which provides a place for microbial growth and reproduction, and improves the efficiency of biological degradation. Compared with traditional large oxidation tanks, low construction costs, and are easy to operate and maintain. They are suitable for small-scale wastewater treatment.

4.1.2 Constructed Wetland Combination Process

The constructed wetland combined process is a wastewater treatment technology that integrates constructed wetlands with other treatment processes. This process utilizes the synergistic effects plants, soil, and microorganisms in constructed wetlands to provide advanced treatment of wastewater. After preliminary treatment, the sewage enters the constructed wetland, where removal of pollutants is achieved through plant absorption, soil filtration, and microbial degradation. The constructed wetland combined process offers advantages such as good treatment performance, low operating, and ease of maintenance, making it particularly suitable for treating wastewater with high organic content [10].

4.2 Anaerobic Combined Treatment Process

The anaerobic combined treatment process decomposes organic pollutants in wastewater through the metabolic action of anaerobic microorganisms, while combining with other units to achieve comprehensive removal of pollutants. This process is suitable for treating high-concentration organic wastewater, characterized by high treatment efficiency and low energy consumption. The anaerobic combined treatment process includes improvements in anaerobic tank design, combination of anaerobic processes with constructed wetlands, and combination of anaerobic processes with filters, which can significantly improve wastewater treatment effectiveness [11].

4.2.1 Improvement in Anaerobic Tank Design

The improvement of anaerobic tank design mainly involves optimizing the structure and operating parameters of the tank to enhance the efficiency of anaerobic treatment Specific measures include increasing the effective volume of the anaerobic tank, setting up baffle plates and fillers to enhance the mixing and contact efficiency of the sl. Additionally, by controlling the temperature, pH value, and sludge retention time of the anaerobic tank, the growth environment of anaerobic microorganisms optimized, further improving the removal rate of organic matter. The improved anaerobic tank has advantages such as good treatment effect and stable operation, as shown in Fig. 1.

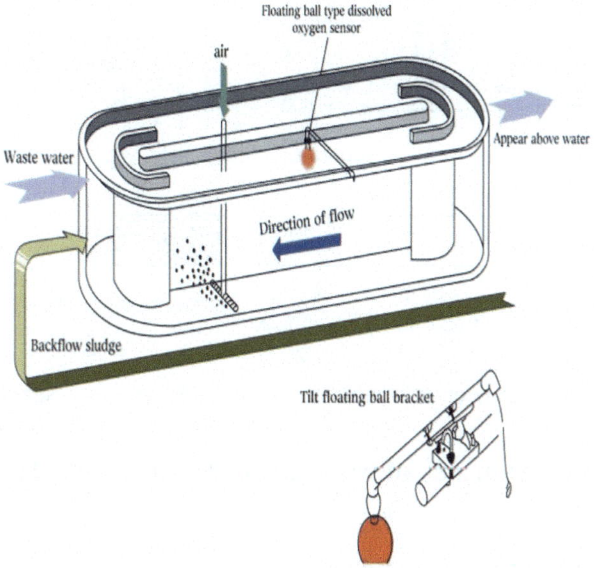

Fig. 1. Improved design of anaerobic tank

4.2.2 Combination of Anaerobic Process and Constructed Wetland

The combination of anaerobic process with constructed wetland is a wastewater treatment technology that organically combines anaerobic treatment with constructed wetland. The anaerobic process removes most of the organic pollutants at the front end, and the effluent is then further treated through the constructed wetland. The wetland uses plant absorption, soil filtration, and microbial degradation to deeply purify the sewage. This combined process not only improves the effectiveness of wastewater treatment but also operating costs. It is suitable for treating wastewater from substations with small treatment volumes and large fluctuations in pollutant concentrations [12].

4.2.3 Anaerobic Process Combined with Biological Filter

The combination of anaerobic process with biological filter is a wastewater treatment technology that combines anaerobic treatment with biological filter treatment. The anaerobic process treats most of the organic pollutants in the sewage at the front end, and the effluent enters the biological filter, where the microorganisms on the filter further degrade the residual pollutants [13]. The biological filter uses the adsorption, filtration, and biodegradation effects of the filter media to carry out advanced of sewage. This combined process has the advantages of high treatment efficiency, good effluent water quality, and low operating costs, and is especially suitable for treating highcentration organic wastewater, as shown in Fig. 2.

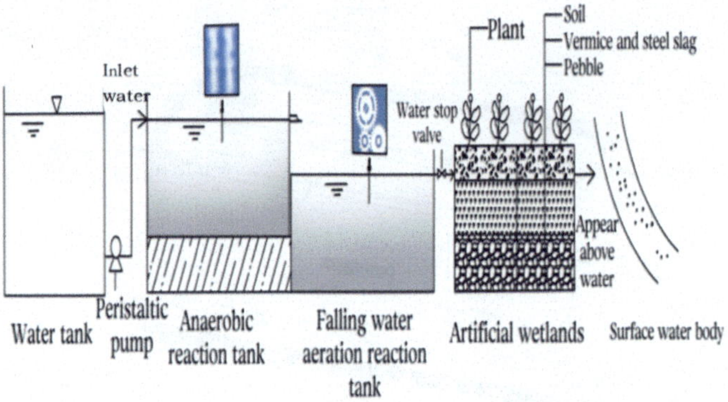

Fig. 2. Schematic diagram of the combination of anaerobic process and artificial wetland

5 Application Examples of Substation Wastewater Treatment Technology

5.1 Septic Tank Treatment Examples

The septic tank treatment technology, with its simple structure and low cost, is widely used in the primary stage of wastewater treatment in subst. In a 220 kV substation, septic tanks are used to treat domestic sewage. The substation adopts a three-chamber septic tank which removes most of the organic matter and suspended solids through the sedimentation of sewage and anaerobic fermentation. However, with the improvement of wastewater treatment standards the effluent quality after septic tank treatment is difficult to meet the new discharge standards. The substation faces frequent septic tank cleaning and large maintenance workload during, and needs to regularly clean and maintain the septic tank to ensure its treatment effect, as shown in Tables 1 and 2.

Table 1. Typical emission standard limits for domestic sewage (unit: mg/L)

Emission standards	suspended solids (mg/L)	BOD_5 (mg/L)	COD_{cr} (mg/L)	Total phosphorus (mg/L)	Ammonia nitrogen (mg/L)
GB/T 8978–1996 *comprehensive wastewater discharge standard*					
Class I	70	20	100	0.5	15
Class II	150	30	150	1	25
Class III	200	60	200	1.5	35
GB/T 18918–2002 *pollutant discharge standards for urban sewage treatment plants*					
Class A	10	10	50	0.5	5
Class B	20	20	60	1	8
Secondary	30	30	120	1.5	15

Table 2. Comparison of results before and after wastewater treatment at the substation

	Suspended solids (mg/L)	BOD_5 (mg/L)	COD_{cr} (mg/L)	Total phosphorus (mg/L)	Ammonia nitrogen (mg/L)
Data 1 (before processing)	300	350	250	8	45
Data 1 (processed)	8	9	45	0.4	5
Data 2 (before processing)	189	263	246	9	42
Data 2 (processed)	7	8	38	0.7	4
Data 3 (before processing)	266	323	278	10	51
Data 3 (processed)	7	10	48	0.8	7
Data 4 (before processing)	192	350	132	10	31
Data 4 (processed)	6	9	32	0.7	4

The wastewater treated by the substation wastewater treatment technology meets the national standards of GB/T 8978–1996 "Com Wastewater Discharge Standard" and GB/T 18918–2002 "Pollutant Discharge Standard for Urban Waste Treatment Plants."

5.2 Example of a Powered Integrated Sewage Treatment Device

The application of integrated wastewater treatment equipment in substations is becoming increasingly common. In a 500 kV substation, an wastewater treatment equipment of type WSZ-A-1 was installed to treat the domestic wastewater within the station. This equipment uses the biological contact oxidation method, effectively removing pollutants from the wastewater through multi-stage treatment including a sand settling tank, a contact oxidation tank, and a secondary settling tank. The operation results show that the equipment stably meet the national wastewater discharge standards, with good effluent quality. However, some problems have also occurred during the operation of the equipment, such as a equipment failure and poor operation of the sludge system, which require regular equipment maintenance and process adjustments [14].

5.3 Case Studies of New Processing Technology Application

The application of new treatment processes in substation wastewater treatment has provided new solutions to improve treatment efficiency and effluent quality. For example, application of a combination process of magnetic flocculation and electrocatalytic oxidation has achieved good results in the treatment of wastewater from an accident oil pit at a station. This process removes suspended solids from the wastewater through magnetic flocculation, and then degrades organic pollutants using electrocatalytic oxidation, ensuring the treatedluent meets the national first-class discharge standard. Additionally, some substations have adopted a combination of anaerobic treatment processes and artificial wetland treatment processes, further the effectiveness of wastewater treatment, as shown in Fig. 3. These new treatment processes not only improve treatment efficiency but also reduce operating costs, making them widely.

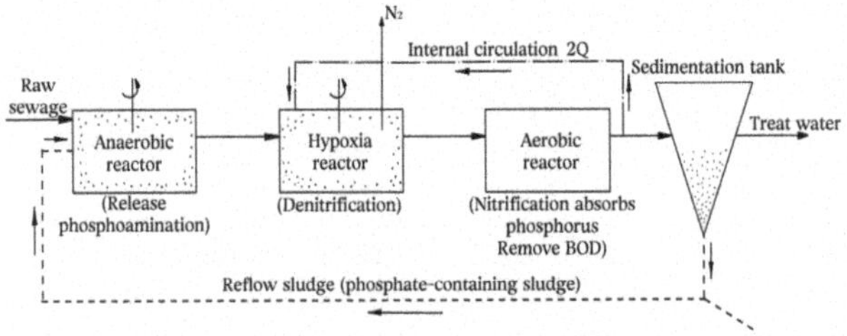

Fig. 3. Principle of the new treatment process

6 Conclusion

Research and application of wastewater treatment technology in substations is of great significance to improving environmental protection and maintaining ecological environment. This paper analyzes the and disadvantages of existing treatment technologies, and proposes an optimized solution based on anaerobic combined treatment process and dynamic oxidation treatment process. Practical application examples show that the treatment process performs excellently in improving treatment efficiency, reducing operating costs and maintenance difficulties. However, further optimization and promotion of the technology still need continuous exploration and research In the future, combining the characteristics of substation wastewater discharge, developing more efficient and economical treatment technologies will provide a solid technical guarantee for the green development ofations.

Acknowledgments. This paper is one of the research achievement funded by the State Grid Heilongjiang Electric Power Co., Ltd. The project name is "Research and Application of Wastewater Treatment Technology in Substations"(52243724000 W).

References

1. Kang, C., Jiaqi, W., Wendong, Z.: Research on green renovation of 220kV substations [J]. Energy Conserv. Environ. Protect. **22**(02), 78–80 (2023)

2. Tao, W., Yao, C., Heng, L., et al.: Treatment of accident oil pool wastewater in substations based on magnetic coagulation-electrocatalytic oxidation combined process [J]. Hunan Electr. Power. **42**(05), 69–74 (2022)
3. Zihan, Q., Xuan, C., Ying, Z., et al.: Current status and prospects of sewage treatment technology in manned substations [J]. Hubei Electr. Power. **44**(04), 104–110 (2020)
4. Xiaohua, L., Hui, L., Zhenguo, G., et al.: Research on mobile integrated oil-containing wastewater treatment device for substations [J]. Green Technol. **45**(02), 109–110 (2020)
5. Libin, P.: Discussion on the design of water supply and drainage system for substations [J]. Constr. Supervis. Testing Cost. **12**(03), 54–55+72 (2019)
6. Tiqiang, Z.: Current status and prospects of substation sewage treatment [J]. Environ. Sci. Manage. **42**(06), 83–86 (2017)
7. Nianyong, Y.: A brief discussion on harmless treatment methods for rural substation sewage [J]. Build. Mater. Decorat. **33**(22), 180–181 (2017)
8. Huahua, H., Zijian, Z.: Survey on the current status of domestic sewage treatment in substations in China [J]. Environ. Protect. Technol. **19**(06), 42–45 (2013)
9. Zhen, C., Jiazheng, L., Qunce, Z., et al.: Analysis of substation wastewater discharge and environmental protection assessment [J]. Electr. Power Energy. **34**(01), 62–65 (2013)
10. Chengyi, Y.: Electrical and Automation Design of Small Urban Sewage Treatment Plants [D]. Southwest Jiaotong University (2013)
11. Xingchu, C.A.I., Jie, Z.H.A.N.G.: Ecological treatment process for domestic sewage in substations [J]. Electr. Power Surv. Des. **54**(06), 51–53 (2011)
12. Minglan, C., Hui, L.: A brief discussion on the application of substation sewage treatment equipment [J]. Guangdong Transm. Substat. Technol. **12**(06), 36–38 (2010)
13. Yan, S.: Research on water-saving and emission reduction Technologies for Environmentally Friendly Substations [J]. Sci. Technol. Wind. **28**(24), 49–52 (2013)
14. Fanling, L.I.: Renovation of the substation in urban sewage treatment plants [J]. J. Instrum. **63**(01), 463–464 (2001)

Risk Assessment of Sudden Water Pollution Accidents in the Inflow Navigation Channels of the Taihu Lake Basin

Fang Zhang, Shiyue Wang(✉), Fang Zhou, Xianchao Yang, Feng Xiao, and Xiaoguang You

China Waterborne Transport Research Institute, Beijing, China
wangshiyue@wti.ac.cn

Abstract. The issue of water environmental pollution caused by sudden water pollution accidents in inland waterways has garnered significant attention in recent years. With the rapid development of inland waterway transportation in the Taihu Lake basin, the scale of oil and chemical transportation has been continuously expanding, leading to an upward trend in the probability of sudden pollution accidents. The relatively complex river network of the Taihu Lake basin has increased the difficulty of studying the impact of accidents. This paper takes the inland waterways leading into the Taihu Lake basin as an example, by establishing hydrodynamic models, oil spill prediction models, and chemical spill models, to predict potential sudden oil spills and chemical spills during ship transportation, assess the response time for handling sudden pollution accidents, and propose suggestions on risk prevention and emergency measures based on simulation results. This provides a scientific basis for the prediction and emergency handling of sudden accidents under complex river network conditions.

Keywords: Channel · Water pollution accidents · Risk assessment

1 Introduction

Taihu Lake is the third largest freshwater lake in China and the most important drinking water source in the Yangtze River Delta region. Strengthening the protection of the Taihu Lake basin is of great significance for ensuring water security and ecological safety in the lower reaches of the Yangtze River and the Yangtze River Delta region, promoting the integrated development of the Yangtze River Delta, and jointly protecting the Yangtze River Economic Belt. In recent years, China has vigorously regulated river channels and developed inland waterway transportation, leading to rapid development in the industry with a navigable mileage of approximately 128,200 km. Inland waterway transportation is an important mode of transport for the river channels entering Taihu Lake, with dense cargo ships traveling the waterways and a wide variety of transported goods, including a large amount of oil and chemicals. In the event of an accident involving these ships on the waterways leading to the lake, it could cause severe pollution to the water environment of the river channels, threatening the water ecological environment of Taihu Lake.

V. Achal et al. (Eds.): EPPC 2024, ESE, pp. 292–301, 2025.
https://doi.org/10.1007/978-3-031-92123-0_31

Inland waterways are relatively narrow, with rapid water flow and rapid changes in discharge, which means that the speed at which harm becomes apparent from a risk accident is faster. Tian Wei [1] and others established an EFDC model, combined with the Lagrangian particle tracking method, to predict the drifting behavior of oil films in the tidal river section downstream of Jingjiang in the lower reaches of the Yangtze River. Guo Danyang [2] and others took the main river channel of the Pearl River where the water resource allocation project of the Pearl River Delta is located as the study area, and used an oil spill model to simulate the scenario of a sudden oil spill pollution incident. It can be seen that previous studies on river oil spill issues have mainly focused on large rivers, with relatively few studies on oil spill issues in complex river networks. Cao Shengwen [3] and others established an environmental risk assessment model for marine hazardous chemical mobile risk sources based on AIS data. Zhang Jun [4] took a petrochemical enterprise in the Dalian Dagushan chemical port as the research object and predicted the influence range of the drifting path of a chemical spill accident in the nearby sea area. Currently, there are relatively few studies on inland chemical spills, with research objects mainly focusing on coastal ports.

This study sets up two scenarios of sudden water pollution incidents, oil spill and chemical leakage, and selects three sections of the waterways on the river channels entering Taihu Lake as the spill points. Through the oil particle model and material transport equations, it analyzes and predicts the impact of sudden oil spill and chemical leakage incidents on Taihu Lake, which is of great significance for the construction of a risk prevention and control system for the river channels entering Taihu Lake under complex river network conditions.

2 Study Area

The Taihu Lake basin spans Jiangsu Province, Zhejiang Province, and Shanghai City, with a dense network of rivers and numerous lakes. The water surface area accounts for 15% of the total area, the total length of the river channels is approximately 120,000 km, and the river density reaches 3.3 km per square kilometer. Taihu Lake is a typical saucer-shaped lake with an average water depth of only 1.9 m, which means its environmental capacity is very limited. The plain areas within the basin have an interwoven network of rivers with slow water flow and low self-purification ability. The multi-year average annual water surface evaporation of the Taihu Lake basin is 821.7 mm, with a variation range of 750–900 mm, and the spatial distribution is greater in the east than in the west and greater in the plains than in the mountainous areas. The multi-year average annual natural runoff of the Taihu Lake basin is 16.01 billion m^3, equivalent to an annual runoff depth of 438 mm, and the multi-year average annual runoff coefficient is 0.37.

3 Scenario Setting

3.1 Accident Point Setting

The water system in the plain river network area of the Taihu Lake basin is relatively complex, with back-and-forth flow phenomena, and there are as many as 215 rivers entering and exiting the lake. Based on the investigation of navigation infrastructure, river runoff, and wharf operations, this study selects three rivers entering Taihu Lake in the upstream Zhejiang region as the sites of the accidents. Accident site A is located 10.7 km upstream from the mouth of Taihu Lake, in the vicinity of an oil transportation dock. The river at this location has a reduced flow and a narrow channel, yet it is equipped with a sluice gate at the lake inlet. Accident site B is situated 16.5 km upstream from the mouth of Taihu Lake, where the river exhibits a greater flow and a wider channel; no sluice gate is installed, and an oil transportation dock is present in the area. Accident site C is positioned 11 km upstream from the mouth of Taihu Lake, featuring a substantial river flow and a broad channel, and is adjacent to a chemical transportation dock. The geographical locations of the accident sites are shown in Fig. 1.

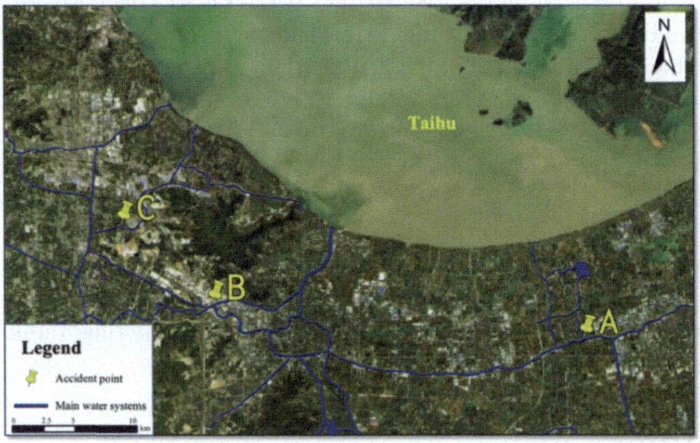

Fig. 1. Location of sudden water pollution incidents

3.2 Setting of Sudden Water Pollution Accidents

There are many environmental risk factors distributed around the rivers entering the Taihu Lake Lake. According to the nature and location of the risk factors, this study assumes two types of sudden water pollution events.

- Oil spill accident: A collision accident occurred between a 1000ton refined oil vessel (with a density of 850 kg·m^{-3} at accident sites A and B, resulting in a complete leakage of oil from a single cargo hold (85m^3) for 10 min.
- Chemical leakage accident: At accident point C, a 1000 ton ethylene glycol ship's bulkhead was severely damaged, resulting in a complete leakage of ethylene glycol from a single cargo hold (85 m^3) for 10 min. The physicochemical properties and hazardous characteristics of ethylene glycol chemical products are shown in Table 1.

Table 1. Physical and chemical properties and hazardous characteristics of ethylene glycol chemicals.

Name	CAS number	Appearance	Solubility	Hazard category
Ethylene glycol	107-21-1	Transparent viscous liquid	Soluble	Acute toxicity, oral (category 4), H302 Specific target organ systemic toxicity (repeated exposure), Oral (category 2), kidney, H373

4 Model Building

4.1 Hydrodynamic Model

This study employs the MIKE21 software to establish a hydrodynamic model, upon which subsequent oil spill and chemical leak simulations are based [5]. The two-dimensional Hydrodynamic module is selected according to the hydrological characteristics of the inland river, which is based on the three-dimensional incompressible and Reynolds-averaged Navier-Stokes equations, complying with the Boussinesq assumption and the hydrostatic pressure assumption. This module utilizes the Alternating Direction Implicit (ADI) method to handle the control equations of water flow, expressing the parameters of the equations through the central difference method. The Thomas algorithm is used to solve these discretized matrix equations, effectively preventing calculation errors that may arise from discretization. The control equations of the two-dimensional hydrodynamic model are shown in Eq. (1).

$$\frac{\partial \zeta}{\partial t} + \frac{\partial p}{\partial x} + \frac{\partial q}{\partial y} = \frac{\partial h}{\partial t} \tag{1}$$

In the formula (1): ζ is the water level; p is the flow flux in the x-direction; q is in the y direction Flow flux; h is the water depth.

The momentum equation in the x-direction is given by formula (2).

$$\begin{aligned} &\frac{\partial p}{\partial t} + \frac{\partial}{\partial x}\left(\frac{p^2}{H}\right) + \frac{\partial}{\partial y}\left(\frac{pq}{H}\right) + gH\frac{\partial \zeta}{\partial x} + \frac{gq\sqrt{p^2+q^2}}{C^2H^2} \\ &-\frac{1}{\rho}\left[\frac{\partial}{\partial x}(H\tau_{xx}) + \frac{\partial}{\partial y}(H\tau_{xy})\right] - fq - f_w|W|W_x = 0 \end{aligned} \tag{2}$$

The momentum equation in the y-direction is given by formula (3).

$$\begin{aligned} &\frac{\partial q}{\partial t} + \frac{\partial}{\partial y}\left(\frac{p^2}{H}\right) + \frac{\partial}{\partial x}\left(\frac{pq}{H}\right) + gH\frac{\partial \zeta}{\partial y} + \frac{gq\sqrt{p^2+q^2}}{C^2H^2} \\ &-\frac{1}{\rho}\left[\frac{\partial}{\partial y}(H\tau_{yy}) + \frac{\partial}{\partial x}(H\tau_{xy})\right] - fq - f_w|W|W_y = 0 \end{aligned} \tag{3}$$

In the formula (2) and (3): $H = h+\zeta$; C is the Xie Cai coefficient; ρ is the fluid density; τ_{xx}, $\tau_{xy}$and τ_{yy} are effective shear force components; f is the Coriolis force coefficient; f_w is Wind resistance coefficient; W_x and W_y are the wind speeds in the x and y directions, respectively weight.

4.2 Oil Leakage Model

The accident simulation prediction employs an oil particle model [6] for calculations, taking into account the movement of particles caused by wind, currents, physical dispersion, and Stokes drift.

4.2.1 Push Flow and Random Walk Diffusion Processes

The model assumes that the leaked oil can be generalized as independent Lagrangian particles with known mass. The position vector of oil particles at time t is represented by formula 4. Δt is time step, s; X is the position of surface oil particles; $\overrightarrow{U_{oil}}$ is the oil film drift rate, m/s.

$$\overrightarrow{X_t} = \overrightarrow{X_{t-1}} + \Delta t\overrightarrow{U_{oil}} \tag{4}$$

The model incorporates a random walk diffusion process, and the dispersion distance of the oil film is calculated using formulas (5) and (6). D_x and D_y are the horizontal dispersion coefficients in the x and y directions, m^2/s, respectively; Δt is the time step, γ is a random number, −1 to 1.

$$x_{dd} = \gamma\sqrt{6D_x\Delta t} \tag{5}$$

$$y_{dd} = \gamma\sqrt{6D_y\Delta t} \tag{6}$$

4.3 Destination Model

The extension process determines the expansion of the surface floating oil, which further affects the evaporation, dissolution, diffusion, and photo oxidation of the oil film on the water surface. The calculation formula for the rate of change in floating oil area (m^2/s) caused by the extension of thick oil film is shown in formula (7). A_{tk} is the surface area of the floating oil, m^2; K_1 is the elongation rate constant, 1/s; V_m is the volume of floating oil, m^3; t is time, s.

$$\tilde{A}_{tk} = \frac{dA_{tk}}{dt} - K_1 A_{tk}^{1/3}\left(\frac{V_m}{A_{tk}}\right)^{4/3} \tag{7}$$

The evaporation process can cause 20–40% of the floating oil to enter the atmosphere from the water surface, depending on the type of oil. The calculation formula for oil evaporation rate is shown in formula (8). T_0 is the initial boiling point of the corrected distillation curve, K; T_G is the gradient of the corrected distillation curve; T is the ambient temperature, K; A, B are dimensionless constants; t is time, s; θ is evaporation capacity.

$$F_V = \ln\left[1 + B(T_G/T)\theta\exp(A - BT_0/T)\right][T/(BT_G)] \tag{8}$$

4.4 Chemical Leakage Model

The material transport equation is used to predict the diffusion of chemicals, and the vertically averaged material transport equation is formula (9). P is the concentration of pollutants; u,v represents the flow velocity components in the x and y directions, respectively; Dx and Dy are the diffusion coefficients in the x and y directions; In the simulation, a triangular unstructured grid is used, and there is a significant difference in the time step and spatial step of each grid, resulting in a large difference in its diffusion coefficient. In the model, autonomous calculation and allocation are carried out through the set time step and spatial step; S is the emission rate of pollutants per unit time; k is the attenuation coefficient.

$$\frac{\partial(\mathrm{HP})}{\partial t}+\frac{\partial(\mathrm{HP_u})}{\partial x}+\frac{\partial(\mathrm{HP_v})}{\partial y}-\frac{\partial}{\partial x}\left(\mathrm{HD_x}\frac{\partial P}{\partial x}\right)-\frac{\partial}{\partial y}\left(\mathrm{HD_y}\frac{\partial P}{\partial y}\right)=\mathrm{HS}-\mathrm{kp} \quad (9)$$

5 Analysis of Simulated Result

5.1 Oil Spill Prediction Results

The simulation results of the oil spill model are shown in Figs. 2, 3 and Table 2. After an oil spill accident occurs at the leakage point A, it can reach the nearest sluice dam within 19 h. After an oil leak occurs at the leakage point B, it can reach the Taihu Lake waters within 38 h. Xu Qian et al. [7] used MIKE software to simulate the oil spill risk caused by collisions of inland river ships in Jiaxing during the operation period of the waterway, based on the river network of the Taihu Lake basin. The results show that the model reflects the impact pattern of sudden pollution accidents in inland rivers quite well.

Fig. 2. Oil trajectory at different time points at accident site A

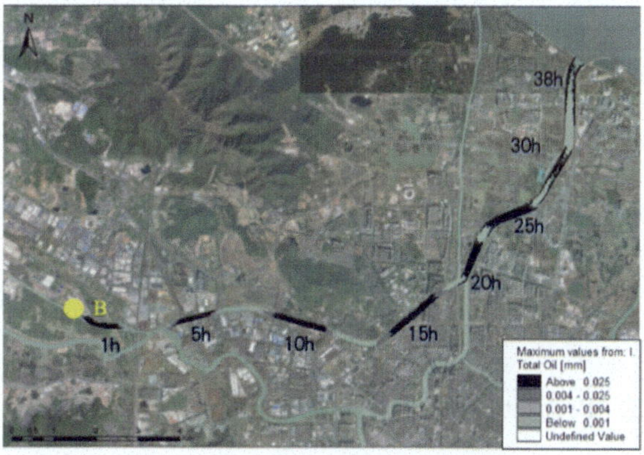

Fig. 3. Oil trajectory at different time points at accident site B

Table 2. Oil film overflow area at different times

Accident site	Time (h)	Oil film area (km^2)
A	1	0.007
	5	0.010
	10	0.022
	16	0.029
	19	0.097
B	1	0.075
	5	0.11
	10	0.15
	15	0.19
	20	0.20
	25	0.22
	30	0.16
	38	0.15

The diffusion area of the oil film at different times is shown in Table 2. With the passage of time, the maximum diffusion area of the oil film at accident point A is 0.097 km^2. The maximum diffusion area of the oil film at accident point B is 0.22 km^2.The harm caused by oil spill accidents is comprehensive [8]. After oil pollutants enter the river, they form an oil film on the water surface, which hinders the gas exchange between the atmosphere and the water body, leading to a decrease in dissolved oxygen in the water and affecting the self-purification effect of the water body. Oil products will partially sink after being affected by various external environmental factors. Although there may be no residual oil on the water surface, the underwater oil pollution can still be toxic to aquatic life and disrupt the normal ecological environment of the water body. The oil pollutants contain harmful substances, such as polycyclic aromatic hydrocarbons and heavy metals, which can endanger the safety of water source water and human health. The oil film moves to different positions at different times, and the area of the oil film varies. Based on the time it takes for the oil film to reach Taihu Lake, full preparations can be made within the emergency response time to activate the emergency plan in a timely manner. The appropriate emergency plan can be selected according to the area of the oil film to minimize the adverse effects of oil spill diffusion.

5.2 Prediction Results of Chemical Leakage

The simulation results of the chemical spill model are shown in Fig. 4. After a leak at the accident site C, it can reach the Taihu Lake waters in as fast as 31 h. As the leaked material drifts and diffuses, the concentration gradually decreases, with the material concentration at Taihu Lake being approximately 0.0002–0.0004 mg/L. Xu Cundong et al. [9] utilized MIKE software to establish a two-dimensional hydrodynamic-water quality coupling model, simulating the hydrodynamic-water quality coupling in the river network of Nanxun District, Zhejiang Province. The results indicated that the model outcomes align with the dynamic patterns of plain river networks.

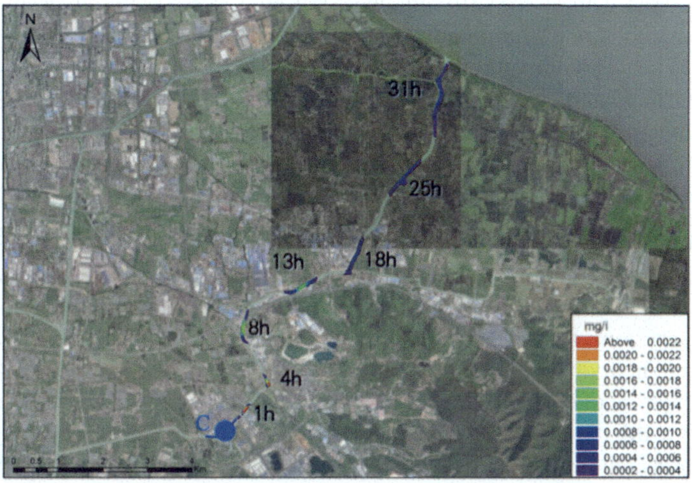

Fig. 4. Chemical trajectory of accident site C

The diffusion area of chemicals at different times is shown in Table 3. Over time, the maximum diffusion area of chemical leakage at accident point C is 0.231 km^2. After a chemical leak, chemicals may volatilize into the atmosphere, float on the water surface, dissolve in the water, or settle at the bottom, contaminating the nearby water quality and air. Rivers have a self-purification function, but only some types of chemicals will rapidly reduce in concentration after entering the water. Different types of chemicals also vary in the degree of toxicity they cause to surrounding organisms, with severe cases capable of damaging the cellular structure of organisms or even causing direct death [10].

Table 3. Oil film overflow area at different times

Accident site	Time (h)	Chemical diffusion area (km^2)
C	1	0.02
	4	0.041
	8	0.067
	13	0.069
	18	0.124
	25	0.129
	31	0.231

6 Conclusion

Oil spills and chemical leaks can cause severe pollution to the aquatic environment and even pose a threat to human health. Emergency incidents themselves are highly random and unpredictable. Due to the different characteristics of the water environment, many emergency methods and equipment used in coastal areas are not suitable for inland waterways. However, compared to maritime emergencies, emergency control measures for accidents in inland waterways are relatively simple and effective. Based on the simulation and analysis results mentioned above, combined with real-world situations, this study proposes the following recommendations for risk prevention and emergency measures for the navigation channels entering Taihu Lake:

- Risk prevention should be prioritized from the source to avoid accidents. The ship-to-shore traffic management system should be improved, and communication between the ship and shore should be enhanced. Additionally, the navigation and assistive navigation systems should be perfected. Strengthening the construction, maintenance, and management of waterway channels and assistive navigation facilities is crucial. Regular clearing of obstacles in the waterways should be carried out to ensure that the depth and width of the channels meet the requirements for safe vessel navigation. For river sections entering Taihu Lake where sluice gates have been installed, the gates can be closed to effectively control the entry of oil and hazardous chemicals into the Taihu Basin.
- To address oil spill incidents, emergency equipment and materials should be strengthened and managed effectively. Currently, the emergency supplies of port and terminal enterprises in the study area mainly consist of oil booms and absorbent pads. In accordance with the "Emergency Equipment Allocation Requirements for Oil Spills at Ports and Terminals" (JT/T451–2017) and "Guidelines for Ship Oil Spill Emergency Response Capability Assessment" (JT/T877–2013), it is recommended to supplement the existing equipment with emergency unloading devices, oil spill containment equipment, mechanical recovery equipment, oil-absorbing materials, storage and transportation devices, and other supporting equipment.
- In response to chemical spill accidents, prior to the berthing and loading/unloading operations of chemical tanker ships, the pipelines, valves, and other related equipment should be inspected to ensure they are in good condition, and the mooring status should be checked for stability. During the loading/unloading operations, close attention should be paid to the condition of the pipelines to prevent leaks, spills, drips, or any other leakage incidents. It is recommended to equip areas prone to leakage, such as loading/unloading pipelines, with toxic and flammable gas monitoring and alarm systems.
- Promote the establishment of a joint emergency response mechanism for waterway pollution in port and terminal enterprises, and develop a collaborative pollution emergency response system with neighboring regions. Build an environmental pollution forecasting, early warning, and emergency decision support system to enhance the capacity for rapid emergency response and the ability to address major pollution incidents. Strengthen emergency training and drills to improve response efficiency and emergency capability, while incorporating relevant emergency response measures

and procedures into the emergency plans for sudden environmental incidents at port enterprises, ensuring that the emergency response time meets the required standards.

Acknowledgments. Support for this work was provided by China Waterborne Transport Research Institute for Researchers.

References

1. Wei, T., Li, Q., Yiping, L.: Risk prediction of oil spill accidents in tidal river sections based on EFDC model [J]. Water Resour. Protect. **31**(6), 5 (2015)
2. Danyang, G., Silong, Z., Dong, W.: Simulation study on sudden water pollution events at intake of water resources allocation project in the Pearl River Delta [J]. J. Beijing Normal Univ. (Na. Sci. Ed.). **56**(3), 7 (2020)
3. Shengwen, C., Xiaoyi, J., Lili, S., et al.: Research and application of environmental risk assessment model for maritime hazardous chemicals Mobile risk sources based on AIS Data [J]. Mar. Inform. Technol. Appl. **36**(4), 10 (2021)
4. Jun, Z.: Research on Marine Environmental Pollution Risk Assessment and Emergency Countermeasures of Dalian Dagushan Chemical Terminal [D]. Dalian Maritime University (2015)
5. Li, X., Huang, M., Wang, R.: Numerical simulation of Donghu Lake hydrodynamics and water quality based on remote sensing and MIKE 21 [J]. ISPRS Int. J. Geo Inf. **9**(2), 94 (2020)
6. Xing, Z., Xianxin, W., Liang, W., et al.: Simulation of oil spill risk in the Weitouwan Sea area and study on its impact on sensitive areas of the surrounding environment [J]. Marine Forecast. **40**(1), 57–64 (2023)
7. Qian, X., Yong, P., Weiwei, S.: Risk prediction of the impact of Jiaxing River network oil spill accident on sensitive targets [J]. Environ. Technol. **006**, 033 (2020)
8. Cuiping, K., Jilong, C., Jie, W., et al.: Effect of wind-wave-current interaction on oil spill in the Yangtze River Estuary [J]. J. Mar. Sci. Eng. **11**(3), 494 (2023)
9. Cundong, X., Zihao, R., Zhirui, L., et al.: Research on coupled simulation of hydrodynamics and water quality in Nanxun District river network based on MIKE21 [J]. Environ. Sci. Technol. (2022)
10. Baoyou, S.: Solution to water pollution: reducing chemical input from the source [J]. J. Environ. Eng. **8**, 2 (2019)

Study on the Performance of Electrocatalytic Treatment of High-Salinity Ammonia-Nitrogen Wastewater

Zheng Chen[1,2](✉), Peng Li[3], Liming Xing[1,2], Zhaoyang Wang[1,2], Ying Tian[1,2], Zhumin Wu[1,2], Zhengkun Li[1,2], Zhenhua Ren[1,2], Ruozheng Li[1,2], Jilin Teng[1,2], Zilong Zhou[3], and Xiaoxuan Bai[3]

[1] Beijing Guodian Futong Science and Technology Development Co., Ltd., Beijing, China
chen_zh@126.com

[2] Nanrui Group (State Grid Electric Power Research Institute) Co., Ltd., Nanjing, China

[3] State Grid Jibei Electric Power Co., Ltd., Electric Power Science Research Institute, Beijing, China

Abstract. With the acceleration of industrialization and urbanization, wastewater pollution is becoming increasingly serious, threatening human health and the environment. High-salinity ammonia-nitrogen wastewater, generated from pharmaceutical wastewater, smelting wastewater, landfill leachate, aquaculture wastewater, etc., is a type of wastewater that is difficult to treat and poses a serious threat to the environment. Recently, with the rapid development of electrocatalytic technology, it has demonstrated excellent performance in treating High-salinity ammonia-nitrogen wastewater. However, there are few studies on their treatment performance on high-salinity ammonia-nitrogen wastewater. Therefore, the performance of $Ti/IrO_2–Ta_2O_5$, Ti/IrO_2-RuO_2, and boron-doped diamond in the treatment of high-salinity ammonia-nitrogen wastewater is investigated. Meanwhile, the operation factors are optimized, such as plate spacing, current density, pH, cathode materials, and the energy consumption is evaluated. The results show that under optimized operating parameter conditions, the ammonia nitrogen removal rate can reach 100% appling the three anodes in the treatment of wastewater with 600 mg/L ammonia-nitrogen and 10,000 TDS, and the energy consumption is 0.0375 kW·h/g, 0.0749 kW·h/g, and 0.0300 kW·h/g respectively.

Keywords: Electrocatalytic · High-salinity ammonia-nitrogen wastewater · Energy consumption

1 Introduction

High-salinity ammonia-nitrogen (NH_4^+-N) wastewater is a significant environmental concern, primarily originating from industries such as petrochemicals, textiles, and aquaculture. The elevated concentrations of ammonia and salts in these effluents pose substantial challenges to conventional wastewater treatment processes [1, 2]. Conventional methods, such as biological and physicochemical processes, struggle to effectively

V. Achal et al. (Eds.): EPPC 2024, ESE, pp. 302–309, 2025.
https://doi.org/10.1007/978-3-031-92123-0_32

remove NH_4^+-N from high-salinity wastewater due to reduced microbial efficiency and high operational costs [3–5]. Consequently, there is an urgent need for alternative treatment technologies capable of effectively addressing the complexities associated with high-salinity NH_4^+-N wastewater.

Electrocatalytic oxidation has emerged as a promising approach for the treatment of such challenging wastewaters [6, 7]. Electrocatalytic oxidation involves the anodic oxidation of ammonia on electrode surfaces, converting NH_4^+ directly into nitrogen gas (N_2), thereby mitigating the formation of harmful by-products like nitrites and nitrates. The process is particularly advantageous in high-salinity environments, where the presence of chloride ions (Cl^-) can facilitate the generation of active chlorine species, enhancing the oxidation efficiency [8–11].

Recent advancements in electrocatalytic oxidation have focused on the development of robust and efficient electrode materials. Boron-doped diamond (BDD) electrodes have garnered significant attention due to their high overpotential for oxygen evolution, chemical inertness, and long operational life [12, 13]. Studies have demonstrated that BDD electrodes can achieve high ammonia removal efficiencies even in the presence of high chloride concentrations, making them suitable for saline wastewater applications. In addition to BDD, mixed metal oxide (MMO) electrodes, such as those composed of titanium substrates coated with oxides of ruthenium, iridium, or tin, have been investigated for their electrocatalytic properties [14, 15]. These materials offer a balance between cost and performance, exhibiting good conductivity and catalytic activity. Research indicates that MMO electrodes can effectively oxidize ammonia in high-salinity conditions, with the added benefit of being more economically feasible compared to noble metal-based electrodes.

The operational parameters of the electrocatalytic process, including current density, electrode configuration, and electrolyte composition, play crucial roles in determining the efficiency of ammonia removal. Optimizing these parameters is essential to minimize energy consumption and maximize treatment efficacy. For instance, studies have shown that higher current densities can enhance ammonia oxidation rates but may also lead to increased energy usage and potential electrode degradation [16, 17]. Therefore, a balance must be struck to achieve sustainable operation.

Despite the promising potential of electrocatalytic oxidation, several challenges remain. The formation of disinfection by-products (DBPs), such as chloramines and trihalomethanes, due to the reaction between active chlorine species and organic matter in wastewater, is a concern that needs to be addressed [18, 19]. Additionally, the long-term stability and fouling resistance of electrode materials under harsh operating conditions require further investigation to ensure the viability of this technology for large-scale applications.

In conclusion, electrocatalytic oxidation presents a viable and effective method for treating high-salinity NH_4^+-N wastewater. Ongoing research into electrode material development, process optimization, and mitigation of potential by-products is essential to advance this technology towards practical implementation. Addressing these challenges will contribute significantly to sustainable wastewater management and environmental protection.

2 Methods and Materials

The experiment of electrocatalytic degradation of high-salinity ammonia-nitrogen wastewater was conducted on a small-scale test device, including an electrocatalytic reaction tank (1), a power supply, and a stirrer (3) as shown in Fig. 1. 1 L of experimental water sample was filled in the reaction tank (Fig. 1). Then the anode and cathode for the experiment are respectively connect to the power supply via tabs 3 and 4, fixed by a connecting rod passing through hole 5. The stir was set to a speed of 100 rpm/min.

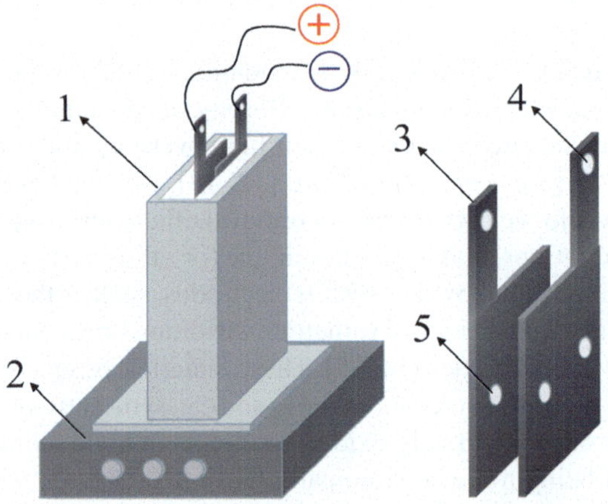

Fig. 1. Schematic of the experimental equipment

Above all, the optimization experiment of current density was conducted under different parameters of 5.76 A, 8.64 A, 11.52 A, and 14.4 A, corresponding to current densities of 200 A/m^2, 300 A/m^2, 400 A/m^2, and 500 A/m^2, applying commercial electrodes of $Ti/IrO_2–Ta_2O_5$ (Ir-Ta), Ti/IrO_2-RuO_2 (Ir-Ru), boron-doped diamond (BDD) as the anode, and Ti as the cathode, using 5 mm electrode spacing, respectively. The ammonia-nitrogen concentration of the water sample was tested at intervals of 5–10 min within 30 min of reaction, using the Nessler spectrophotometry. As a result, the parameters that showed the maximum amount of ammonia nitrogen degraded per unit time can be determined as the optimal conditions. Then, the optimization experiments of electrode spacing, pH, and cathode were investigated one by one under the optimal conditions determined previously, using different parameters of scales (5 mm, 10 mm, 15 mm), pH (3, 7, 11), and cathodes (Ti, Carbon fiber) respectively, as demonstrated above. The pH was adjusted with aqueous sulfuric acid and sodium hydroxide and tested using a pH meter.

Finally, the energy consumption required for degradation of unit mass of ammonia nitrogen using different electrodes was calculated and evaluated, under the optimal conditions.

3 Results and Discussions

3.1 Optimal Current Density

Current density is one of the most important factors affecting the performance of electrocatalytic degradation of pollutants. Therefore, the ammonia nitrogen degradation effect of different electrode plates at virous current densities was first investigated to determine the optimal current density parameters.

As shown in Fig. 2, while the current density increases, the degradation capabilities of the three electrodes (a) Ir-Ta, (b) Ir-Ru, and (c) BDD also increase. When the current density reaches 500 A/m^2, the Ir-Ta and Ir-Ru electrodes can completely degrade ammonia nitrogen in high-salinity wastewater within 30 min of reaction. In contrast, the BDD anode can completely degrade ammonia nitrogen under 400 A/m^2 or 500 A/m^2 within 30 min, and 500 A/m^2 within 20 min.

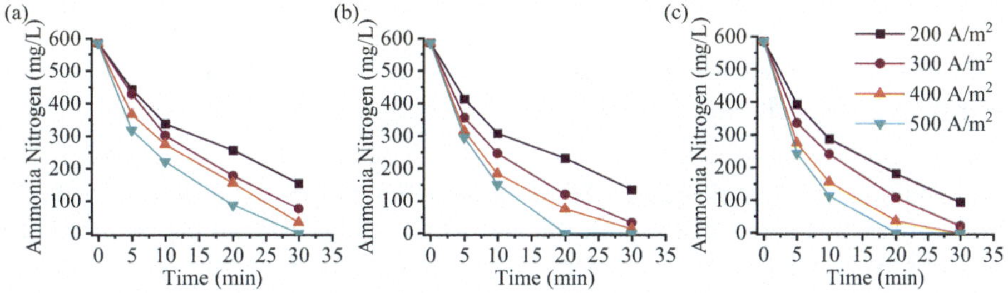

Fig. 2. The ammonia nitrogen degradation effect of different electrode plates at virous current densities. (a) Ir-Ta, (b) Ir-Ru, (c) BDD

Therefore, it can be concluded that among the high-salinity wastewater ammonia nitrogen degradation performance, BDD is the strongest, followed by iridium tantalum, and ruthenium iridium is relatively weak. However, it is worth noting that the performance of the three electrodes shows a large gap only at high density (300–500 A/m^2), which is not obvious at low current density (200 A/m^2). This suggests that when degradation speed is not pursued, one of the three electrodes can be selected for the treatment of high-salinity ammonia-nitrogen wastewater based on cost investment of electrodes. However, the operating cost is still closely related to the performance of the electrode plate.

In summary, the optimal current density of Ir-Ta, Ir-Ru, and BDD is determined as 500 A/m^2, 500 A/m^2, and 400 A/m^2 respectively.

3.2 Optimal Electrode Spacing

In addition to current density, the distance between the electrodes is another key factor affecting the electrocatalytic performance and energy consumption. Therefore, after determining the above optimal current density parameters, the degradation performance of different electrode plates under different electrode plate distances was studied.

The experimental results are shown in Fig. 3. It can be observed that the performance of all the three electrodes in degrading ammonia nitrogen in high-salinity wastewater decreases slightly with the increase of the electrode distance, during the reaction process (0–10 min). The three electrodes have a higher degradation rate under the condition of 5 mm plate spacing rather than 10 mm and 15 mm. Since the water sample conductivity is as high as 20 ms/cm, the performance is close either under the conditions of 10 mm or 15 mm.

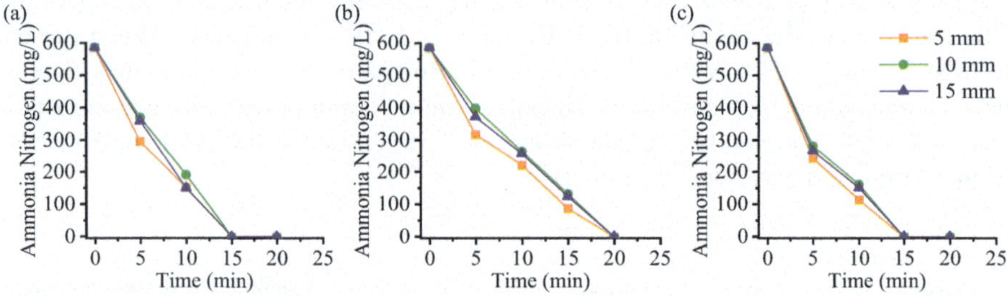

Fig. 3. The ammonia nitrogen degradation effect of different electrode plates at virous plate distances. (a) Ir-Ta, (b) Ir-Ru, (c) BDD.

In addition, the complete degradation of ammonia nitrogen can be achieved after 15 min of reaction using BDD and Ir-Ta, and 20 min using Ir-Ru. It can be concluded that in this experiment, the performance of BDD and Ir-Ta electrodes are relatively close, and both are better than Ir-Ru electrodes.

However, a larger electrode plate spacing will result in a higher cell voltage, which means that more energy is required to degrade the same amount of ammonia nitrogen. Therefore, according to the experimental results, it can be determined that the optimal plate spacing parameters of the three electrode plates are 5 mm.

3.3 Optimal pH

There are two forms of reaction in the electrocatalytic degradation of ammonia nitrogen. One is that the electrode material directly catalyzes the oxidation of ammonia nitrogen into N_2, and the other is that the active chlorine produced by Cl^- indirectly oxidized ammonia nitrogen into N_2. In the indirect reaction process, the pH of the reaction system has an important influence on the reaction rate of ammonia nitrogen degradation. Therefore, the optimal pH of different anodes is investigated via electrocatalytic experiments under virous adjusted pH of the water samples.

The experiment results are shown in Fig. 4. According to that, under acidic to neutral conditions (pH = 3, 7), the amount of ammonia nitrogen degraded by each electrode in the same time is not much different, and the performance difference in degrading ammonia nitrogen is not obvious. However, each electrode has a higher ammonia nitrogen degradation performance under alkaline conditions (pH = 11). At pH = 11, BDD and Ir-Ta can completely degrade all ammonia nitrogen within 15 min, while Ir-Ru requires 20 min. BDD and Ir-Ta showed better ammonia nitrogen degradation performance than Ir-Ru.

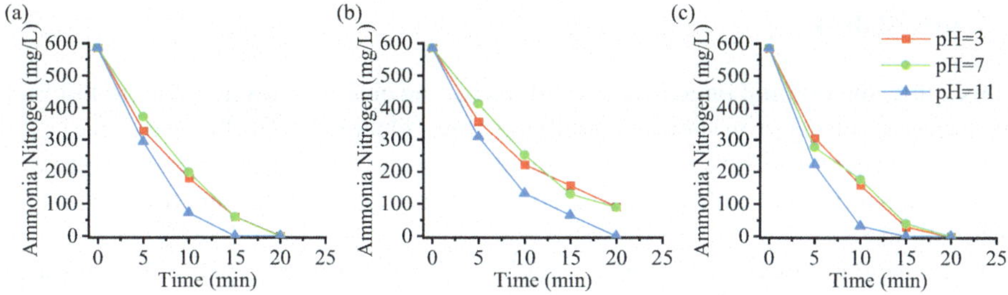

Fig. 4. The ammonia nitrogen degradation effect of different electrode plates at virous pH. (a) Ir-Ta, (b) Ir-Ru, (c) BDD.

In summary, based on the above results, it can be determined that the optimal pH of the three electrode application systems is 11. In addition, the difference of performance between BDD and Ir-Ta is close, which means that Ir-Ta is a better anode choice than BDD, considering the perspective of electrode cost.

3.4 Optimal Cathode and Energy Consumption

In the process of electrocatalytic water treatment, cathode material is also a key factor affecting the performance of electrocatalytic treatment. During the electrocatalytic degradation of ammonia nitrogen, part of the ammonia nitrogen may be converted into nitrate nitrogen, which is then oxidized to ammonia nitrogen and finally degraded to N_2. Therefore, the performance of Ti and carbon fiber cathodes were investigated, and the energy consumption is evaluated under all the determined optimal conditions.

According to the results shown in Fig. 5, the performance difference of each electrode in ammonia nitrogen degradation under the condition of titanium and carbon fiber as cathode plate is not obvious. The performance of using Ti cathode is slightly better than that of using carbon fiber cathode. The average energy consumption of the three electrodes in completely degrading ammonia nitrogen is 0.0375 kW·h/g, 0.0749 kW·h/g, and 0.0300 kW·h/g respectively. Under all the above optimal conditions, the ammonia nitrogen degradation performance of BDD and Ir-Ta is still better than that of Ir-Ru.

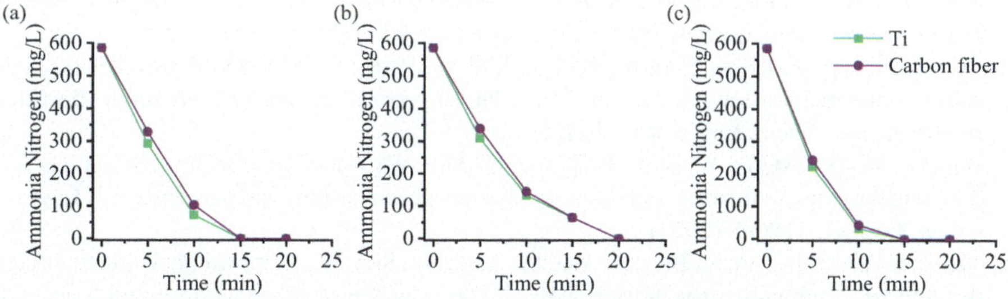

Fig. 5. The ammonia nitrogen degradation effect of different electrode plates using different cathodes. (a) Ir-Ta, (b) Ir-Ru, (c) BDD.

4 Conclusion

In this study, the optimal operating conditions for the electrocatalytic treatment of high-salt ammonia nitrogen wastewater applying three electrodes (Ir-Ta, Ir-Ru, and BDD anodes) were determined through experiments. Furthermore, the treatment performance of different anodes and cathodes was compared based on degradation effect and energy consumption.

At a current density of 500 A/m^2, a plate spacing of 5 mm, pH = 11, and using Ti cathode, the Ir-Ta and Ir-Ru electrodes have the best ammonia nitrogen degradation effect. Under the same conditions, BDD has the best ammonia nitrogen degradation effect at a current density of 400 A/m^2. However, using Ir-Ta and BDD has close energy consumption. Therefore, in high-salt ammonia nitrogen wastewater treatment, Ir-Ta is a type of appropriate electrodes with obvious material cost and energy consumption advantages.

Acknowledgments. This work was supported by the State Grid technology project [Research and application of key technologies for power grid full-process water management based on refined management and control objectives, 5108-202218280A-2-326-XG].

References

1. Srivastava, A., Parida, V.K., Majumder, A., Gupta, B., Gupta, A.K.: Treatment of saline wastewater using physicochemical, biological, and hybrid processes: insights into inhibition mechanisms, treatment efficiencies and performance enhancement. J. Environ. Chem. Eng. **9**(4), 105775 (2021)
2. Marmanis, D., Thysiadou, A., Diamantis, V., Christoforidis, A., Dermentzis, K.: Performance of electrocoagulation processes for the removal of COD and ammonia from high salinity landfill-leachate using iron or aluminum electrodes. J. Eng. Sci. Technol. Rev. **14**(4), 105–109 (2021)
3. Ahmad, N.N.R., Ang, W.L., Leo, C.P., Mohammad, A.W., Hilal, N.: Current advances in membrane technologies for saline wastewater treatment: a comprehensive review. Desalination. **517**, 115170 (2021)
4. Marathe, D., Singh, A., Raghunathan, K., Thawale, P., Kumari, K.: Current available treatment technologies for saline wastewater and land-based treatment as an emerging environment-friendly technology: a review. Water Environ. Res. **93**(11), 2461–2504 (2021)
5. Li, W., Zhang, Q., Li, J., Gao, R., Kao, C., Li, X., Peng, Y.: Nitrogen removal from high-saline municipal wastewater via anammox-based process driven by both **nitritation and denitratation**. Chem. Eng. J. **462**, 142269 (2023)
6. Bao, H., Wu, M., Meng, X., Han, H., Zhang, C., Sun, W.: Application of electrochemical oxidation technology in treating high-salinity organic ammonia-nitrogen wastewater. J. Environ. Chem. Eng. **11**, 110608 (2023)
7. Meng, X., Zeng, P., Lin, S., Bao, H., Wu, M., Yang, L., Sun, W.: Removal of chemical oxygen demand and ammonia nitrogen from high salinity tungsten smelting wastewater by one-step electrochemical oxidation: from bench-scale test, pilot-scale test, to industrial test. J. Environ. Manag. **340**, 117983 (2023)
8. Oyekunle, D.T., Cai, J., Gendy, E.A., Chen, Z.: Impact of chloride ions on activated persulfates based advanced oxidation process (AOPs): a mini review. Chemosphere. **280**, 130949 (2021)

9. Li, Q., Liu, G.H., Qi, L., Wang, H., Xian, G.: Chlorine-mediated electrochemical advanced oxidation process for ammonia removal: mechanisms, characteristics and expectation. Sci. Total Environ. ***896***, 165169 (2023)
10. Li, Q., Liu, G.H., Chen, S., Qi, L., Wang, H.: Clarifying the contributor of ammonia oxidation in chlorine-mediated electrochemical process. Chem. Eng. J. **473**, 145120 (2023)
11. Choi, S., Choi, W.I., Lee, J.S., Lee, C.H., Balamurugan, M., Schwarz, A.D., Nam, K.T.: A reflection on sustainable anode materials for electrochemical chloride oxidation. Adv. Mater. **35**(43), 2300429 (2023)
12. Martínez-Huitle, C.A., Brillas, E.: A critical review over the electrochemical disinfection of bacteria in synthetic and real wastewaters using a boron-doped diamond anode. Curr. Opin. Solid State Mater. Sci. **25**(4), 100926 (2021)
13. Herrera-Chávez, S., Pacheco-Álvarez, M., Kadier, A., Brillas, E., Peralta-Hernández, J.M.: Efficient electrochemical advanced degradation of red CL and red WB dyes from the tanning industry using a boron-doped diamond anode. Chemosphere. **363**, 142825 (2024)
14. Snowdon, M.R., Rathod, S., Fattahi, A., Khan, A., Bragg, L.M., Liang, R., Servos, M.R.: Water purification and electrochemical oxidation: meeting different targets with BDD and MMO anodes. Environments. **9**(11), 135 (2022)
15. Kumar, D., Gupta, S.K.: Electrochemical oxidation of direct blue 86 dye using MMO coated Ti anode: modelling, kinetics and degradation pathway. Chem. Eng. Process. **181**, 109127 (2022)
16. Almomani, F., Saad, M.A.H.S.: Electrochemical oxidation of ammonia (NH4+/NH3) ON synthesized nickel-cobalt oxide catalyst. Int. J. Hydrog. Energy. **46**(6), 4678–4690 (2021)
17. Feng, H., Chen, Z., Wang, X., Chen, S., Crittenden, J.: Electrochemical advanced oxidation for treating ultrafiltration effluent of a landfill leachate system: impacts of organics and inorganics and economic evaluation. Chem. Eng. J. **413**, 127492 (2021)
18. Weng, C., Hinkle, M.M., Mitch, W.A.: Electrochemical dichlorination of municipal wastewater effluent. Environ. Sci. Technol. **57**(3), 1425–1432 (2023)
19. İskurt, Ç., Aliyev, E., Gengec, E., Kobya, M., Khataee, A.: Electrochemical oxidation of pretreated landfill leachate nanofiltration concentrate in terms of pollutants removal and formation of by-products. Chemosphere. **307**, 135954 (2022)

Author Index

V. Achal et al. (Eds.): EPPC 2024, ESE, pp. 311–312, 2025.
https://doi.org/10.1007/978-3-031-92123-0

The manufacturer's authorised representative in the EU is Springer Nature Customer Service Centre GmbH, Europaplatz 3, 69115 Heidelberg, Germany. If you have any concerns regarding our products, please contact ProductSafety@springernature.com

Printed and bound by CPI Group (UK) Ltd, Croydon, CR0 4YY
30/09/2026
02225596-0007